T0136855

Springer Proceedings in Mathematics & Statistics

Volume 283

Springer Proceedings in Mathematics & Statistics

This book series features volumes composed of selected contributions from workshops and conferences in all areas of current research in mathematics and statistics, including operation research and optimization. In addition to an overall evaluation of the interest, scientific quality, and timeliness of each proposal at the hands of the publisher, individual contributions are all refereed to the high quality standards of leading journals in the field. Thus, this series provides the research community with well-edited, authoritative reports on developments in the most exciting areas of mathematical and statistical research today.

More information about this series at http://www.springer.com/series/10533

Peter Friz · Wolfgang König ·
Chiranjib Mukherjee · Stefano Olla
Editors

Probability and Analysis in Interacting Physical Systems

In Honor of S.R.S. Varadhan,
Berlin, August, 2016

 Springer

Editors
Peter Friz
Institute of Mathematics
Technische Universität Berlin
Berlin, Germany

Wolfgang König
Weierstrass Institute for Applied
Analysis and Stochastics
Berlin, Germany

Chiranjib Mukherjee
Institute for Mathematical Stochastics
University of Münster
Münster, Germany

Stefano Olla ⓘ
CEREMADE, UMR CNRS 7534
Université Paris Dauphine - PSL
Paris, France

ISSN 2194-1009   ISSN 2194-1017   (electronic)
Springer Proceedings in Mathematics & Statistics
ISBN 978-3-030-15340-3   ISBN 978-3-030-15338-0   (eBook)
https://doi.org/10.1007/978-3-030-15338-0

Library of Congress Control Number: 2019934517

Mathematics Subject Classification (2010): 05C80, 60B10, 60B12, 60B20, 60F10, 60F17, 60H15, 60K35, 60K37, 60H05, 82A42, 82B21, 82B44, 82C22

This Springer imprint is published by the registered company Springer Nature Switzerland AG
The registered company address is: Gewerbestrasse 11, 6330 Cham, Switzerland

Preface

This festschrift marks the occasion of the 75th birthday of S. R. S. Varadhan, one of the most influential researchers in the field of probability for the last 50 years. This volume contains ten research articles authored by several of Varadhan's former PhD students and/or close collaborators. The topics of the papers are more or less closely linked with some of his deepest interests over the decades: large deviations, Markov processes, interacting particle systems, motions in random media and homogenisation, reaction–diffusion equations and directed last-passage percolation. This diverse span of subjects illustrates the wide range of Varadhan's research.

S. R. S. Varadhan was a pioneer in many important developments in the understanding of the asymptotic behaviour of random processes, such as Brownian motions, or interacting particle systems, in random and in non-random environments. Many of his contributions are fundamental and have attracted entire generations of researchers, giving them material to work on for decades. Consider, for example, the idea of martingale characterisations of Markov processes that he developed with Daniel Stroock around 1970, the series of papers from the early 1970s with Monroe Donsker that laid the foundations of a large-deviation analysis of exponential functionals of Markov chains, the work on homogenisation with George Papanicolaou and the joint paper with Claude that introduced a method for proving central limit theorems for motions in random media (which is still the basis of most research in this area), his work on hydrodynamic limits of large interacting particle systems, and many more groundbreaking developments that he initiated with his co-authors.

His worldwide esteem has been enormous over the decades, and he has been the recipient of a number of prestigious prizes, of which we mention here only the Abel Prize (2007) for his *fundamental contributions to the theory of probabilities, in particular the creation of a unified theory of large deviations.*

S. R. S. Varadhan has inspired and motivated many young gifted students, and he has attracted a great number of strong early-career PhD researchers. This is reflected in the total of 37 successfully finished PhD projects over the decades. Furthermore, a large percentage of these students have developed respectable academic careers themselves, spreading Varadhan's ideas and favourite subjects to their own PhD students and colleagues.

Let us give a short survey of the scientific content of this festschrift.

Chatterjee introduces and surveys the probabilistic theory and open questions of the Euclidean version of the Yang–Mills theory and corresponding lattice gauge theories, in particular their continuum limits. He formulates in probabilistic terms the questions that theoretical physicists ask and gives a brief survey of the probabilistic literature.

Chevyrev, Friz, Korepanov, Melbourne and Zhang review the origins of the convergence of fast-slow deterministic systems to stochastic differential equations and revisit and improve a proof of Kelly and Melbourne using recent progress on p-variation and càdlàg rough-path analysis.

Kosioris, Loulakis and Souganidis study the shallow lake problem from economics and identify the welfare function as a viscosity solution of the associated Bellman equation. They then derive several properties of the solution, including its asymptotic behaviour at infinity, and conclude with a numerical scheme.

Joseph, Rassoul-Agha and Seppäläinen study the motion of independent particles in a certain kind of dynamical random environment in the d-dimensional discrete space, where the distribution of the environment has a product structure. They characterize the class of spatially ergodic invariant measures, study their correlation structure and draw conclusions about the convergence of the particle distribution to equilibrium in dimensions one and two.

Reaction–diffusion equations, more precisely, the heat equation in random (here Weibull distributed) potential, are considered by Ben Arous, Molchanov and Ramirez; they concentrate on approximation in boxes that are so large that the mean over them is a kind of interpolation between the moment asymptotics (ergodic theorem) and the quenched (i.e., almost-sure) asymptotics; stable limiting distributions are obtained.

Bröker and Mukherjee consider a mollified version of the stochastic heat equation (random Brownian polymer in time-space white noise) in dimension ≥ 3 and prove the convergence of the rescaled polymer in distribution to a Gaussian distribution.

Bisi and Zygouras consider point-to-line and point-to-half-line directed last-passage percolation with exponentially distributed waiting times. They derive Sasamoto's Fredholm determinant formula for the Tracy–Widom GOE distribution and the one-point marginal distribution of the $\text{Airy}_{2\to1}$ process, which was originally derived by Borodin, Ferrari and Sasamoto.

Landim, Chang and Lee prove an energy estimate for the polar empirical measure of the two-dimensional symmetric simple exclusion process. They deduce from this estimate, and from their earlier results, large deviations principles for the polar empirical measure and for the occupation time of the origin.

Sethuraman and Venkataramani consider a time-dependent growing random-graph model of preferential-attachment type, where new nodes are attached to existing ones according to some superlinear function of their degrees. From earlier work, the emergence of condensation is known. Here, they establish laws of large number and fluctuation results for the number of nodes at a given time with a given degree and recover the emergence of the condensate in greater detail.

Pinsky considers the distribution of a certain random polynomial of order N of the prime numbers, whose powers are independent geometric random variables with parameter equal to one minus the reciprocal of the basis (the prime). He shows that the logarithm of this random quantity, when divided by log N, converges in distribution to the Buchstab distribution. As a corollary, Merten's theorem from multiplicative number theory is recovered.

This festschrift grew out of a birthday workshop on the occasion of Varadhan's 75th birthday, which took place at TU Berlin on 15–19 August 2016. It was a great honour and pleasure for us to organize this event.

We wish you, dear Raghu, many further years of much joy of doing mathematical research and a most stable health to carry through all your plans that you have!

Berlin, Münster and Rome   Peter Friz
December 2018   Wolfgang König
Chiranjib Mukherjee
Stefano Olla

List of Participants of the workshop at TU Berlin, 15–19 August 2016

Inez Armendariz, Universidad de Buenos Aires, Buenos Aires, Argentina
Sigurd Assing, University of Warwick, Coventry, UK
Peter Bank, TU Berlin, Berlin, Germany
Christian Bayer, WIAS Berlin, Berlin, Germany
Gérard Ben Arous, Courant Institute, New York, New York, USA
Erwin Bolthausen, University of Zurich, Zurich, Switzerland
Yvain Bruned, University of Warwick, Coventry, UK
Catriona Byrne, Springer, Heidelberg, Germany
Jiawei Cheng, University of Oxford, Oxford, UK
Ilya Chevyrev, TU Berlin, Berlin, Germany
Alessandra Cipriani, WIAS Berlin, Berlin, Germany
David Criens, TU München München, Germany
Apostolos Damialis, De Gruyter, Berlin, Germany
Jean Dominique, Deuschel, TU Berlin, Berlin, Germany
Joscha Diehl, Max-Planck-Institut, Leipzig, Germany
Alexander Drewitz, Universität zu Köln, Köln, Germany
Leif Döring, Universität Mannheim, Mannheim, Germany
Peter Eichelsbacher, Universität Bochum, Bochum, Germany
Shima Elesaely, TU Berlin, BMS, Berlin, Germany
Benjamin Fehrman, Max-Planck-Institut, Leipzig, Germany
Franziska Flegel, WIAS Berlin, Berlin, Germany
Klaus Fleischmann, WIAS Berlin, Berlin, Germany
Peter Friz, WIAS Berlin/TU Berlin, Berlin, Germany
Nina Gantert, TU München, München, Germany
Adrian Gonzalez Casanova, WIAS Berlin, Berlin, Germany
Jürgen Gärtner, TU Berlin, Berlin, Germany
Onur Gün, WIAS Berlin, Berlin, Germany
Matthias Hammer, TU Berlin, Berlin, Germany
Martin Hairer, University of Warwick, Warwick, UK
Christian Hirsch, WIAS Berlin, Berlin, Germany
Antoine Hocquet, TU Berlin, Berlin, Germany

Antoine Jacquier, Imperial College London, London, UK
Benedikt Jahnel, WIAS Berlin, Berlin, Germany
Marvin Kettner, TU Darmstadt, Darmstadt, Germany
Frederik Klement, Universität Mainz, Mainz, Germany
Tom Klose, HU Berlin, Berlin, Germany
Elena Kosygina, Baruch College/CUNY, Graduate Center, New York, New York, USA
Richard Kraaij, TU Delft, Delft, The Netherlands
Christof Külske, Universität Bochum, Bochum, Germany
Wolfgang König, WIAS Berlin/TU Berlin, Berlin, Germany
Claudio Landim, IMPA, Rio de Janeiro, Brasilien
Michail Loulakis, National Technical University of Athens, Athens, Greece
Hoang Duc Luu, Max-Planck-Institut, Leipzig, Germany
Terry Lyons, Oxford University, Oxford, UK
Matthias Löwe, Universität Münster, Münster, Germany
Vlad Margarint, University of Oxford, Oxford, UK
Mario Maurelli, WIAS Berlin/TU Berlin, Berlin, Germany
Georg Menz, University of California, Los Angeles, California, USA
Markus Mittnenzweig, WIAS Berlin, Berlin, Germany
Marvin Mueller, ETH Zurich, Zurich, Switzerland
Chiranjib Mukherjee, Courant Institute, New York, New York, USA/WIAS Berlin, Berlin, Germany
Christian Mönch, TU Darmstadt, Darmstadt, Germany
Jan Nagel, TU München, München, Germany
Sina Nerjad, University of Oxford, Oxford, UK
Stefano Olla, Université Paris Dauphine, Paris, France
Nicolas Perkowski, HU Berlin, Berlin, Germany
Ross Pinsky, Technion-Israel Institute of Technology, Haifa, Israel
David Prömel, ETH Zurich, Zurich, Switzerland
Alejando Ramírez, Pontificia Universidad Catolica de Chile, Santiago, Chile
Jose Ramírez, Universidad de Costa Rica, San Pedro, Costa Rica
Firas Rassoul-Agha, University of Utah, Salt Lake City, Utah, USA
Martin Redmann, WIAS Berlin, Berlin, Germany
Max von Renesse, Universität Leipzig, Leipzig, Germany
Michiel Renger, WIAS Berlin, Berlin, Germany
Fraydoun Rezakhanlou, University of California, Berkeley, California, USA
Sebastian Riedel, TU Berlin, Berlin, Germany
Michael Röckner, Universität Bielefeld, Bielefeld, Germany
Renato Soares dos Santos, WIAS Berlin, Berlin, Germany
Alexandros Saplaouras, TU Berlin, Berlin, Germany
Michael Scheutzow, TU Berlin, Berlin, Germany
Massimo Secci, TU Berlin, Berlin, Germany
Vitalii Senin, TU Berlin, Berlin, Germany
Insuk Seo, University of California, Berkeley, California, USA
Sunder Sethuraman, University of Arizona, Tucson, Arizona, USA

Atul Shekhar, Indian Statistical Institute, Bangalore, India
Sergio Simonella, TU München, München, Germany
Martin Slowik, TU Berlin, Berlin, Germany
Herbert Spohn, TU München, München, Germany
Claudia Strauch, Universität Heidelberg, Heidelberg, Germany
Tat Dat Tran, Max-Planck-Institut, Leipzig, Germany
Srinivasa Varadhan, Courant Institute, New York, USA
Isabell Vorkastner, TU Berlin, Berlin, Germany
Moritz Voss, TU Berlin, Berlin, Germany
Florian Völlering, Universität Münster, Münster, Germany
Stefan Walter, TU Darmstadt, Darmstadt, Germany
Yilin Wang, ETH Zurich, Zurich, Switzerland
Martin Weidner, Imperial College London, London, UK
Heinrich von Weizsäcker, TU Kaiserslautern, Kaiserslautern, Germany
Bo Xia, Université Paris Sud, Paris, France
Danyu Yang, University of Oxford, Oxford, UK
Horng-Tzer Yau, Harvard University, Cambridge, UK
Atilla Yilmaz, Koc University, Istanbul, Turkey
Ofer Zeitouni, Weizmann Institute, Rehovot, Israel
Deng Zhang, Universität Bielefeld, Bielefeld, Germany
Nikos Zygouras, University of Warwick, Coventry, UK

Contents

Yang–Mills for Probabilists

Sourav Chatterjee

Dedicated to friend and teacher Raghu Varadhan on the occasion of his 75th birthday.

Abstract The rigorous construction of quantum Yang–Mills theories, especially in dimension four, is one of the central open problems of mathematical physics. Construction of Euclidean Yang–Mills theories is the first step towards this goal. This article presents a formulation of some of the core aspects this problem as problems in probability theory. The presentation begins with an introduction to the basic setup of Euclidean Yang–Mills theories and lattice gauge theories. This is followed by a discussion of what is meant by a continuum limit of lattice gauge theories from the point of view of theoretical physicists. Some of the main issues are then posed as problems in probability. The article ends with a brief review of the mathematical literature.

Keywords Lattice gauge theory · Yang–Mills theory · Wilson loop variable · Continuum limit · Area law · Quark confinement

2010 Mathematics Subject Classification. 70S15 · 81T13 · 81T25 · 82B20

1 Introduction

Four-dimensional quantum Yang–Mills theories are the building blocks of the Standard Model of quantum mechanics. The Standard Model encapsulates the sum total of all that is currently known about the basic particles of nature (see [48] for more

Research partially supported by NSF grant DMS-1608249.

S. Chatterjee (✉)
Department of Statistics, Stanford University Sequoia Hall, 390 Serra Mall, Stanford, CA 94305, USA
e-mail: sourave@stanford.edu

© Springer Nature Switzerland AG 2019
P. Friz et al. (eds.), *Probability and Analysis in Interacting Physical Systems*, Springer Proceedings in Mathematics & Statistics 283, https://doi.org/10.1007/978-3-030-15338-0_1

1

details about the physics). Unfortunately, in spite of their incredible importance, quantum Yang–Mills theories have no rigorous mathematical foundation. In fact, the mathematical foundation is so shaky that we do not even know for certain the right spaces on which these theories should be defined, or the right observables to look at.

Quantum Yang–Mills theories are defined in Minkowski spacetime. Euclidean Yang–Mills theories are 'Wick-rotated' quantum Yang–Mills theories that are defined in Euclidean spacetime. These should, at least in principle, be easier to understand and analyze than their Minkowski counterparts. Although these theories are not rigorously defined either, they are formally probability measures on spaces of connections on certain principal bundles (more details to follow). Moreover, they have lattice analogs, known as lattice gauge theories or lattice Yang–Mills theories, that are rigorously defined probabilistic models. Therefore, the construction of Euclidean Yang–Mills theories, when viewed as a problem of taking scaling limits of lattice gauge theories, reveals itself as a problem in probability theory.

The problem of rigorously constructing Euclidean Yang–Mills theories, and then extending the definition to Minkowski spacetime via Wick rotation, is the problem of Yang–Mills existence, posed as a 'millennium prize problem' by the Clay Institute [54].

Quantum Yang–Mills theories are certain kinds of quantum field theories. A standard approach to the rigorous construction of quantum field theories is via the program of constructive quantum field theory, as outlined in the classic monograph of [46]. One of the important objectives of constructive quantum field theory is to define Euclidean quantum field theories as probability measures on appropriate spaces of generalized functions, and then show that these probability measures satisfy certain axioms (the Wightman axioms, or the Osterwalder–Schrader axioms), which would then imply that the theory can be 'quantized' to obtain the desired quantum field theories in Minkowski spacetime. However, as noted in the monograph of [73], there is a fundamental problem in following this path for Yang–Mills theories: the key observables in these theories do not take values at points, but at curves. Thus, it is not clear how to describe these theories as probability measures on spaces of generalized functions on manifolds—they should, rather, be probability measures on spaces of generalized functions on spaces of curves. These considerations led [73] to pose the problem as a problem of constructing an appropriate random function on a suitable space of closed curves. This is the route that we will adopt in this section, partly because it gives the most straightforward way of stating the problem.

That said, however, we still need a bit of preparation. After presenting quick introductions to Euclidean Yang–Mills theories, lattice gauge theories and Wilson loops in the first three sections, the physicist's definition of a continuum limit of lattice gauge theories is discussed and well-defined mathematical problems are formulated. The last section contains a brief review of the mathematical literature.

2   Euclidean Yang–Mills Theories

An Euclidean Yang–Mills theory involves a dimension n, and a 'gauge group' G, usually a compact Lie group. For simplicity, let us assume that G is a closed subgroup of the group of unitary matrices of some order N. Examples are $G = U(1)$ for quantum electrodynamics, $G = SU(3)$ for quantum chromodynamics, and $G = SU(3) \times SU(2) \times U(1)$ for the Standard Model, with dimension $n = 4$ in each case.

The Lie algebra $\mathfrak{g}$ of the Lie group G is a subspace of the vector space of all $N \times N$ skew-Hermitian matrices. A G connection form on $\mathbb{R}^n$ is a smooth map from $\mathbb{R}^n$ into $\mathfrak{g}^n$. If A is a G connection form, its value $A(x)$ at a point x is an n-tuple $(A_1(x), \ldots, A_n(x))$ of skew-Hermitian matrices. In the language of differential forms,

$$A = \sum_{j=1}^{n} A_j dx_j.$$

The curvature form F of a connection form A is the $\mathfrak{g}$-valued 2-form

$$F = dA + A \wedge A.$$

In coordinates, this means that at each point x, $F(x)$ is an $n \times n$ array of skew-Hermitian matrices of order N, whose (j, k)th entry is the matrix

$$F_{jk}(x) = \frac{\partial A_k}{\partial x_j} - \frac{\partial A_j}{\partial x_k} + [A_j(x), A_k(x)],$$

where $[B, C] = BC - CB$ denotes the commutator of B and C.

Let $\mathcal{A}$ be the space of all G connection forms on $\mathbb{R}^n$. The Yang–Mills action on this space is the function

$$S_{\text{YM}}(A) := -\int_{\mathbb{R}^n} \text{Tr}(F \wedge *F),$$

where F is the curvature form of A and $*$ denotes the Hodge star operator, assuming that this integral is finite. Explicitly, this is

$$S_{\text{YM}}(A) = -\int_{\mathbb{R}^n} \sum_{j,k=1}^{n} \text{Tr}(F_{jk}(x)^2) \, dx. \qquad (2.1)$$

The Euclidean Yang–Mills theory with gauge group G on $\mathbb{R}^n$ is formally described as the probability measure

$$d\mu(A) = \frac{1}{Z} \exp\left(-\frac{1}{4g^2} S_{\text{YM}}(A)\right) dA,$$

where A belongs to the space $\mathcal{A}$ of all $U(N)$ connection forms, S_{YM} is the Yang–Mills functional defined above,

$$dA = \prod_{j=1}^{n} \prod_{x \in \mathbb{R}^n} d(A_j(x))$$

is infinite-dimensional Lebesgue measure on $\mathcal{A}$, g is a parameter called the coupling strength, and Z is the normalizing constant that makes this a probability measure.

The above description of Euclidean Yang–Mills theory with gauge group G is not directly mathematically meaningful because any simple-minded way of defining infinite-dimensional Lebesgue measure on $\mathcal{A}$ would yield $Z = \infty$ and make it impossible to define μ as a probability measure. While it has been possible to circumvent this problem in roundabout ways and give rigorous meanings to similar descriptions of Brownian motion and various quantum field theories in dimensions two and three, Euclidean Yang–Mills theories have so far remained largely intractable.

3 Lattice Gauge Theories

In 1974, [78] proposed a discretization of Euclidean Yang–Mills theories. These are now known as lattice gauge theories or lattice Yang–Mills theories. Let G be as in the previous section. The lattice gauge theory with gauge group G on a finite set $\Lambda \subseteq \mathbb{Z}^n$ is defined as follows. Suppose that for any two adjacent vertices $x, y \in \Lambda$, we have a unitary matrix $U(x, y) \in G$, with the constraint that $U(y, x) = U(x, y)^{-1}$. Let us call any such assignment of matrices to edges a 'configuration'. Let $G(\Lambda)$ denote the set of all configurations. A square bounded by four edges is called a plaquette. Let $P(\Lambda)$ denote the set of all plaquettes in Λ. For a plaquette $p \in P(\Lambda)$ with vertices x_1, x_2, x_3, x_4 in anti-clockwise order, and a configuration $U \in G(\Lambda)$, define

$$U_p := U(x_1, x_2)U(x_2, x_3)U(x_3, x_4)U(x_4, x_1). \tag{3.1}$$

The Wilson action of U is defined as

$$S_\Lambda(U) := \sum_{p \in P(\Lambda)} \mathrm{Re}(\mathrm{Tr}(I - U_p)), \tag{3.2}$$

where I is the identity matrix of order N. Let σ_Λ be the product Haar measure on $G(\Lambda)$. Given $\beta > 0$, let $\mu_{\Lambda,\beta}$ be the probability measure on $G(\Lambda)$ defined as

$$d\mu_{\Lambda,\beta}(U) := \frac{1}{Z} e^{-\beta S_\Lambda(U)} d\sigma_\Lambda(U),$$

where Z is the normalizing constant. This probability measure is called the lattice gauge theory on Λ for the gauge group G, with inverse coupling strength β.

Often, it is convenient to work with an infinite volume limit of the theory, that is, a weak limit of the above probability measures as $\Lambda \uparrow \mathbb{Z}^n$. The infinite volume limit may or may not be unique. Indeed, the uniqueness (or non-uniqueness) is in general unknown for lattice gauge theories in dimensions higher than two when β is large.

Lattice gauge theories in several variations are a huge computational engine for numerical approximations for the Standard Model. They are used to make very accurate predictions of quantities like the masses of hadrons [16]. For readers who want to learn more about the physics of lattice gauge theories, two textbook references are [45, 69]. There are also two rather extensive scholarpedia articles on lattice gauge theories and lattice quantum field theory, respectively.

The passage from a lattice gauge theory to an Euclidean Yang–Mills theory is heuristically justified as follows. First, discretize the space $\mathbb{R}^n$ as the scaled lattice $\epsilon \mathbb{Z}^n$ for some small ϵ. Next, take a G connection form

$$A = \sum_{j=1}^n A_j dx_j.$$

Let $e_1, \ldots, e_n$ denote the standard basis vectors of $\mathbb{R}^n$. For a directed edge $(x, x + \epsilon e_j)$ of $\epsilon \mathbb{Z}^n$, define

$$U(x, x + \epsilon e_j) := e^{\epsilon A_j(x)},$$

and let $U(x + \epsilon e_j, x) := U(x, x + \epsilon e_j)^{-1}$. This defines a configuration of unitary matrices assigned to directed edges of $\epsilon \mathbb{Z}^n$. For a plaquette p in $\epsilon \mathbb{Z}^n$, let U_p be defined as in (3.1). Then a formal calculation using the Baker–Campbell–Hausdorff formula for products of matrix exponentials shows that when ϵ is small,

$$\sum_p \mathrm{Re}(\mathrm{Tr}(I - U_p)) \approx \frac{\epsilon^{4-n}}{4} S_{\mathrm{YM}}(A),$$

where S_{YM} is the Yang–Mills action defined in (2.1). The calculation goes as follows. Take any $x \in \epsilon \mathbb{Z}^n$ and any $1 \le j < k \le n$, and let

$$x_1 = x, \quad x_2 = x + \epsilon e_j, \quad x_3 = x + \epsilon e_j + \epsilon e_k, \quad x_4 = x + \epsilon e_k.$$

Let p be the plaquette formed by the vertices x_1, x_2, x_3, x_4. Let U_p be defined as in (3.1). Then

$$U_p = e^{\epsilon A_j(x_1)} e^{\epsilon A_k(x_2)} e^{-\epsilon A_j(x_4)} e^{-\epsilon A_k(x_1)}.$$

Recall the Baker–Campbell–Hausdorff formula for products of matrix exponentials:

$$e^B e^C = \exp\left(B + C + \frac{1}{2}[B, C] + \text{higher commutators} \right).$$

Iterating this gives, for any m and any $B_1, \ldots, B_m$,

$$e^{B_1} \cdots e^{B_m} = \exp\left(\sum_{a=1}^{m} B_a + \frac{1}{2} \sum_{1 \le a < b \le m} [B_a, B_b] + \text{higher commutators} \right).$$

Recall that the eigenvalues of a skew-Hermitian matrix are all purely imaginary, and that the commutator of two skew-Hermitian matrices is skew-Hermitian. Consequently, the term within the exponential on the right side of the above display is skew-Hermitian and therefore has a purely imaginary trace. This implies that if N is the order of the matrices, if the entries of $B_1, \ldots, B_m$ are of order ϵ and if the entries of $B_1 + \cdots + B_m$ are of order ϵ^2, then

$$\mathrm{Re}(\mathrm{Tr}(I - e^{B_1} \cdots e^{B_m})) = -\frac{1}{2} \mathrm{Tr}\left[\left(\sum_{a=1}^{m} B_a + \frac{1}{2} \sum_{1 \le a < b \le m} [B_a, B_b] \right)^2 \right]$$
$$+ O(\epsilon^5),$$

where the real part of the trace was replaced by the trace on the right because the square of a skew-Hermitian matrix has real eigenvalues. Writing

$$A_k(x_2) = A_k(x + \epsilon e_j) = A_k(x) + \epsilon \frac{\partial A_k}{\partial x_j} + O(\epsilon^2)$$

and using a similar Taylor expansion for $A_j(x_4)$, we get

$$A_j(x_1) + A_k(x_2) - A_j(x_4) - A_k(x_1) = \epsilon \left(\frac{\partial A_k}{\partial x_j} - \frac{\partial A_j}{\partial x_k} \right) + O(\epsilon^2).$$

Combining the above observations gives

$$\mathrm{Re}(\mathrm{Tr}(I - U_p)) = -\frac{1}{2} \epsilon^4 \mathrm{Tr}\left[\left(\frac{\partial A_k}{\partial x_j} - \frac{\partial A_j}{\partial x_k} + [A_j(x), A_k(x)] \right)^2 \right] + O(\epsilon^5)$$
$$= -\frac{1}{2} \epsilon^4 \mathrm{Tr}(F_{jk}(x)^2) + O(\epsilon^5).$$

This gives the formal approximation

$$S(U) = \sum_p \mathrm{Re}(\mathrm{Tr}(I - U_p))$$
$$\approx -\frac{1}{4} \sum_{x \in \epsilon \mathbb{Z}^n} \sum_{j,k=1}^{n} \epsilon^4 \mathrm{Tr}(F_{jk}(x)^2)$$
$$\approx -\frac{\epsilon^{4-n}}{4} \int_{\mathbb{R}^n} \sum_{j,k=1}^{n} \mathrm{Tr}(F_{jk}(x)^2) \, dx = \frac{\epsilon^{4-n}}{4} S_{\mathrm{YM}}(A).$$

The above heuristic was used by Wilson to justify the approximation of Euclidean Yang–Mills theory by lattice gauge theory, scaling the inverse coupling strength β like ϵ^{4-n} as the lattice spacing $\epsilon \to 0$. The most important dimension is $n = 4$, because spacetime is four-dimensional. In the above formulation, β does not scale with ϵ at all when $n = 4$. Currently, however, the general belief in the physics community is that β should scale like $\log(1/\epsilon)$ in dimension four, although there are doubts about this belief and the question remains an open mathematical problem.

The problem of constructing continuum limits of lattice gauge theories, from the point of view of theoretical physicists, is discussed in greater detail in the following sections. Most of this is 'common knowledge' in the theoretical physics community, but not formalized in the sense of rigorous mathematics. I do not have references, but I have found [59, 60, 73] helpful. In particular, [73] proposes a formulation of continuum limits in terms of Wilson loop expectations, from which I borrow.

4 Wilson Loop Variables and Quark Confinement

Any physical theory should have observables of interest. For Yang–Mills theories, the most important observables are Wilson loop variables. These are defined as follows. Suppose that we have an Euclidean Yang–Mills theory on $\mathbb{R}^n$ with gauge group G, as defined in Sect. 2. Given a piecewise smooth closed path γ in $\mathbb{R}^n$ and a G connection A, the Wilson loop variable for γ is defined as

$$W_\gamma := \mathrm{Tr}\left(\mathcal{P}\exp\left(\int_\gamma \sum_{j=1}^n A_j dx_j\right)\right),$$

where $\mathcal{P}$ is the path-ordering operator. In differential geometric terminology, the term inside the trace in the above display is the holonomy of A along the closed path γ. Alternatively, it is the parallel transport of the identity matrix along γ by the connection A. If the reader is unfamiliar with these concepts, there is nothing to worry. A simple definition of Wilson loop variables for lattice gauge theories is given below.

The physical importance of Wilson loop variables stems in part from their connection with the static quark potential. It was argued by [78] that the potential between a static quark and antiquark separated by distance R is given by the formula

$$V(R) = -\lim_{T\to\infty} \frac{1}{T} \log\langle W_{\gamma_{R,T}}\rangle,$$

where $\gamma_{R,T}$ is the boundary of a rectangle of length T and breadth R, and $\langle \cdot \rangle$ denotes expectation with respect to a suitable Yang–Mills theory. If $V(R)$ grows to infinity as $R \to \infty$, the quark-antiquark pair cannot separate beyond a fixed distance. This is the phenomenon of quark confinement, observed in experiments but currently lacking a satisfactory theoretical understanding (much less proof) due to the uncertainty about the existence of a continuum limit (although extensive numerical work in the lattice community points to a positive answer). In fact, it is believed that $V(R)$ grows like a multiple of R for non-Abelian Yang–Mills theories in dimension four. This is known as Wilson's area law. If the area law holds, then the quantity

$$\lim_{R \to \infty} \frac{V(R)}{R}$$

has physical significance. It is called the 'string tension' of the continuum theory, and represents the energy density per unit length in the theory.

For lattice gauge theories, the definition of a Wilson loop variable is very simple. Suppose that we have a lattice gauge theory on $\Lambda \subseteq \mathbb{Z}^n$ with gauge group G, as in Sect. 3. A loop in $\mathbb{Z}^n$ is simply a directed path in the lattice which ends where it started. Given a loop γ with directed edges $e_1, \ldots, e_m$, the Wilson loop variable W_γ is defined as

$$W_\gamma := \mathrm{Tr}(U(e_1)U(e_2) \cdots U(e_m)).$$

The rationale for this definition is as follows. Let A be a smooth G connection on $\mathbb{R}^n$. Take some small ϵ and define a configuration of group elements assigned to directed edges of $\epsilon \mathbb{Z}^n$ using the connection A, as in Sect. 3. Let γ be a smooth closed path in $\mathbb{R}^n$ and let γ_ϵ be a loop in $\epsilon \mathbb{Z}^n$ that approximates this path. Then, as $\epsilon \to 0$, the discrete Wilson loop variable W_{γ_ϵ} approaches the continuous Wilson loop variable W_γ.

Wilson's original motivation for investigating lattice gauge theories was to gain a theoretical proof of quark confinement. Since the expected values of Wilson loop variables are mathematically well-defined for lattice gauge theories, one can hope to give a rigorous proof of the area law in the discrete setting. In fact, the following upper bound suffices.

Problem 4.1 (*Area law*) Take any compact non-Abelian Lie group $G \subseteq U(N)$ for some $N \geq 2$ and consider any infinite volume limit of four-dimensional lattice gauge theory with gauge group G at inverse coupling strength β. Let $\gamma_{R,T}$ be a rectangular loop of breadth R and length T in the lattice. Prove that

$$|\langle W_{\gamma_{R,T}} \rangle| \leq C(\beta) e^{-c(\beta)RT},$$

where $C(\beta)$ and $c(\beta)$ are positive constants that depend only on the inverse coupling strength β and the gauge group.

Soon after the appearance of Wilson's paper, physicists realized that the area law holds for any lattice gauge theory at sufficiently small β. A rigorous proof was given by [71]. However, this also implied that the area law at small β cannot be evidence for quark confinement, because there are certain Yang–Mills theories that should not be confining quarks. An example is four-dimensional $U(1)$ Yang–Mills theory, which is the theory of electromagnetism. It is a fact of nature that there are no confined quarks in electromagnetism.

This apparent paradox was resolved by [53], who showed that four-dimensional $U(1)$ lattice gauge theory fails to satisfy the area law at large β. A fully rigorous proof of Guth's theorem was given by [44]. This result suggested that to prove quark confinement in a Yang–Mills theory, the area law has to be proved for the corresponding lattice theory at arbitrarily large β. For non-Abelian theories, this is currently known only in dimension two, where it is not very hard to prove. For $U(1)$ theory, there is a remarkable result of [47], who proved the area law at arbitrary β in dimension three. The solution of Problem 4.1 for four-dimensional non-Abelian lattice gauge theories at large β remains elusive.

5 The Problem of Defining the Continuum Limit

Take a lattice gauge theory as in Sect. 3, and consider an infinite volume limit of this theory on $\mathbb{Z}^n$ obtained by taking a weak limit of the theories on finite cubes. Let γ_1 and γ_2 be two Wilson loops of fixed length, such as two plaquettes. The correlation between W_{γ_1} and W_{γ_2} is defined as the quantity

$$\langle W_{\gamma_1} W_{\gamma_2} \rangle - \langle W_{\gamma_1} \rangle \langle W_{\gamma_2} \rangle. \tag{5.1}$$

Let $d(\gamma_1, \gamma_2)$ denote the Euclidean distance between the two loops. If the logarithm of the above correlation behaves like $-d(\gamma_1, \gamma_2)/\xi$ for some $\xi > 0$ as $d(\gamma_1, \gamma_2) \to \infty$, then the number ξ is called the correlation length of the model. We have to take the logarithm because there may be polynomial correction terms to the exponential decay in the actual correlation [35, 72].

Physicists say that the model has a continuum limit if there is a critical point $\beta_c \in [0, \infty]$ such that as $\beta \to \beta_c$, the correlation length tends to infinity. The reason for saying this is that if such a critical point exists, then it is possible to define the model on the scaled lattice $\epsilon \mathbb{Z}^n$ instead of $\mathbb{Z}^n$, and send $\epsilon \to 0$ in an appropriate manner as $\beta \to \beta_c$ such that the correlation length tends to a finite nonzero limit. In other words, it is possible to define correlations in the continuum.

It is believed that in dimension four (which, as stated earlier, is the dimension of greatest physical significance since spacetime is four-dimensional), many of the non-Abelian lattice models of interest have $\beta_c = \infty$. That is, one needs to take $\beta \to \infty$ while sending the lattice spacing ϵ to zero, to obtain a nontrivial correlation function in the limit.

The following is a possible formulation of the above discussion as a concrete mathematical problem.

Problem 5.1 (*Mass gap*) Take any compact non-Abelian Lie group $G \subseteq U(N)$ for some $N \geq 2$ and consider any infinite volume limit of four-dimensional lattice gauge theory with gauge group G at inverse coupling strength β. For each $x \in \mathbb{R}^4$, let p_x

be the plaquette that is closest to x (breaking ties by some arbitrary rule). Let $f_\beta(x)$ denote the correlation between W_{p_0} and W_{p_x}, as defined in (5.1). Show that for any $\beta > 0$, there exists some $\xi(\beta) \in (0, \infty)$ such that

$$\lim_{|x| \to \infty} \frac{\log f_\beta(x)}{|x|} = -\frac{1}{\xi(\beta)}.$$

Moreover, prove that

$$\lim_{\beta \to \infty} \xi(\beta) = \infty.$$

The correlation length ξ has a physical meaning. Any lattice gauge theory contains information of an associated class of elementary particles called 'glueballs' or 'gluonballs'. The existence of glueballs is considered to be one of the most important predictions of the Standard Model, but has not yet been experimentally verified. The number ξ represents the reciprocal of the mass of the lightest glueball in the theory.

One approach to the construction of continuum limits of lattice gauge theories is via Wilson loops. This is the approach that is advocated by [73, Chap. 8], from which we draw inspiration. While [73] gives a detailed description of the desired properties of the continuum limit that would presumably facilitate the quantization of the theory, we will restrict attention to the most basic question that needs to be solved before making any further progress.

Let β_c be as above. The problem of constructing a continuum limit at this critical point in terms of Wilson loop expectations can be stated as follows. As $\beta \to \beta_c$, one would like to show that the lattice spacing ϵ can be taken to 0 in such a way that if γ_ϵ is any sequence of lattice loops converging to a loop γ in $\mathbb{R}^n$, then $\langle W_{\gamma_\epsilon} \rangle$ converges to a nontrivial limit after some appropriate renormalization.

Since ∞ is believed to be a critical point of compact non-Abelian lattice gauge theories in dimension four, one way to formulate the above question for the simple case of rectangular loops is the following.

Problem 5.2 (*Continuum limit*) Take any compact non-Abelian Lie group $G \subseteq U(N)$ for some $N \geq 2$ and consider any infinite volume limit of four-dimensional lattice gauge theory with gauge group G at inverse coupling strength β. Let $\gamma_{R,T}$ denote a rectangular loop of length T and breadth R. Prove that as $\beta \to \infty$, there are sequences $\epsilon = \epsilon(\beta) \to 0$ and $c = c(\beta) \to \infty$, and a nonzero constant d, such that for any R and T,

$$\log\langle W_{\gamma_{R/\epsilon,T/\epsilon}} \rangle = -c(R + T) - dRT + o(1). \tag{5.2}$$

Note that $R + T$ is the limiting perimeter of the rectangular loops after scaling by ϵ, and RT is the limiting area. Since $c \to \infty$, the above conjecture says that Wilson's area law holds in the continuum only after subtracting off ('renormalizing away') the first term when taking the limit. That is, the logarithm of the Wilson loop expectation $\langle W_{\gamma_{R,T}} \rangle$ in the continuum should be defined as

$$\log\langle W_{\gamma_{R,T}}\rangle := \lim_{\beta\to\infty}(\log\langle W_{\gamma_{R/\epsilon,T/\epsilon}}\rangle + c(R+T)).$$

With this definition, the string tension of the continuum theory (as defined in Sect. 4) is the number d in (5.2).

Since none of the above has been proved, it is not clear to me whether the renormalization term $c(R+T)$ is indeed necessary. It seems entirely possible that the limit in (5.2) holds without the renormalization term.

The next section gives a brief summary of existing rigorous results on the problem of constructing continuum limits of lattice gauge theories in various dimensions.

6 Review of the Mathematical Literature

There is a long and quite successful development of two-dimensional Yang–Mills theories in the mathematical literature. The two-dimensional Higgs model, which is $U(1)$ Yang–Mills theory with an additional Higgs field, was constructed by [20–22] and further developed by [18]. Building on an idea of [19, 52] formulated a rigorous mathematical approach to performing calculations in two-dimensional Yang–Mills theories via stochastic calculus. Different ideas leading to the same goal were implemented by [30, 31, 58]. The papers of [30, 31] made precise the idea of using objects called lassos to define the continuum limit of Yang–Mills theories. Explicit formulas for Yang–Mills theories on compact surfaces were obtained by [42, 43, 79, 80]. All of these results were generalized and unified by [74–76] using a stochastic calculus approach.

Yet another approach was introduced by [61, 62], who constructed two-dimensional Yang–Mills theories as random holonomy fields. A random holonomy field is a stochastic process indexed by curves on a surface, subject to boundary conditions, and behaving under surgery as dictated by a Markov property. Lévy's framework allows parallel transport along more general curves than the ones considered previously, and makes interesting connections to topological quantum field theory. A relatively non-technical description of this body of work is given in [63].

Recently, [70] has established the mathematical validity of the perturbative approach to 2D Yang–Mills theory by comparing its predictions with rigorous results obtained by the approaches outlined above. Another important body of recent work consists of the papers of [32–34, 64], who establish the validity of the Makeenko–Migdal equations for 2D Yang–Mills theories [68] by a number of different approaches.

Euclidean Yang–Mills theories in dimensions three and four have proved to be more challenging to construct mathematically. At sufficiently strong coupling (that is, small β), a number of conjectures about lattice gauge theories in arbitrary dimensions—such as quark confinement and the existence of a positive self-adjoint transfer matrix—were rigorously proved by [71]. An expansion of partition functions of lattice gauge theories as asymptotic series in the dimension of the gauge group was proposed by [77], leading to a large body of work. The papers of [15, 26, 28, 55] contain some recent advances on 't Hooft type expansions and connections with gauge-string duality at strong coupling. Confinement and deconfinement in three- and four-dimensional lattice gauge theories at weak coupling were investigated by [17, 44, 47, 53].

None of the above techniques, however, help in constructing the continuum limit. The problems posed in Sect. 5 have not been mathematically tractable. An alternative route is the method of phase cell renormalization. In this approach, one starts with a lattice gauge theory on $\epsilon\mathbb{Z}^n$ for some small ϵ. Choosing an integer L, the theory is then 'renormalized' to yield an 'effective field theory' on the coarser lattice $L\epsilon\mathbb{Z}^n$, which is just another lattice gauge theory but with a more complicated action. A survey of the various ways of carrying out this renormalization step is given in [46, Chap. 22]. The process is iterated to produce effective field theories on $L^k\epsilon\mathbb{Z}^n$ for $k = 2, 3, \ldots$, until the lattice spacing $L^k\epsilon$ attains macroscopic size (for example, becomes greater than 1). Note that the macroscopic effective field theory obtained in this way is dependent on ϵ. The goal of phase cell renormalization is to show that the effective field theory at the final macroscopic scale converges to a limit as $\epsilon \to 0$. Usually, convergence is hard to prove, so one settles for subsequential convergence by a compactness argument. The existence of a convergent subsequence is known as ultraviolet stability, for the following reason. The approximation of an Euclidean Yang–Mills theory by a lattice gauge theory on a lattice with spacing ϵ is analogous to truncating a Fourier series at a finite frequency, which grows as $\epsilon \to 0$. In this sense, a lattice approximation is an ultraviolet cutoff (ultraviolet = high frequency), and the compactness of the effective field theories is ultraviolet stability, that is, stability with respect to the cutoff frequency.

A notable success story of phase cell renormalization is the work of [56, 57], who established the existence of the continuum limit of the three-dimensional Higgs model. The continuum limit of pure $U(1)$ Yang–Mills theory (that is, without the Higgs field) was established earlier by [49], but with a different notion of convergence. Gross's approach was later used by [29] to construct a continuum limit of 4D $U(1)$ lattice gauge theory.

Ultraviolet stability of three- and four-dimensional non-Abelian lattice gauge theories by phase cell renormalization, as outlined above, was famously established by [1–14] in a long series of papers spanning six years. A somewhat different approach, again using phase cell renormalization, was pursued by [36–41].

Phase cell renormalization is not the only approach to constructing Euclidean Yang–Mills theories. [67] formulated a program of directly constructing Yang–Mills theories in the continuum instead of using lattice theories. The main idea in [67] was to regularize the continuum theory by introducing a quadratic term in the Hamiltonian. The problem with this regularization is that it breaks gauge invariance. The problem is taken care of by showing that it is possible to remove the quadratic term and restore gauge invariance by taking a certain kind of limit of the regularized theories.

In spite of the remarkable achievements surveyed above, there is yet no construction of a continuum limit of a lattice gauge theory in any dimension higher than two where Wilson loop variables have been shown to have nontrivial behavior. The standard approach of regularizing Wilson loop variables by phase cell renormalization has not yielded definitive results. In a recent series of papers, [23–25, 50, 51] have proposed a method of regularizing connection forms in $\mathbb{R}^3$ by letting them flow for a small amount of time according to the Yang–Mills heat flow (see also the papers of [65, 66] for a similar idea). It will be interesting to see whether this new approach, in combination with some ideas developed in the paper [27], can lead to the construction of nontrivial three-dimensional Euclidean Yang–Mills theories with non-Abelian gauge groups.

Acknowledgements I thank Erik Bates, David Brydges, Persi Diaconis, Len Gross, Jafar Jafarov, Erhard Seiler, Scott Sheffield, Steve Shenker, Tom Spencer and Edward Witten for many valuable conversations and comments.

References

1. Bałaban, T.: Regularity and decay of lattice Green's functions. Comm. Math. Phys. **89**(4), 571–597 (1983)
2. Bałaban, T.: Renormalization Group Methods in Non-abelian Gauge Theories, p. B134. Harvard preprint, HUTMP (1984a)
3. Bałaban, T.: Propagators and renormalization transformations for lattice gauge theories. I. Comm. Math. Phys. **95**(1), 17–40 (1984b)
4. Bałaban, T.: Propagators and renormalization transformations for lattice gauge theories. II. Comm. Math. Phys. **96**(2), 223–250 (1984c)
5. Bałaban, T.: Recent results in constructing gauge fields. Phys. A **124**(1–3), 79–90 (1984d)
6. Bałaban, T.: Averaging operations for lattice gauge theories. Comm. Math. Phys. **98**(1), 17–51 (1985a)
7. Bałaban, T.: Spaces of regular gauge field configurations on a lattice and gauge fixing conditions. Comm. Math. Phys. **99**(1), 75–102 (1985b)
8. Bałaban, T.: Propagators for lattice gauge theories in a background field. Comm. Math. Phys. **99**(3), 389–434 (1985c)
9. Bałaban, T.: Ultraviolet stability of three-dimensional lattice pure gauge field theories. Comm. Math. Phys. **102**(2), 255–275 (1985d)
10. Bałaban, T.: The variational problem and background fields in renormalization group method for lattice gauge theories. Comm. Math. Phys. **102**(2), 277–309 (1985e)
11. Bałaban, T.: Renormalization group approach to lattice gauge field theories. I. Generation of effective actions in a small field approximation and a coupling constant renormalization in four dimensions. Comm. Math. Phys. **109**(2), 249–301 (1987)
12. Bałaban, T.: Convergent renormalization expansions for lattice gauge theories. Comm. Math. Phys. **119**(2), 243–285 (1988)
13. Bałaban, T.: Large field renormalization. I. The basic step of the R operation. Comm. Math. Phys. **122**(2), 175–202 (1989a)
14. Bałaban, T.: Large field renormalization. II. Localization, exponentiation, and bounds for the R operation. Comm. Math. Phys. **122**(3), 355–392 (1989b)
15. Basu, R., Ganguly, S.: $SO(N)$ Lattice gauge theory, planar and beyond. Comm. Pure Appl, Math. (To appear in) (2016)
16. Bazavov, A., Toussaint, D., Bernard, C., Laiho, J., DeTar, C., Levkova, L., Oktay, M.B., Gottlieb, S., Heller, U.M., Hetrick, J.E., Mackenzie, P.B., Sugar, R., Van de Water, R.S.: Nonperturbative QCD simulations with $2 + 1$ flavors of improved staggered quarks. Rev. Mod. Phys. **82**(2), 1349–1417 (2010)

17. Borgs, C.: Confinement, deconfinement and freezing in lattice Yang-Mills theories with continuous time. Comm. Math. Phys. **116**(2), 309–342 (1988)
18. Borgs, C., Seiler, E.: Lattice Yang-Mills theory at nonzero temperature and the confinement problem. Comm. Math. Phys. **91**(3), 329–380 (1983)
19. Bralić, N.E.: Exact computation of loop averages in two-dimensional Yang-Mills theory. Phys. Rev. D (3) **22**(12), 3090–3103 (1980)
20. Brydges, D., Fröhlich, J., Seiler, E.: On the construction of quantized gauge fields. I. General results. Ann. Phys. **121**(1–2), 227–284 (1979)
21. Brydges, D., Fröhlich, J., Seiler, E.: Construction of quantised gauge fields. II. Convergence of the lattice approximation. Comm. Math. Phys.**71**(2), 159–205 (1980)
22. Brydges, D., Fröhlich, J., Seiler, E.: On the construction of quantized gauge fields. III. The two-dimensional abelian Higgs model without cutoffs. Comm. Math. Phys. **79**(3), 353–399 (1981)
23. Charalambous, N., Gross, L.: The Yang-Mills heat semigroup on three-manifolds with boundary. Comm. Math. Phys. **317**(3), 727–785 (2013)
24. Charalambous, N., Gross, L.: Neumann domination for the Yang–Mills heat equation. J. Math. Phys. **56**(7), 073505, 21 (2015)
25. Charalambous, N., Gross, L.: Initial behavior of solutions to the Yang-Mills heat equation. J. Math. Anal. Appl. **451**(2), 873–905 (2017)
26. Chatterjee, S.: Rigorous solution of strongly coupled $SO(N)$ lattice gauge theory in the large N limit. Comm. Math. Phys. (To appear in) (2015)
27. Chatterjee, S.: The leading term of the Yang-Mills free energy. J. Funct. Anal. **271**, 2944–3005 (2016)
28. Chatterjee, S., Jafarov, J.: The $1/N$ expansion for $SO(N)$ lattice gauge theory at strong coupling. Preprint. arXiv:1604.04777 (2016)
29. Driver, B.K.: Convergence of the $U(1)_4$ lattice gauge theory to its continuum limit. Comm. Math. Phys. **110**(3), 479–501 (1987)
30. Driver, B.K.: Classifications of bundle connection pairs by parallel translation and lassos. J. Funct. Anal. **83**(1), 185–231 (1989a)
31. Driver, B.K.: YM_2: continuum expectations, lattice convergence, and lassos. Comm. Math. Phys. **123**(4), 575–616 (1989b)
32. Driver, B.K.: A Functional Integral Approaches to the Makeenko-Migdal equations. Preprint. arXiv:1709.04041 (2017)
33. Driver, B.K., Gabriel, F., Hall, B. C., Kemp, T.: The Makeenko–Migdal equation for Yang-Mills theory on compact surfaces. Preprint. arXiv:1602.03905 (2016)
34. Driver, B.K., Hall, B.C., Kemp, T.: Three proofs of the Makeenko–Migdal equation for Yang–Mills theory on the plane. Preprint. arXiv:1601.06283 (2016)
35. Faria da Veiga, P.A., O'Carroll, M., Schor, R.: Existence of baryons, baryon spectrum and mass splitting in strong coupling lattice QCD. Comm. Math. Phys. **245**(2), 383–405 (2004)
36. Federbush, P.: A phase cell approach to Yang-Mills theory. I. Modes, lattice-continuum duality. Comm. Math. Phys. **107**(2), 319–329 (1986)
37. Federbush, P.: A phase cell approach to Yang-Mills theory. III. Local stability, modified renormalization group transformation. Comm. Math. Phys. **110**(2), 293–309 (1987a)
38. Federbush, P.: A phase cell approach to Yang-Mills theory. VI. Nonabelian lattice-continuum duality. Ann. Inst. H. Poincaré Phys. Théor. **47**(1), 17–23 (1987b)
39. Federbush, P.: A phase cell approach to Yang-Mills theory. IV. The choice of variables. Comm. Math. Phys. **114**(2), 317–343 (1988)
40. Federbush, P.: A phase cell approach to Yang-Mills theory. V. Analysis of a chunk. Comm. Math. Phys. **127**(3), 433–457 (1990)
41. Federbush, P., Williamson, C.: A phase cell approach to Yang-Mills theory. II. Analysis of a mode. J. Math. Phys. **28**(6), 1416–1419 (1987)
42. Fine, D.S.: Quantum Yang-Mills on the two-sphere. Comm. Math. Phys. **134**(2), 273–292 (1990)

43. Fine, D.S.: Quantum Yang-Mills on a Riemann surface. Comm. Math. Phys. **140**(2), 321–338 (1991)
44. Fröhlich, J., Spencer, T.: Massless phases and symmetry restoration in abelian gauge theories and spin systems. Comm. Math. Phys. **83**(3), 411–454 (1982)
45. Gattringer, C., Lang, C.B.: Quantum Chromodynamics on the Lattice. An Introductory Presentation. Springer, Berlin (2010)
46. Glimm, J., Jaffe, A.: Quantum Physics. A Functional Integral Point of View. 2 edn Springer, New York (1987)
47. Göpfert, M., Mack, G.: Proof of confinement of static quarks in 3-dimensional $U(1)$ lattice gauge theory for all values of the coupling constant. Comm. Math. Phys. **82**(4), 545–606 (1982)
48. Griffiths, D.: Introduction to Elementary Particles. Wiley (2008)
49. Gross, L.: Convergence of $U(1)_3$ lattice gauge theory to its continuum limit. Comm. Math. Phys. **92**(2), 137–162 (1983)
50. Gross, L.: The Yang-Mills heat equation with finite action. Preprint. arXiv:1606.04151 (2016)
51. Gross, L.: Stability of the Yang-Mills heat equation for finite action. Preprint. arXiv:1711.00114 (2017)
52. Gross, L., King, C., Sengupta, A.: Two-dimensional Yang-Mills theory via stochastic differential equations. Ann. Phys. **194**(1), 65–112 (1989)
53. Guth, A.H.: Existence proof of a nonconfining phase in four-dimensional $U(1)$ lattice gauge theory. Phys. Rev. D **21**(8), 2291–2307 (1980)
54. Jaffe, A., Witten, E.: Quantum Yang–Mills Theory. The Millennium Prize Problems, pp. 129–152. Clay Math. Inst., Cambridge, MA (2006)
55. Jafarov, J.: Wilson loop expectations in $SU(N)$ lattice gauge theory. Preprint. arXiv:1610.03821 (2016)
56. King, C.: The $U(1)$ Higgs model. I. The continuum limit. Comm. Math. Phys. **102**(4), 649–677 (1986a)
57. King, C.: The $U(1)$ Higgs model. II. The infinite volume limit. Comm. Math. Phys. **103**(2), 323–349 (1986b)
58. Klimek, S., Kondracki, W.: A construction of two-dimensional quantum chromodynamics. Comm. Math. Phys. **113**(3), 389–402 (1987)
59. Kogut, J.B.: An introduction to lattice gauge theory and spin systems. Rev. Mod. Phys. **51**(4), 659–713 (1979)
60. Kogut, J.B.: The lattice gauge theory approach to quantum chromodynamics. Rev. Mod. Phys. **55**(3), 775–836 (1983)
61. Lévy, T.: Yang-Mills measure on compact surfaces. Mem. Amer. Math. Soc. **166**(790) (2003)
62. Lévy, T.: Two-dimensional Markovian holonomy fields. Astérisque, No 329 (2010)
63. Lévy, T.: Topological quantum field theories and Markovian random fields. Bull. Sci. Math. **135**(6–7), 629–649 (2011)
64. Lévy, T.: The master field on the plane. Preprint. arXiv:1112.2452 (2011)
65. Lüscher, M.: Trivializing maps, the Wilson flow and the HMC algorithm. Comm. Math. Phys. **293**(3), 899–919 (2010)
66. Lüscher, M.: Properties and uses of the Wilson flow in lattice QCD. J. High Energy Phys. **2010**(8), 071, 18 (2010)
67. Magnen, J., Rivasseau, V., Sénéor, R.: Construction of YM$_4$ with an infrared cutoff. Comm. Math. Phys. **155**(2), 325–383 (1993)
68. Makeenko, Y.M., Migdal, A.A.: Exact equation for the loop average in multicolor QCD. Phys. Lett. B **88**(1), 135–137 (1979)
69. Montvay, I., Münster, G.: Quantum fields on a lattice. Reprint of the 1994 original. Cambridge University Press, Cambridge (1997)
70. Nguyen, T.: Quantum Yang–Mills Theory in Two Dimensions: exact versus Perturbative. Preprint. arXiv:1508.06305 (2015)
71. Osterwalder, K., Seiler, E.: Gauge field theories on a lattice. Ann. Phys. **110**(2), 440–471 (1978)
72. Paes-Leme, P.J.: Ornstein-Zernike and analyticity properties for classical lattice spin systems. Ann. Phys. **115**(2), 367–387 (1978)

73. Seiler, E.: Gauge Theories as a Problem of Constructive Quantum Field Theory and Statistical Mechanics. Springer, Berlin (1982)
74. Sengupta, A.: The Yang-Mills measure for S^2. J. Funct. Anal. **108**(2), 231–273 (1992)
75. Sengupta, A.: Quantum gauge theory on compact surfaces. Ann. Phys. **221**(1), 17–52 (1993)
76. Sengupta, A.: Gauge theory on compact surfaces. Mem. Amer. Math. Soc. **126**(600) (1997)
77. t'Hooft, G.: A planar diagram theory for strong interactions. Nucl. Phys. B **72**(3), 461–473 (1974)
78. Wilson, K.G.: Confinement of quarks. Phys. Rev. D **10**(8), 2445–2459 (1974)
79. Witten, E.: On quantum gauge theories in two dimensions. Comm. Math. Phys. **141**(1), 153–209 (1991)
80. Witten, E.: Two-dimensional gauge theories revisited. J. Geom. Phys. **9**(4), 303–368 (1992)

Multiscale Systems, Homogenization, and Rough Paths

Ilya Chevyrev, Peter K. Friz, Alexey Korepanov, Ian Melbourne and Huilin Zhang

Dedicated to Professor S.R.S Varadhan on the occasion of his 75th birthday

Abstract In recent years, substantial progress was made towards understanding convergence of fast-slow deterministic systems to stochastic differential equations. In contrast to more classical approaches, the assumptions on the fast flow are very mild. We survey the origins of this theory and then revisit and improve the analysis of Kelly-Melbourne [Ann. Probab. Volume 44, Number 1 (2016), 479–520], taking into account recent progress in p-variation and càdlàg rough path theory.

Keywords Fast-slow systems · Homogenization · Rough paths

I. Chevyrev
Mathematical Institute, University of Oxford, Andrew Wiles Building, Radcliffe Observatory Quarter, Woodstock Road, Oxford OX2 6GG, United Kingdom
e-mail: chevyrev@maths.ox.ac.uk

P. K. Friz (✉)
Institut für Mathematik, Technische Universität Berlin, and Weierstraß–Institut für Angewandte Analysis und Stochastik, Berlin, Germany
e-mail: friz@math.tu-berlin.de

A. Korepanov · I. Melbourne
Mathematics Institute, University of Warwick, Coventry CV4 7AL, United Kingdom
e-mail: a.korepanov@warwick.ac.uk

I. Melbourne
e-mail: i.melbourne@warwick.ac.uk

H. Zhang
Institute of Mathematics, Fudan University, and Technische Universität Berlin, Shanghai 200433, China
e-mail: huilinzhang2014@gmail.com

© Springer Nature Switzerland AG 2019
P. Friz et al. (eds.), *Probability and Analysis in Interacting Physical Systems*, Springer Proceedings in Mathematics & Statistics 283,
https://doi.org/10.1007/978-3-030-15338-0_2

MSC Codes 60H10 · 39A50 · 37D20 · 37D25 · 37A50

1 Introduction

The purpose of this article is to survey and improve several recent developments in the theories of homogenization and rough paths, and the interaction between them. From the side of homogenization, we are interested in the programme initiated by [56] and continued in [32] of studying fast-slow systems without mixing assumptions on the fast flow. From the side of rough paths, we are interested in surveying recent extensions of the theory to the discontinuous setting [18, 27, 30] (see also [17, 21, 39, 73] for related results); the continuous theory, for the purposes of this survey, is well-understood [29]. The connection between the two sides first arose in [40, 41] in which the authors were able to employ rough path techniques (in the continuous and discontinuous setting) to study systems widely generalising those considered in [32, 56].

In this article we address both continuous and discrete systems. The continuous fast-slow systems take the form of the ODEs

$$\frac{d}{dt}x_\varepsilon = a(x_\varepsilon, y_\varepsilon) + \varepsilon^{-1}b(x_\varepsilon, y_\varepsilon), \qquad \frac{d}{dt}y_\varepsilon = \varepsilon^{-2}g(y_\varepsilon). \tag{1}$$

The equations are posed on $\mathbb{R}^d \times M$ for some compact Riemannian manifold M, and $g : M \to TM$ is a suitable vector field. We assume a fixed initial condition $x_\varepsilon(0) = \xi$ for some fixed $\xi \in \mathbb{R}^d$, while the initial condition for y_ε is drawn randomly from (M, λ), where λ is a Borel probability measure on M.

For the discrete systems, we are interested in dynamics of the form

$$X_{j+1}^{(n)} = X_j^{(n)} + n^{-1}a(X_j^{(n)}, Y_j) + n^{-1/2}b(X_j^{(n)}, Y_j), \quad Y_{j+1} = TY_j. \tag{2}$$

The equations are again posed on $\mathbb{R}^d \times M$, and $T : M \to M$ is an appropriate transformation. As before, $X_0^{(n)} = \xi \in \mathbb{R}^d$ is fixed and Y_0 is drawn randomly from a probability measure λ on M.

Let $x_\varepsilon : [0, 1] \to \mathbb{R}^d$ denote either the solution to (1), or the piecewise constant path $x_\varepsilon(t) = X_{\lfloor t/\varepsilon^2 \rfloor}^{(\lfloor 1/\varepsilon^2 \rfloor)}$, where $X_j^{(n)}$ is the solution to (2). The primary goal of this article is to show convergence in law $x_\varepsilon \to X$ in the uniform (or stronger) topology as $\varepsilon \to 0$. Here X is a stochastic process, which in our situation will be the solution to an SDE.

Throughout this note we shall focus on the case where $a(x, y) \equiv a(x)$ depends only on x and $b(x, y) \equiv b(x)v(y)$, where $v : M \to \mathbb{R}^m$ is an observable of y and $b : \mathbb{R}^d \to L(\mathbb{R}^m, \mathbb{R}^d)$ This is precisely the situation considered in [40]. One restriction of the method in [40] is the use of Hölder rough path topology which necessitates moment conditions on the fast dynamics which are suboptimal from the point of view of homogenization. Our main insight is that switching from α-Hölder to p-variation rough path topology allows for optimal moment assumptions on the fast dynamics. The non-product case was previously handled, also with suboptimal moment assumptions, in [41] and also [7], using infinite-dimensional and flow-based rough paths respectively. We briefly discuss this and some other extensions in Sect. 5, leaving a full analysis of the general (non-product) case, under equally optimal moment assumptions, to a forthcoming article [19].

An Example of Fast Dynamics. There are many examples to which the results presented here apply, however we feel it is important to have a concrete (and simple to state) example in mind from the very beginning. In this regard, Pomeau and Manneville [63] introduced a class of maps that exhibit *intermittency* as part of their study of turbulent bursts. The most-studied example [48] is the one-dimensional map $T : M \to M$, $M = [0, 1]$, given by

$$Ty = \begin{cases} y(1 + 2^\gamma y^\gamma) & y < \frac{1}{2} \\ 2y - 1 & y \geq \frac{1}{2} \end{cases}. \tag{3}$$

Here $\gamma \geq 0$ is a parameter. When $\gamma = 0$ this is the doubling map $Ty = 2y \bmod 1$ which is uniformly expanding (see Sect. 2.3). For $\gamma > 0$, there is a neutral fixed point at 0 ($T'(0) = 1$) which has more and more influence as γ increases. For each value of $\gamma \in [0, 1)$, there is a unique absolutely continuous invariant probability measure μ. This measure is ergodic and equivalent (in fact equal when $\gamma = 0$) to the Lebesgue measure.

Suppose that $v : M \to \mathbb{R}^m$ is Hölder continuous and $\int v \, d\mu = 0$. Let

$$v_n = \sum_{0 \leq j < n} v \circ T^j.$$

By [48, 75], for $\gamma \in [0, \frac{1}{2})$, the random variable $n^{-1/2} v_n$, defined on the probability space (M, μ), converges in law to a normal distribution. (Convergence in law also holds on (M, Leb).) In other words, the *central limit theorem* (CLT) holds. However, by [34], the CLT fails for $\gamma > \frac{1}{2}$ (instead there is convergence to a stable law of index γ^{-1}); for $\gamma = \frac{1}{2}$ the CLT holds but with non-standard normalization $(n \log n)^{-1/2}$. Hence from now on we restrict to $\gamma \in [0, \frac{1}{2})$.

Define

$$S_n = \sum_{0 \leq i \leq j < n} (v \circ T^i) \otimes (v \circ T^j).$$

The approach to homogenization of fast-slow systems in [40] requires convergence of the pair of stochastic processes $\left(n^{-1/2} v_{\lfloor nt \rfloor}, n^{-1} S_{\lfloor nt \rfloor}\right)$ to an enhanced Brownian motion, which is established for all $\gamma \in [0, \frac{1}{2})$. (See Sects. 2.3 and 4.1.) Further, the approach based on Hölder rough path theory requires that $\|v_n\|_{2q} = O(n^{1/2})$ and $\|S_n\|_q = O(n)$ for some $q > 3$. These estimates are established in [40] for $\gamma \in [0, \frac{2}{11})$. An improvement in [44] covers $\gamma \in [0, \frac{1}{4})$ and this is known to be sharp [52,

55]. Hence the parameter regime $\gamma \in [\frac{1}{4}, \frac{1}{2})$ is beyond the Hölder rough path theory. In contrast, the p-variation rough path theory described here requires the moment estimates only for some $q > 1$ and [44] applies for all $\gamma \in [0, \frac{1}{2})$. Hence we are able to prove homogenization theorems in the full range $\gamma \in [0, \frac{1}{2})$.

The remainder of this article is organized as follows. In Sect. 2, we discuss the WIP and chaotic dynamics, and several situations of homogenization where rough path theory is not required. In Sect. 3, we introduce the parts of rough path theory required in the Brownian motion setting of this paper. This is applied to fast-slow systems in Sect. 4. In Sect. 5, we mention extensions and related work.

2   Emergence of Randomness in Deterministic Dynamical Systems

In this section, we review a simplified situation where the ordinary weak invariance principle (see below) suffices, and rough path theory is not required.

2.1   The Weak Invariance Principle

Consider a family of stochastic processes indexed by $\varepsilon \in (0, 1)$, say $W_\varepsilon = W_\varepsilon(t, \omega)$ with values in $\mathbb{R}^m$. We are interested in convergence of the respective laws. In the case of continuous sample paths (including smooth or piecewise linear) we say that the *weak invariance principle* (WIP) holds if

$$W_\varepsilon \to_w W \text{ in } C([0, 1], \mathbb{R}^m) \text{ as } \varepsilon \to 0,$$

where W is an m-dimension Brownian motion with covariance matrix Σ; in the case of càdlàg sample paths (including piecewise constant) we mean

$$W_\varepsilon \to_w W \text{ in } D([0, 1], \mathbb{R}^m) \text{ as } \varepsilon \to 0,$$

where C resp. D denotes the space of continuous resp. càdlàg paths, equipped with the uniform topology.[1] For notational simplicity only, assume (W_ε) are defined on a common probability space (Ω, F, λ); we then write $W_\varepsilon \to_\lambda W$ to indicate convergence in law, i.e. $\mathbb{E}_\lambda[f(W_\varepsilon)] \to E_\lambda[f(W)]$ for all bounded continuous functionals.

In many cases, one has convergence of second moments. This allows to compute the covariance of the limiting Brownian motion,

$$\Sigma = \mathbb{E}(W(1) \otimes W(1)) = \lim_{\varepsilon \to 0} \mathbb{E}_\lambda(W_\varepsilon(1) \otimes W_\varepsilon(1)). \tag{4}$$

---

[1]Since our limit processes here—a Brownian motion—is continuous, there is no need to work with the Skorokhod topology on D.

The WIP is also known as the *functional central limit theorem*, with the CLT for finite-dimensional distributions as a trivial consequence. Conversely, the CLT for f.d.d. together with tightness gives the WIP.

Donsker's invariance principle [23] is the prototype of a WIP: consider a centered *m*-dimensional random walk $Z_n := \xi_1 + \cdots + \xi_n$, with $\mathbb{R}^m$-valued IID increments of zero mean and finite covariance Σ. Extend to either a continuous piecewise linear process or càdlàg piecewise constant process $(Z_t : t \geq 0)$. Then the WIP holds for the rescaled random walk

$$Z_t^\varepsilon := \varepsilon Z_{t/\varepsilon^2},$$

and the limiting Brownian motion has covariance Σ. This result has an important generalization to a *(functional) martingale CLT*: using similar notation, assume (Z_n) is a zero mean L^2-martingale with stationary and ergodic increments (ξ_i). Then, with the identical rescaling, the WIP holds true, with convergence of second moments [14] (or e.g. [10, Theorem 18.3]).

Another interesting example is given by *physical Brownian motion* with positive mass $\varepsilon^2 > 0$ and friction matrix M, where the trajectory is given by

$$X_t^\varepsilon := \varepsilon \int_0^{t/\varepsilon^2} Y_s \, ds$$

and Y follows an *m*-dimensional OU process, $dY = -MY dt + dB$, $Y_0 = y_0$. Here, M is an $m \times m$-matrix whose spectrum has positive real part, and B is an *m*-dimensional standard Brownian motion. One checks without difficulties [25, 62] that a WIP holds, even in the sense of weak convergence in the Hölder space $C^\alpha([0, 1], \mathbb{R}^m)$ with any $\alpha < 1/2$. The covariance matrix of the limiting Brownian is given by $\Sigma = M^{-1}(M^{-1})^T$, as can be seen from the Newton dynamics $\varepsilon^2 \ddot{X}^\varepsilon = -M\dot{X}^\varepsilon + \dot{B}$ with white noise $\dot{B}$.

Finally, *sufficiently chaotic deterministic dynamical systems* are a rich source of WIPs. To fix ideas, consider a compact Riemannian manifold M with a Lipschitz vector field g and corresponding flow g_t, for which there is an ergodic, invariant Borel probability measure μ on M. We regard (g_t) as an M-valued stochastic process, given by $g_t(y_0)$ with initial condition y_0 distributed according to λ, another Borel probability measure on M. (It is possible but not necessary to have $\lambda = \mu$.) Consider further a suitable observable $v : M \to \mathbb{R}^m$ with $\mathbb{E}_\mu v = 0$. A family of C^1-processes $(W_\varepsilon)_{\varepsilon > 0}$, with values in $\mathbb{R}^m$, is then given by

$$W_\varepsilon(t) = \varepsilon \int_0^{t\varepsilon^{-2}} v \circ g_s \, ds.$$

As will be reviewed in Sect. 2.3, also in a discrete time setting, in many situations a WIP holds. That is,

$$W_\varepsilon \to_\lambda W \text{ in } C([0, 1], \mathbb{R}^m) \text{ as } \varepsilon \to 0.$$

Typically one also has convergence of second moments, so that W is a Brownian motion with covariance Σ given by (4). Under (somewhat restrictive) assumptions on the decay of correlations, this can be simplified to a Green-Kubo type formula

$$\Sigma = \int_0^\infty \mathbb{E}_\mu \{ v \otimes (v \circ g_s) + (v \circ g_s) \otimes v \} ds.$$

2.2   First Applications to Fast-Slow Systems

In the setting of deterministic, sufficiently chaotic dynamical systems discussed in the previous paragraph, Melbourne–Stuart [56] consider the fast-slow system posed on $\mathbb{R}^d \times M$ (with $m = d$),

$$\dot{x}_\varepsilon = a(x_\varepsilon, y_\varepsilon) + \varepsilon^{-1} v(y_\varepsilon), \qquad \dot{y}_\varepsilon = \varepsilon^{-2} g(y_\varepsilon),$$

with deterministic initial data $x_\varepsilon(0) = x_0$ and $y_\varepsilon(0)$ sampled randomly with probability λ. We wish to study the limiting dynamics of the slow variable x_ε. Assuming for simplicity $a(x, y) = a(x)$, the basic observation is to rewrite

$$\dot{x}_\varepsilon = a(x_\varepsilon) + \dot{W}_\varepsilon.$$

We see that the noise W_ε enters the equation in an additive fashion and one checks without difficulty that the "Itô-map" $W_\varepsilon \mapsto x_\varepsilon$ extends continuously (w.r.t. uniform convergence) to any continuous noise path. Now assume validity of a WIP, i.e. $W_\varepsilon \to_\lambda W$. Then, together with continuity of the Itô-map, one obtains the desired limiting SDE dynamics of the slow variable as

$$dX = dW + a(X)dt.$$

In the general case when a depends on x_ε and y_ε, the drift term is given by $\bar{a}(x) = \int_M a(x, y) \, d\mu(y)$

In subsequent work, Gottwald–Melbourne [32] consider the one-dimensional case $d = m = 1$ with

$$\dot{x}_\varepsilon = a(x_\varepsilon, y_\varepsilon) + \varepsilon^{-1} b(x_\varepsilon) v(y_\varepsilon), \qquad \dot{y}_\varepsilon = \varepsilon^{-2} g(y_\varepsilon).$$

Again, taking $a(x, y) = a(x)$ for simplicity, the limiting SDE turns out to be of Stratonovich form

$$dX = a(X)dt + b(X) \circ dW .$$

The essence of the proof is a robust representation of such SDEs. Indeed, taking $a \equiv 0$ for notational simplicity, an application of the (first order) Stratonovich chain rule exhibits the explicit solution as $X_t = e^{W_t b}(X_0)$, where $e^{W_t b}$ denotes the flow at "time" $W_t \in \mathbb{R}$ along the vector field b; this clearly depends continuously on X_0 and W w.r.t. uniform convergence. Hence, as in the additive case, the problem is reduced to having a WIP. This line of reasoning can be pushed a little further, namely to the case $\dot{x}_\varepsilon = a(x_\varepsilon, y_\varepsilon) + \varepsilon^{-1} V(x_\varepsilon) v(y_\varepsilon)$ with commuting vector fields $V = (V_1, ..., V_m)$, a.k.a. the Doss–Sussmann method, but fails for general vector fields, not to mention the non-product case when $V(x)v(y)$ is replaced by $b(x, y)$. This is a fundamental problem which is addressed by Lyons' theory of rough paths.

Gottwald-Melbourne [32] consider also discrete time fast-slow systems posed on $\mathbb{R}^d \times M$,

$$X_{j+1}^{(n)} = X_j^{(n)} + n^{-1} a(X_j^{(n)}, Y_j) + n^{-1/2} b(X_j^{(n)}) v(Y_j), \quad Y_{j+1} = T Y_j,$$

Again we suppose for notational simplicity that $a(x, y) = a(x)$. We continue to suppose that μ is an ergodic T-invariant probability measure on M and that $\mathbb{E}_\mu v = 0$. Also, λ is another probability measure on M. Recall that $x_\varepsilon(t) = X_{\lfloor t/\varepsilon^2 \rfloor}^{\lfloor 1/\varepsilon^2 \rfloor}$ and assume validity of a WIP, i.e. $W_\varepsilon \to_\lambda W$. When $b \equiv 1$, it is shown in [32] that $x_\varepsilon \to_\lambda X$ where $dX = a(X) \, dt + dW$. For $d = m = 1$, under additional mixing assumptions it is shown that $x_\varepsilon \to_\lambda X$ where $dX = \tilde{a}(X) \, dt + b(X) \, dW$ with

$$\tilde{a}(x) = a(x) + b(x)b'(x) \sum_{n=1}^{\infty} \mathbb{E}_\mu (v \, v \circ T^n).$$

2.3 Chaotic Dynamics: CLT and the WIP

In this subsection, we describe various classes of dynamical systems with good statistical properties, focusing attention on the CLT and WIP.

Somewhat in contrast to rough path theory, the ergodic theory of smooth dynamical systems is much simpler for discrete time than for continuous time—indeed the continuous time theory proceeds by reducing to the discrete time case. Also, the simplest examples are noninvertible. The reasons behind this are roughly as follows. Since the papers of Anosov [4] and Smale [71], it is has been understood that the way to study dynamical systems is to exploit expansion and contraction properties. The simplest systems are uniformly expanding; these are necessarily discrete time and noninvertible. Anosov and Axiom A (uniformly hyperbolic) diffeomorphisms have uniformly contracting and expanding directions. Anosov and Axiom A (uniformly hyperbolic) flows have a neutral time direction and are uniformly contracting and expanding in the remaining directions. The neutral direction makes flows much harder to study. The mixing properties of uniformly hyperbolic flows are still poorly understood (see for example the review in [53]); fortunately the CLT and WIP do not rely on mixing.

Accordingly, we consider in turn expanding maps, hyperbolic diffeomorphisms, and hyperbolic flows, in Sects. 2.3.1, 2.3.2 and 2.3.3 respectively. This includes the uniform cases mentioned in the previous paragraph, but also dynamical systems that are nonuniformly expanding/hyperbolic, which is crucial for incorporating large classes of examples.

2.3.1 Uniformly and Nonuniformly Expanding Maps

The CLT and WIP are proved in [36, 38] for large classes of dynamical systems (in fact, they prove a stronger statistical property, known as the almost sure invariance principle). For recent developments in this direction, see [20, 43] and references therein. In particular, the CLT and WIP hold for smooth uniformly expanding maps and for systems modelled by Young towers with summable decay of correlations [75], which provide a rich source of examples including the intermittent maps (3).

Here we review the results in various situations, focusing on various issues that are of importance for fast-slow systems: CLT, WIP, covariance matrices, nondegeneracy. Also, we mention the notions of spectral decomposition, mixing up to a finite cycle, basins of attraction, and strong distributional convergence, which are necessary for understanding how the theory is applied.

Smooth Uniformly Expanding Maps. The simplest chaotic dynamical system is the doubling map $T : M \to M$, $M = [0, 1]$, given by $Ty = 2y$ mod 1. More generally, let $T : M \to M$ be a C^2 map on a compact Riemannian manifold M and let $\mathcal{B}$ be the σ-algebra of Borel sets. The map is *uniformly expanding* if there are constants $C > 0$, $L > 1$ such that $\|DT^n|_y z\| \ge CL^n \|z\|$ for all $y \in M$, $z \in T_y M$. By [45], there is a unique ergodic T-invariant Borel probability measure μ on M equivalent to the volume measure. (Recall that μ is T-invariant if $\mu(T^{-1}B) = \mu(B)$ for all $B \in \mathcal{B}$, and is ergodic if $\mu(B) = 0$ or $\mu(B) = 1$ for all $B \in \mathcal{B}$ with $TB \subset B$.) By Birkhoff's ergodic theorem [11] (an extension of the strong law of large numbers) the sum $v_n = \sum_{0 \le j < n} v \circ T^j$ satisfies $n^{-1} v_n \to \int_M v \, d\mu$ a.e. for all $v \in L^1$.

To make further progress it is necessary to impose some regularity on the observable v; the CLT fails in general for continuous observables. Hence, we suppose that v is Hölder. Specifically, fix $\kappa \in (0, 1)$ and let $C_0^\kappa(M)$ be the space of C^κ observables $v : M \to \mathbb{R}$ with $\int_M v \, d\mu = 0$. It is well known [12, 66, 70] that there are constants $\gamma \in (0, 1)$ and $C > 0$ depending only on T and κ such that

$$\left| \int_M v \, w \circ T^n \, d\mu \right| \le C\gamma^n \|v\|_{C^\kappa} \|w\|_1 \quad \text{for all } v \in C_0^\kappa(M), w \in L^1, n \ge 1. \quad (5)$$

An immediate consequence is that the limit

$$\sigma^2 := \lim_{n \to \infty} n^{-1} \int_M v_n^2 \, d\mu,$$

exists and that

$$\sigma^2 = \int_M v^2 \, d\mu + 2 \sum_{n=1}^{\infty} \int_M v \, v \circ T^n \, d\mu.$$

For $1 \leq p \leq \infty$, we recall that the Koopman operator $U : L^p \to L^p$ is given by $Uv = v \circ T$ and that the transfer operator $P : L^q \to L^q$ is given by $\int_M Pv \, w \, d\mu = \int_M v \, w \circ T \, d\mu$ for $v \in L^q$, $w \in L^p$, where $p^{-1} + q^{-1} = 1$. These operators satisfy $\|U\|_p = 1$ and $\|P\|_p \leq 1$ for $1 \leq p \leq \infty$. In addition, $PU = I$ and $UP = \mathbb{E}(\cdot \,|T^{-1}\mathcal{B})$. Note that property (5) is equivalent to

$$\|P^n v\|_\infty \leq C\gamma^n \|v\|_{C^\kappa} \quad \text{for all } v \in C_0^\kappa(M), n \geq 1. \tag{6}$$

Following the classical approach of Gordin [31], we define $\chi = \sum_{j=1}^\infty P^j v$ and $m = v - \chi \circ T + \chi$. By (6), $m, \chi \in L^\infty$. It follows from the definitions that $m \in \ker P$ and hence that $n^{-1} \int_M m_n^2 \, d\mu = \int_M m^2 \, d\mu$ for all n. Since

$$v_n - m_n = \chi \circ T^n - \chi \in L^\infty,$$

it follows that $\int_M m^2 \, d\mu = \sigma^2$.

Moreover, $\mathbb{E}(m|T^{-1}\mathcal{B}) = UPm = 0$, so $\{m \circ T^n, n \geq 0\}$ is an L^∞ stationary ergodic reverse martingale difference sequence. Hence, standard martingale limit theorems apply. In particular, by [9, 51] we obtain the CLT for m and thereby v:

$$n^{-1/2} v_n \to_\mu N(0, \sigma^2) \quad \text{as } n \to \infty.$$

We refer to the decomposition $v = m + \chi \circ T - \chi$ as an L^∞ *martingale-coboundary decomposition*, since m is a reverse martingale increment. The coboundary term $\chi \circ T - \chi \in L^\infty$ telescopes under iteration and therefore is often negligible. Next, define the process $W_n \in C[0, 1]$ by setting $V_n(t) = n^{-1/2} v_{nt}$ for $t = 0, 1/n, 2/n \dots$ and linearly interpolating. By [14, 51] we obtain the WIP:

$$W_n \to_\mu W \quad \text{in } C[0, 1] \text{ as } n \to \infty,$$

where W is a Brownian motion with variance σ^2.

The CLT and WIP are said to be *degenerate* if $\sigma^2 = 0$. We now show that this is extremely rare. Since $v = m + \chi \circ T - \chi$ and $\sigma^2 = \int_M m^2 \, d\mu$, we obtain that $\sigma^2 = 0$ if and only if $v = \chi \circ T - \chi$ where $\chi \in L^\infty$. Moreover, the series $\chi = \sum_{n=1}^\infty P^n v$ converges in C^κ (see for example [67]), so in particular χ is continuous. Let $C_{\deg}^\kappa$ consist of observables $v \in C_0^\kappa(M)$ with $\sigma^2 = 0$.

Proposition 2.1 $C_{\deg}^\kappa$ *is a closed, linear subspace of infinite codimension in* $C_0^\kappa(M)$.

Proof Suppose that $v \in C_{\deg}^\kappa$ so $v = \chi \circ T - \chi$ where χ is continuous. Iterating, we obtain $v_k = \chi \circ T^k - \chi$. Hence if $y \in M$ is a period k point, i.e. $T^k y = y$, then $v_k(y) = 0$. Since periodic points are dense in M [71] we obtain infinitely many linear constraints on v.

$\square$

Analogous results hold for vector-valued observables $v : M \to \mathbb{R}^m$. Let $v \in C_0^\kappa(M, \mathbb{R}^m)$. Then

$$\lim_{n\to\infty} n^{-1} \int_M v_n \otimes v_n \, d\mu = \Sigma, \tag{7}$$

where $\Sigma \in \mathbb{R}^m \otimes \mathbb{R}^m$ is symmetric and positive semidefinite, and

$$\Sigma = \int_M v \otimes v \, d\mu + \sum_{n=1}^{\infty} \int_M \{v \otimes (v \circ T^n) + (v \circ T^n) \otimes v\} \, d\mu. \tag{8}$$

Define $v_n \in \mathbb{R}^m$ and $W_n \in C([0, 1], \mathbb{R}^m)$ as before. By the above results, $n^{-1/2} c^T v_n$ converges in distribution to a normal distribution with variance $c^T \Sigma c$ for each $c \in \mathbb{R}^m$, and hence by Cramer-Wold we obtain the multi-dimensional CLT $n^{-1/2} v_n \to_d N(0, \Sigma)$. Similarly, $W_n \to_w W$ in $C([0, 1], \mathbb{R}^m)$ where W is m-dimensional Brownian motion with covariance Σ. Finally $c^T \Sigma c = 0$ for $c \in \mathbb{R}^m$ if and only if $c^T v \in C_{\text{deg}}^\kappa$. Hence the degenerate case $\det \Sigma = 0$ occurs only on a closed subspace of infinite codimension.

Since the CLT is a consequence of the WIP, generally we only mention the WIP in the remainder of this subsection.

Returning to the ergodic theorem, if $v \in L^1$, then $n^{-1} v_n(y_0) \to \int_M v \, d\mu$ for μ almost every initial condition $y_0 \in M$. Since μ is equivalent to volume, we could equally choose the initial condition y_0 randomly with respect to volume, which is perhaps more natural since volume is the intrinsic measure on M. Similar considerations apply to the WIP. Based on ideas of [24], it follows from [78, Corollary 2] that if $v \in C_0^\kappa(M, \mathbb{R}^m)$ then $W_n \to_\lambda W$ in $C([0, 1], \mathbb{R}^m)$ for every absolutely continuous Borel probability measure λ (including μ and volume as special cases). This property is often called *strong distributional convergence* [78]. Of course $C_0^\kappa(M, \mathbb{R}^m)$ is defined using μ regardless of the choice of λ.

Piecewise Expanding Maps. There are numerous extensions of the above arguments in various directions. For example, Keller [38, Theorem 3.5] considers piecewise $C^{1+\varepsilon}$ transformations $T : M \to M$, $M = [0, 1]$, with finitely many monotone branches and $|T'| \geq L$ for some $L > 1$. There exists an ergodic T-invariant absolutely continuous probability measure (acip) μ. Let $\Lambda = \text{supp} \, \mu$. Recall that Λ is *mixing* if $\lim_{n\to\infty} \mu(T^{-n} A \cap B) = \mu(A)\mu(B)$ for all measurable sets $A, B \subset \Lambda$. In this case, by [38, Theorem 3.3], condition (5) holds (with M replaced by Λ). Hence we obtain the WIP for all $v \in C_0^\kappa(\Lambda, \mathbb{R}^m)$ with Σ given as in (7) and (8). Also $\det \Sigma = 0$ if and only if there exists $c \in \mathbb{R}^m$ such that $c^T v = \chi \circ T - \chi$ for some $\chi : \Lambda \to \mathbb{R}$ in L^∞.

If Λ is not mixing, then condition (5) fails. Nevertheless, by [38, Theorem 3.3] Λ is *mixing up to a finite cycle*: we can write Λ as a disjoint union $\Lambda = A_1 \cup \cdots \cup A_k$ for some $k \geq 2$ such that T permutes the A_j cyclically and $T^k : A_j \to A_j$ is mixing with respect to $\mu|A_j$ for each j. Moreover, condition (5) holds for the map $T^k : A_j \to A_j$. It is easily verified that the WIP goes through for $T : \Lambda \to \Lambda$ and that the limit formula (7) for Σ remains valid. (Of course in the nonmixing case, (8) no longer makes sense.)

The *basin of attraction* of the ergodic probability measure μ is defined as

$$B_\mu = \{y \in M : \lim_{n\to\infty} n^{-1} v_n(y) = \int_M v \, d\mu \ \text{ for all } v : M \to \mathbb{R} \text{ continuous}\}.$$

(The ergodic theorem guarantees that, modulo a zero measure set, $\text{supp}\,\mu \subset B_\mu$, but in general B_μ can be much larger.) The acip μ need not be unique but by [38, Theorem 3.3] there is *a spectral decomposition*: there exist finitely many absolutely continuous ergodic invariant probability measures $\mu_1, \ldots, \mu_k$ such that $\text{Leb}(B_{\mu_1} \cup \cdots \cup B_{\mu_k}) = 1$ and the results described above for μ hold separately for each of $\mu_1, \ldots, \mu_k$.

For related results on C^2 one-dimensional maps with infinitely many branches, we refer to [68]. For higher-dimensional piecewise smooth maps, see for example [15, 69]. Again there is a spectral decomposition into finitely many attractors which are mixing up to a finite cycle. After restricting to an appropriate subset and considering a suitable iterate of T, condition (5) holds and we obtain the WIP etc as described above.

In general, extra work is required to deduce that degeneracy is infinite codimension as in Proposition 2.1. We note that the approach in [15] fits within the Young tower approach of [74, 75] where it is possible to recover Proposition 2.1 as described below.

Nonuniformly Expanding Maps. An important method for studying nonuniformly expanding maps $T : M \to M$ is to construct a *Young tower* as in [75]. This incorporates the maps (3) discussed in the introduction.

Let M be a bounded metric space with finite Borel measure ρ and let $T : M \to M$ be a nonsingular transformation ($\rho(T^{-1}B) = 0$ if and only if $\rho(B) = 0$ for $B \in \mathcal{B}$). Let $Y \subset M$ be a subset of positive measure, and let α be an at most countable measurable partition of Y with $\rho(a) > 0$ for all $a \in \alpha$. We suppose that there is an integrable *return time* function $\tau : Y \to \mathbb{Z}^+$, constant on each a with value $\tau(a) \geq 1$, and constants $L > 1$, $\kappa \in (0, 1)$, $C_0 > 0$, such that for each $a \in \alpha$,

(1) $F = T^\tau$ restricts to a (measure-theoretic) bijection from a onto Y.
(2) $d(Fx, Fy) \geq Ld(x, y)$ for all $x, y \in a$.
(3) $d(T^\ell x, T^\ell y) \leq C_0 d(Fx, Fy)$ for all $x, y \in a$, $0 \leq \ell < \tau(a)$.
(4) $\zeta_0 = \frac{d\rho|_Y}{d\rho|_Y \circ F}$ satisfies $|\log \zeta_0(x) - \log \zeta_0(y)| \leq C_0 d(Fx, Fy)^\kappa$ for all $x, y \in a$.

The *induced map* $F = T^\tau : Y \to Y$ has a unique acip μ_Y.

Remark 2.2 For the intermittent maps (3), we can take $Y = [\frac{1}{2}, 1]$ and we can choose τ to be the first return to Y. In general, it is not required that τ is the first return time to Y.

Define the Young tower [75], $\Delta = \{(y, \ell) \in Y \times \mathbb{Z} : 0 \leq \ell \leq \tau(y) - 1\}$, and the tower map

$$f : \Delta \to \Delta, \qquad f(y, \ell) = \begin{cases} (y, \ell + 1), & \ell \leq \tau(y) - 2 \\ (Fy, 0), & \ell = \tau(y) - 1 \end{cases}. \tag{9}$$

The projection $\pi_\Delta : \Delta \to \Lambda$, $\pi_\Delta(y, \ell) = T^\ell y$, defines a semiconjugacy from f to T. Define the ergodic acip $\mu_\Delta = \mu_Y \times \{\text{counting}\} / \int_Y \tau \, d\mu_Y$ for $f : \Delta \to \Delta$. Then $\mu = (\pi_\Delta)_* \mu_\Delta$ is an ergodic acip for $T : M \to M$ and μ is mixing up to a finite cycle.

Young [75] proved that if μ is mixing and $\mu_Y(y \in Y : \tau(y) > n) = O(n^{-(\beta+1)})$ for some $\beta > 0$, then

$$\left| \int_M v \, w \circ T^n \, d\mu \right| \leq Cn^{-\beta} \|v\|_{C^\kappa} \|w\|_\infty \quad \text{for all } v \in C_0^\kappa(M), \, w \in L^\infty, n \geq 1,$$

(10)

In particular, $\beta > 1$ corresponds to *summable decay of correlations*. (For the maps (3), $\beta = \gamma^{-1} - 1$, so $\beta > 1$ corresponds to $\gamma < \frac{1}{2}$.) Equivalently, $\|P^n v\|_1 \leq Cn^{-\beta} \|v\|_{C^\kappa}$ and by interpolation $\|P^n v\|_p \leq C^{1/p} n^{-\beta/p} \|v\|_{C^\kappa}$ for all $p \geq 1$.

For $\beta > 1$, we have that $\|P^n v\|_1$ is summable for $v \in C_0^\kappa(M, \mathbb{R}^m)$, and a standard calculation shows that formulas (7) and (8) for Σ hold. Also, the series $\chi = \sum_{n=1}^\infty P^n v$ converges in L^p for all $p < \beta$ and we obtain an L^p martingale-coboundary decomposition $v = m + \chi \circ T - \chi$. For $\beta > 2$, we have m, $\chi \in L^2$ and the WIP follows. With extra work it can be shown that the WIP holds for all $\beta > 1$. We refer to [42, 47, 50, 72] for further details. See also [43, 54]. By [54, Rem. 2.11], the degenerate case $\det \Sigma = 0$ is infinite codimension in the sense of Proposition 2.1.

In the case where μ is mixing only up to a finite cycle, the WIP etc go through unchanged, except that formula (8) does not make sense.

2.3.2 Hyperbolic Diffeomorphisms

A WIP for Axiom A diffeomorphisms can be found in [22]. The WIP is also well-known to hold for systems modelled by Young towers with exponential tails [74] as well as those with summable decay of correlations (for an explicit and completely general argument, see [59]). This is a very flexible setting that covers large classes of nonuniformly hyperbolic diffeomorphisms (with singularities).

The results for hyperbolic diffeomorphisms $T : M \to M$ are similar to those in Sect. 2.3.1, subject to two complications. The first complication affects the proofs. Since T is invertible, the transfer operator P is an isometry on L^q for all q. In particular, $\ker P = \{0\}$. Hence the approach in Sect. 2.3.1 cannot be applied directly. The method for getting around this is rather convoluted and is described at the end of this subsection.

The second complication affects the statement of the results. Typically, the invariant measures of interest are supported on zero volume sets and hence there are no acips. We say that μ is a *physical measure* if the basin of attraction B_μ has positive volume. (This is automatic for acips but is an extra assumption now.) Let Vol denote the normalized volume on B_μ.

There is an important class of physical measures μ, known as Sinai-Ruelle-Bowen (SRB) measures [76], for which the WIP with respect to Vol (and hence, by strong distributional convergence, every absolutely continuous probability measure λ on B_μ)

follows from the WIP with respect to μ. Hence it is natural to consider observables v with $\int_\Lambda v \, d\mu = 0$ and to ask that $W_n \to_\lambda W$ for absolutely continuous probability measures λ on B_μ.

Axiom A Diffeomorphisms. Let M be a compact Riemannian manifold. A C^2 diffeomorphism $T : M \to M$ is said to be *Anosov* [4] if there is a continuous DT-invariant splitting $TM = E^s \oplus E^u$ (into stable and unstable directions) where $\|DT^n|E^s\| \leq Ca^n$ and $\|DT^{-n}|E^u\| \leq Ca^n$ for $n \geq 1$. Here $C > 0$ and $a \in (0, 1)$ are constants.

Smale [71] introduced the notion of *Axiom A* diffeomorphism extending the definition in [4]. Since we are interested in SRB measures, we restrict attention to attracting sets, bypassing the full definitions in [71]. Recall that a closed T-invariant set $\Lambda \subset M$ is *attracting* if there is a neighbourhood U of Λ such that $\lim_{n\to\infty} \text{dist}(T^n y, \Lambda) = 0$ for all $y \in U$. An attracting set is called *Axiom A* if there is a continuous DT-invariant splitting $T_\Lambda M = E^s \oplus E^u$ over Λ, again with the properties $\|DT^n|E^s\| \leq Ca^n$ and $\|DT^{-n}|E^u\| \leq Ca^n$ for $n \geq 1$. To avoid trivialities, we suppose that $\dim E^u_y \geq 1$ for all $y \in \Lambda$. (We allow $\dim E^s_y = 0$ though this is just the uniformly expanding case.)

By [71], there is a spectral decomposition of Λ into finitely many attracting sets, called *Axiom A attractors* with the property that none of them can be decomposed further. Moreover, periodic points are dense in Λ.

If Λ is an Axiom A attractor, then by [12, 66, 70] there is a unique ergodic invariant probability measure μ on Λ such that $\text{Leb}(B_\mu) > 0$. Moreover, μ is mixing up to a finite cycle.

All the results described in Sect. 2.3.1 for uniformly expanding maps hold for Axiom A attractors. Specifically, let $C_0^\kappa(\Lambda, \mathbb{R}^m)$ denote the space of C^κ observables $v : \Lambda \to \mathbb{R}^m$ with $\int_\Lambda v \, d\mu = 0$. Then the WIP holds on (Λ, μ) and (B_μ, λ) with Σ satisfying formula (7). Moreover $C_{\text{deg}}^\kappa = \{v \in C_0^\kappa(\Lambda, \mathbb{R}^m) : \det \Sigma = 0\}$ is a closed subspace of infinite codimension in $C_0^\kappa(\Lambda, \mathbb{R}^m)$. If in addition μ is mixing, then formula (8) holds.

Nonuniformly Hyperbolic Diffeomorphisms and Young Towers. A large class of attractors Λ for nonuniformly hyperbolic diffeomorphisms (with singularities) $T : M \to M$ can be modelled by two-sided Young towers with exponential tails [74] and subexponential tails [75]. The Young tower set up covers numerous classes of examples as surveyed in [16, 74, 76, 77] including Axiom A attractors, Lorentz gases, Hénon-like attractors [8], and intermittent solenoids [59]. See also [1–3].

We end this subsection with a very rough sketch of the method of proof of the WIP for Young towers. This includes the Axiom A attractors as a special case for which standard references are [12, 61]. The idea is again to induce to a map $F = T^\tau : Y \to Y$ that is a uniformly hyperbolic transformation with countable partition and full branches, as described in Young [74], with an integrable inducing time $\tau : Y \to \mathbb{Z}^+$ that is constant on partition elements. (Again τ is not necessarily the first return time.) The construction in [74] ensures that there exists an SRB measure μ_Y for F. Starting from F and τ, we construct a "two-sided" Young tower $f : \Delta \to \Delta$ as in (9) with ergodic invariant probability $\mu_\Delta = \mu_Y \times \{\text{counting}\} / \int_Y \tau \, d\mu_Y$. The projection $\pi_\Delta : \Delta \to \Lambda$, $\pi_\Delta(y, \ell) = T^\ell y$, defines a semiconjugacy from f to T, and $\mu = (\pi_\Delta)_* \mu_\Delta$ is the desired SRB measure for $T : M \to M$. Moreover, μ is mixing up to a finite cycle.

Given $v \in C_0^\kappa(M, \mathbb{R}^m)$, we define the lifted observable $\hat{v} = v \circ \pi_\Delta : \Delta \to \mathbb{R}^m$. It suffices to work from now on with $\hat{v}$.

Next, there is a quotienting procedure which projects out the stable directions reducing to an expanding map. Formally, this consists of a "uniformly expanding" map $\bar{F} : \bar{Y} \to \bar{Y}$ and a projection $\pi : Y \to \bar{Y}$ such that $\bar{F} \circ \pi = \pi \circ F$ and such that $\tau(y) = \tau(y')$ whenever $\pi y = \pi y'$. In particular, τ projects to a well-defined return time $\tau : \bar{Y} \to \mathbb{Z}^+$. Using $\bar{F}$ and τ we construct a "one-sided" Young tower $\bar{f} : \bar{\Delta} \to \bar{\Delta}$. The projection π extends to $\pi : \Delta \to \bar{\Delta}$ with $\pi(y, \ell) = (\pi y, \ell)$ and we define $\bar{\mu}_\Delta = \pi_* \mu_\Delta$. The map $\bar{f}$ plays the role of a "nonuniformly expanding map".

As in Sect. 2.3.1, we consider the tails $\mu_Y(\tau > n)$. In the exponential tail setting of [74], $\mu_Y(\tau > n) = O(\gamma^n)$ for some $\gamma \in (0, 1)$ and a version of the "Sinai trick" (see for example [54, Lemma 3.2]) shows that $v \circ \pi_\Delta = \hat{v} + \chi_1 \circ f - \chi_1$ where $\chi_1 \in L^\infty$ and $\hat{v}(y) = \hat{v}(y')$ whenever $\pi y = \pi y'$. In particular, $\hat{v}$ projects to a well-defined observable $\bar{v} : \bar{\Delta} \to \mathbb{R}^m$.

This construction can be carried out so that $\bar{v}$ is sufficiently regular that the analogue of condition (6) holds, where P is the transfer operator on $\bar{\Delta}$. Hence we obtain an L^∞ martingale-coboundary decomposition $\bar{v} = m + \chi_2 \circ \bar{f} - \chi_2$ on $\bar{\Delta}$. This gives the associated decomposition

$$v \circ \pi_\Delta = \hat{v} = m \circ \pi + \chi \circ f - \chi,$$

on Δ where $\chi = \chi_1 + \chi_2 \circ \pi$.

Now the argument is finished, since we can apply the methods from Sect. 2.3.1 to obtain the WIP, etc, for m on $(\bar{\Delta}, \bar{\mu}_\Delta)$, and hence $m \circ \pi$ on (Δ, μ_Δ), $\hat{v}$ on (Δ, μ_Δ), and v on (Λ, μ).

Finally, we consider the case $\mu(\tau > n) = O(n^{-(\beta+1)})$ with $\beta > 1$. In certain situations (nonuniform expansion but uniform contraction) the Sinai trick works as above and reduces to the situation in (10). The general case is more complicated but is covered by [59, Corollary 2.2]. Again, this is optimal since there are many examples with $\beta = 1$ where the CLT with standard scaling does not hold.

2.3.3 Hyperbolic Flows

Let $\dot{y} = g(y)$ be an ODE defined by a C^2 vector field $g : M \to TM$ on a compact Riemannian manifold M. Let $g_t : M \to M$ denote the corresponding flow. Let $X \subset M$ be a codimension one cross-section transverse to the flow and let $\varphi : X \to \mathbb{R}^+$ be a return time function, namely a function such that $g_{\varphi(x)}(x) \in X$ for $x \in X$. The map $T = g_\varphi : X \to X$ is called the *Poincaré map*. We assume (possibly after shrinking X) that $\inf \varphi > 0$. Given an ergodic invariant probability measure μ_X on X and $\varphi \in L^1(X)$, we construct an ergodic invariant probability measure μ on M as follows. Define the suspension

$$X^\varphi = \{(x, u) \in X \times \mathbb{R} : 0 \le u \le \varphi\}/\sim, \qquad (x, \varphi(x)) \sim (Tx, 0).$$

The suspension flow $T_t : X^\varphi \to X^\varphi$ is given by $T_t(x, u) = (x, u + t)$ modulo identifications. The probability measure $\mu^\varphi = (\mu_X \times \text{Leb})/\int_X \varphi \, d\mu_X$ is ergodic and T_t-invariant. Moreover, $\pi : X^\varphi \to M$ given by $\pi(x, u) = T_u x$ is a semiconjugacy from T_t to g_t and $\mu = \pi_* \mu^\varphi$ is the desired ergodic invariant probability measure on M.

Now suppose that $v \in C_0^\kappa(M, \mathbb{R}^m)$ and define the induced observable

$$V : X \to \mathbb{R}^m, \qquad V(x) = \int_0^{\varphi(x)} v(g_u x) \, du.$$

By a purely probabilistic argument [33] (based on [35, 57, 60, 64]), the WIP for $V :$ $X \to \mathbb{R}^m$ with the map T implies a WIP for $v : M \to \mathbb{R}^m$. That is, setting $v_t = \int_0^t v \circ g_s \, ds$ and $W_n(t) = n^{-1/2} v_{nt}$, we obtain $W_n \to_\mu W$ where W is Brownian motion with covariance $\Sigma = \Sigma_X / \int_X \varphi \, d\mu_X$ and Σ_X is the covariance in the WIP for V.

By Bowen [12], Axiom A flows can be realized as suspension flows over uniformly hyperbolic diffeomorphisms, and the above considerations yield the WIP for attractors for Axiom A flows [22]. The same is true for large classes of nonuniformly hyperbolic flows modelled as suspensions over Young towers with summable decay of correlations, including Lorentz gases, Lorenz attractors [6] and singular hyperbolic attractors [5]. In these situations, μ_X and μ are SRB measures on X and M respectively. Also the nondegeneracy property in Proposition 2.1 applies to Σ_X and thereby $\Sigma = \Sigma_X / \int_X \varphi \, d\mu_X$. Moreover,

$$\Sigma = \lim_{t \to \infty} t^{-1} \int_M v_t \otimes v_t \, d\mu,$$

and under extra (rather restrictive) mixing assumptions

$$\Sigma = \int_0^\infty \int_M \{v \otimes (v \circ g_t) + (v \circ g_t) \otimes v\} \, d\mu \, dt.$$

3 General Rough Path Theory

3.1 Limit Theorems from Rough Path Analysis

Consider a (for simplicity only: finite-dimensional) Banach space $(\mathcal{B}, \|\cdot\|)$ and fixed $p \in [2, 3)$. Define the group $G := \mathcal{B} \oplus (\mathcal{B} \otimes \mathcal{B})$ with multiplication $(a, M) \star$ $(b, N) := (a + b, M + a \otimes b + N)$, inverse $(a, M)^{-1} := (-a, -M + a \otimes a)$, and identity $(0, 0)$. A (level-2) p-rough path (over $\mathcal{B}$, on $[0, 1]$) is a path $\mathbf{X} = (\mathbf{X}_t : 0 \le t \le 1)$ with values and increments $\mathbf{X}_{s,t} := \mathbf{X}_s^{-1} \star \mathbf{X}_t := (X_{s,t}, \mathbb{X}_{s,t}) \in G$ of finite p-variation condition, $p \in [2, 3)$, either in the sense ("*homogeneous rough path norm*")

$$\|\mathbf{X}\|_{p\text{-var}} := \|X\|_{p\text{-var}} + \|\mathbb{X}\|_{(p/2)\text{-var}}^{1/2} < \infty, \tag{11}$$

or, equivalently, in terms of the *inhomogeneous rough path norm*

$$\|\mathbf{X}\|_{p\text{-var}} := \|X\|_{p\text{-var}} + \|\mathbb{X}\|_{(p/2)\text{-var}} < \infty ; \tag{12}$$

we used the notation, applicable to any Ξ from $\{0 \le s \le t \le 1\}$ into a normed space, any $q > 0$,

$$\|\Xi\|_{q\text{-var}} := \left(\sup_{\mathcal{P}} \sum_{[s,t] \in \mathcal{P}} |\Xi_{s,t}|^q \right)^{\frac{1}{q}} < \infty. \tag{13}$$

Write $\mathscr{C}^{p\text{-var}}([0, 1], \mathcal{B})$ resp. $\mathscr{D}^{p\text{-var}}([0, 1], \mathcal{B})$ for the space of such (continuous resp. càdlàg) p-rough paths; and also $\mathscr{C}$ resp. $\mathscr{D}$ for the space of continuous resp. càdlàg paths with values in $\mathcal{B} \oplus (\mathcal{B} \otimes \mathcal{B})$. The space of α-Hölder rough paths, $\mathscr{C}^{\alpha\text{-Höl}}$ with $\alpha = 1/p \in (1/3, 1/2]$ forms a popular subclass of $\mathscr{C}^{p\text{-var}}$. A weakly geometric p-rough path, in symbols $\mathbf{X} \in \mathscr{C}_g^{p\text{-var}}$, satisfies a "product rule" of the type $\mathrm{Sym}(\mathbb{X}_t) = (1/2)X_t \otimes X_t$; effectively $\mathbf{X}$ takes values in a sub-group $H \subset G$. (We remark that, when $\mathcal{B} = \mathbb{R}^m$, the (Lie) groups H, G, can be identified with, respectively, the step-2 truncated free nilpotent group with m generators and the step-2 truncated Butcher group with m decorations of its nodes.) Every continuous BV path lifts canonically via

$$\mathbb{X}_t = \int_0^t (X_s - X_0) \otimes dX_s,$$

and gives rise to a (continuous) weakly geometric p-rough path. Conversely, every $\mathbf{X} \in \mathscr{C}_g^{p\text{-var}}$ is the uniform limit of smooth paths, with uniform p-variation bounds. Similarly, every càdlàg BV path X lifts canonically via

$$\mathbb{X}_t = \int_{(0,t]} (X_s^- - X_0) \otimes dX_s$$

to a (càdlàg) p-rough path in $\mathscr{D}^{p\text{-var}}$. We introduce, on $\mathscr{D}^{p\text{-var}}$ (and then by restriction on $\mathscr{C}^{p\text{-var}}$ and $\mathscr{C}_g^{p\text{-var}}$) the *(inhomogeneous) p-rough path distance*[2]

$$\|\mathbf{X}; \tilde{\mathbf{X}}\|_{p\text{-var}} := \|X - \tilde{X}\|_{p\text{-var}} + \|\mathbb{X} - \tilde{\mathbb{X}}\|_{(p/2)\text{-var}}. \tag{14}$$

(A similar Hölder rough path distance can be defined on $\mathscr{C}^{\alpha\text{-Höl}}$ and $\mathscr{C}_g^{\alpha\text{-Höl}}$).

---

[2]In view of the genuine non-linearity of rough path spaces, we refrain from writing $\|\mathbf{X} - \tilde{\mathbf{X}}\|_{p\text{-var},[0,1]}$.

Consider sufficiently regular vector fields $V_0 : \mathbb{R}^d \to \mathbb{R}^d$ and $V : \mathbb{R}^d \to L(\mathcal{B}, \mathbb{R}^d)$. By definition, Y solves the *rough differential equation (RDE)*

$$dY = V_0(Y^-)dt + V(Y^-)d\mathbf{X}$$

if, for all $0 \leq s < t \leq 1$, writing DV for the derivative,[3]

$$Y_t - Y_s = V_0(Y_s)(t - s) + V(Y_s)X_{s,t} + DV(Y_s)V(Y_s)\mathbb{X}_{s,t} + R_{s,t}, \qquad (15)$$

where, for the "remainder term" R, we require, writing $\mathcal{P}(\varepsilon)$ for a partition of $[0, 1]$ with mesh-size less than ε,

$$\sup_{\mathcal{P}(\varepsilon)} \sum_{[s,t] \in \mathcal{P}(\varepsilon)} |R_{s,t}| \to 0 \quad \text{as } \varepsilon \to 0.$$

This definition first encodes that Y is controlled (cf. [27]) by $X \in D^{p\text{-var}}$, with derivative $Y' = V(Y_s) \in D^{p\text{-var}}$ and remainder $Y_{s,t}^{\#} = W(Y_s)\mathbb{X}_{s,t} + R_{s,t}$ with $\|Y^{\#}\|_{(p/2)\text{-var}} < \infty$. As a consequence, Y satisfies a bona fide rough integral equation, for all $t \in (0, 1]$,

$$Y_t = y_0 + \int_{(0,t]} V_0(Y_s^-)ds + \int_{(0,t]} V(Y_s^-)d\mathbf{X}_s.$$

Conversely, (15) is satisfied by every solution to this integral equation. See e.g. [26] for more details on this construction in the Hölder rough path case, and [27, 30] for the càdlàg p-variation case; this contains the discrete Hölder setting of [39]. The following theorem, in the case of continuous p-rough paths, is due to Lyons [49], the recent extension to càdlàg rough paths is taken from [30]. We write C^{p+} to indicate $C^{p+\varepsilon}$, for some $\varepsilon > 0$.

Theorem 3.1 (Continuity of RDE solution map) *Let $p \in [2, 3)$. Consider a càdlàg rough path $\mathbf{X} \in \mathscr{D}^{p\text{-var}}([0, 1], \mathcal{B})$, and assume $V_0 \in C^{1+}$ and $V \in C^{p+}$. Then there exists a unique càdlàg solution $Y \in D([0, 1], \mathbb{R}^d)$ to the rough differential equation*

$$dY_t = V_0(Y_t^-)dt + V(Y_t^-)d\mathbf{X}_t, \quad Y_0 = y_0 \in \mathbb{R}^d,$$

and the solution is locally Lipschitz in the sense that

$$\|Y - \tilde{Y}\|_{p\text{-var}} \lesssim \|\mathbf{X}; \tilde{\mathbf{X}}\|_{p\text{-var}} + |y_0 - \tilde{y}_0|$$

with proportionality constant uniform over bounded classes of driving p-rough paths.

The p-variation rough path distance can be replaced by a p-variation Skorokhod type rough path metric, which adds more flexibility when the limiting (rough) path

---

[3]In coordinates, when $\mathcal{B} = \mathbb{R}^m$, we have $DV(Y_s)V(Y_s)\mathbb{X}_{s,t} = \partial_\alpha V_\gamma(Y_s)V_\beta^\alpha(Y_s)\mathbb{X}_{s,t}^{\beta,\gamma}$ with summation over $\alpha = 1, \ldots, d$ and $\beta, \gamma = 1, \ldots, m$..

has jumps, but we won't need this generality here. Checking p-variation rough path convergence can be done by interpolation: uniform convergence plus uniform p'-variation bounds, for some $p' < p$.

It is known that (càdlàg) semimartingales give rise to càdlàg p-rough paths for any $p \in (2, 3)$ [18]. Solving the resulting random RDE provides exactly a (robust) solution theory for the corresponding SDE. As a consequence, we have the following limit theorems of Stratonovich and Itô type, which cannot be obtained by UCV/UT type argument familiar from stochastic analysis. (Assumptions on V_0, V are as above.) The following theorem applies in particular to sequences of smooth processes, in which case Stratonovich SDEs are simply random ODEs.

Theorem 3.2 (Stratonovich-type limit theorem) *Consider a sequence of continuous semimartingale drivers* (B^n) *with Stratonovich lift* $(\mathbf{B}^{\circ,n})$, *such that* $\mathbf{B}^{\circ,n}$ *converges to* $\mathbf{B} = (B, \mathbb{B}^\circ + \Gamma)$, *for some continuous BV process* Γ, *weakly (resp. in probability, a.s.) in the uniform topology with* $\{\|\mathbf{B}^{\circ,n}\|_{p\text{-}var}(\omega)\}$ *tight, for* $p \in (2, 3)$. (Γ *is necessarily skew-symmetric.*)

(i) *For any* $p' > p$, *it holds that* $\mathbf{B}^{\circ,n} \to \mathbf{B}$ *weakly (resp. in probability, a.s.) in the* p'-*variation rough path topology.*

(ii) *Assume, in the sense of Stratonovich SDEs,*[4]

$$dY_t^n = V_0(Y_t^n)dt + V(Y_t^n) \circ dB_t^n$$

such that $Y_0^n \equiv y_0^n \to y_0$. *Then the Stratonovich SDE solutions* Y^n *converge weakly (resp. in probability, a.s.) to* Y *in the uniform topology, where* $Y_0 = y_0$ *and*

$$dY_t = V_0(Y_t)dt + DV(Y_t)V(Y_t)d\Gamma_t + V(Y_t)dB_t.$$

Moreover, $\{\|Y^n\|_{p\text{-}var}(\omega) : n \geq 1\}$ *is tight and one also has weak (resp. in probability, a.s.) convergence in the* p'-*variation uniform metric for any* $p' > p$.

We now state an analogous Itô-type result. The next theorem in particular applies to sequences of piecewise constant, càdlàg processes, in which case, Itô SDEs are simply stochastic recursions.

Theorem 3.3 (Itô-type limit theorem) *Consider a sequence of càdlàg semimartingale drivers* B^n, *with Itô lift* $\mathbf{B}^n = (B^n, \mathbb{B}^n)$, *such that* $\mathbf{B}^n$ *converges to* $\mathbf{B} = (B, \mathbb{B} + \Gamma)$, *for some càdlàg BV process* Γ, *weakly (resp. in probability, a.s.) in the uniform topology with* $\{\|\mathbf{B}^n\|_{p\text{-}var}(\omega)\}$ *tight, for* $p \in (2, 3)$.

(i) *For any* $p' > p$, *it holds that* $\mathbf{B}^{\circ,n} \to \mathbf{B}$ *weakly (resp. in probability, a.s.) in the* p'-*variation rough path topology.*

---

[4]Often B^n has continuous BV sample paths. Every such process is (trivially) a semimartingale (under its own filtration); the Stratonovich SDE interpretation is the one consistent with the ODE interpretation, in the sense of a Riemann-Stieltjes integral equation.

(ii) Assume, in the sense of Itô SDEs,

$$dY_t^n = V_0(Y_t^{n,-})dt + V(Y_t^{n,-})dB_t^n$$

such that $Y_0^n \equiv y_0^n \to y$. Then Itô SDE solutions Y^n converge weakly (resp. in probability, a.s.) to Y in the uniform topology, where $Y(0) = y$ and

$$dY_t = V_0(Y_t^-)dt + DV(Y_t^-)V(Y_t^-)d\Gamma_t + W(Y_t^-)dB_t.$$

Moreover, $\{\|Y^n\|_{p\text{-}var}(\omega) : n \geq 1\}$ is tight and one also has weak (resp. in probability, a.s.) convergence in the p'-variation uniform metric for any $p' > p$.

Remark 3.4 A minor generalization of Theorem 3.3, which will be convenient later on, states that the drift term $V_0(Y_t^{n,-})dt$ in the approximate problem can be replaced by $V_0(Y_t^{n,-})d\tau^n$ where $\tau^n(t) \to t$ uniformly with uniform 1-variation bounds.

We emphasize that in both Theorems 3.2 and 3.3, the *only purpose* of the semimartingale and adaptedness assumptions is to give a familiar interpretation of what is really a rough differential equation driven by a random rough path. (Consistency with SDEs, in a general semimartingale setting, is established in [18]). The proof is essentially a corollary of interpolation, in a weak convergence setting, with the purely deterministic Theorem 3.1, see [30] for details.

3.2   WIPs in Rough Path Theory

We start with some generalities. A *Brownian rough path* (over $\mathbb{R}^m$) is an $\mathbb{R}^m \oplus (\mathbb{R}^m \otimes \mathbb{R}^m)$-valued continuous process $\mathbf{B} = (B, \mathbb{B})$ with independent increments with respect to the group structure introduced in Sect. 3.1, such that B is centered. (In particular, B is a classical m-dimensional Brownian motion.) It is known that sample paths $\mathbf{B}(\omega)$ are, with probability one, in $\mathscr{C}^{\alpha\text{-Höl}}$ for any $\alpha < 1/2$, and hence also in $\mathscr{C}^{p\text{-var}}$ and $\mathscr{D}^{p\text{-var}}$ for any $p > 2$. We have a full characterization of Brownian rough paths: B is a classical m-dimensional Brownian motion (with some covariance $\Sigma \in \mathbb{R}^m \otimes \mathbb{R}^m$) and

$$\mathbb{B}_{s,t} = \int_s^t B_{s,r} \otimes \circ dB_r + (t-s)\Gamma = \int_s^t B_{s,r} \otimes dB_r + (t-s)(\Gamma + \tfrac{1}{2}\Sigma)$$

for some matrix $\Gamma \in \mathbb{R}^m \otimes \mathbb{R}^m$, which we name *area drift*. (Note that $\mathbf{B}$ is geometric iff Γ is skew-symmetric.) Given a sequence of random rough paths $(\mathbf{B}_\varepsilon)$, we say that the WIP holds *in α-Hölder (resp. p-variation) rough path sense* if, as $\varepsilon \to 0$,

$$\mathbf{B}_\varepsilon \to_w \mathbf{B} \text{ in } \mathscr{C}^{\alpha\text{-Höl}} \text{ (resp. } \mathscr{C}^{p\text{-var}}, \mathscr{D}^{p\text{-var}}) \tag{16}$$

but note that only the regimes $\alpha \in (1/3, 1/2)$ (resp. $p \in (2, 3)$) correspond to a WIP in a bona fide rough path topology. As is implicit in Theorems 3.2 and 3.3, this follows from checking convergence in law in the uniform topology; that is, an *enhanced weak invariance principle* in the sense that[5]

$$\mathbf{B}_\varepsilon \to_w \mathbf{B} \text{ in } \mathscr{C} \text{ (resp. } \mathscr{D}) \text{ as } \varepsilon \to 0,$$

together with tightness of α-Hölder (resp. p-variation) rough path norms, at the expense of replacing α (resp. p) in (16) with $\alpha' < \alpha$ (resp. $p' > p$).

In simple situations, $\mathbf{B}_\varepsilon$ is given as the canonical (Stratonovich or Itô) lift of a *good* sequence of convergent semimartingales. In this case, the limiting area drift is zero and $\{\|\|\mathbf{B}_\varepsilon\|\|_{p\text{-var}} : \varepsilon \in (0, 1]\}$ is automatically tight [18], for any $p > 2$. (This gives a decisive link between classical semimartingale stability theory [37, 46] with rough path analysis.) Immediate applications then include Donsker's theorem in p-variation rough path topology, under identical (finite second) moment assumptions as the classical Donsker theorem. (With piecewise linear interpolation, the limit is the Stratonovich lift $(B, \int B \otimes \circ dB)$, with piecewise constant interpolation the limit is the Itô lift $(B, \int B \otimes dB)$.) As another immediate application, the (functional) CLT for L^2-martingales with stationary, ergodic increments is valid on a rough path level [18]. If interested in the α-Hölder rough path topology, one can use a Kolmogorov-type tightness criterion [29]. Provided $q > 1$, a uniform moment estimate of the form

$$\sup_{\varepsilon \in (0,1]} \mathbb{E}\left[\|\|\mathbf{B}_\varepsilon(s, t)\|\|^{2q}\right]^{1/2q} \lesssim |t - s|^{1/2}, \tag{17}$$

or equivalently,

$$\sup_{\varepsilon \in (0,1]} \mathbb{E}\left[|B_\varepsilon(s, t)|^{2q}\right]^{1/2q} \lesssim |t - s|^{1/2}, \quad \sup_{\varepsilon \in (0,1]} \mathbb{E}\left[|\mathbb{B}_\varepsilon(s, t)|^q\right]^{1/q} \lesssim |t - s|,$$

gives tightness in the α-Hölder topology, for every $\alpha < 1/2 - 1/(2q)$.

Remark 3.5 In the Hölder setting, note that only $\alpha > \frac{1}{3}$ gives a bona fide (level-2) rough path metric under which the Itô map behaves continuously. This leads to the suboptimal moment assumption $q > 3$. To obtain the WIP in the Hölder rough path sense [13], this necessitates increments with 6+ moments. (In contrast, we have the WIP in the p-variation rough path topology under the optimal assumption of 2+ moments.) This is also the main drawback of using the Hölder topology in [40].

Of course, in Gaussian situations such moment assumptions are harmless and can conveniently be reduced to $q = 1$. An instructive example is given by physical Brownian motion X^ε, as introduced in Sect. 2.1. The tightness condition can be seen to be satisfied for all $q < \infty$, giving α-Hölder rough path tightness for any $\alpha < 1/2$. More interestingly, X^ε has a Brownian rough path limit with non-zero area drift [25], provided the particle feels a Lorentz force, expressed through non-symmetry of M.

---

[5] Again it suffices to work with the uniform topology on both $\mathscr{C}$ and $\mathscr{D}$.

(See the notation in Sect. 2.1.) Specifically, this is seen by writing MX^ε as Brownian motion plus a "corrector" which goes uniformly to zero, but leads, in the $\varepsilon \to 0$ limit, to an area contribution.

Remark 3.6 We note that (17) requires no martingale assumptions whatsoever. It is an important observation for the sequel that (17) leads to tightness not only in the α-Hölder rough path topology but also in the p-variation rough path topology: to wit, it follows from the Besov-variation embedding [28] that (17) implies p-variation tightness for any $p > 2$. In this way, for example, one can reprove the WIP in p-variation rough path topology under the almost optimal assumption of $2+$ moments. This argument becomes important when direct martingale arguments are not possible.

4 Applications to Fast-Slow Systems

4.1 Chaotic Dynamics: Enhanced WIP and Moments

In this subsection, we resume the discussion of chaotic dynamical systems from Sect. 2.3 but now focusing on some finer statistical properties, namely an enhanced WIP and moment estimates, that are required for applying rough path theory.

4.1.1 Expanding Maps

Continuing Sect. 2.3.1, we suppose that $T : M \to M$ is a C^2 uniformly expanding map, with unique ergodic absolutely continuous invariant probability measure μ, so conditions (5) and (6) hold. In particular, for any $v \in C_0^\kappa(M, \mathbb{R}^m)$, we have an L^2 martingale-coboundary decomposition decomposition $v = m + \chi \circ T - \chi$. Define the càdlàg processes $W_n \in D([0, 1], \mathbb{R}^m)$, $\mathbb{W}_n \in D([0, 1], \mathbb{R}^m \otimes \mathbb{R}^m)$,

$$W_n(t) = n^{-1/2} \sum_{0 \le j < n} v \circ T^j, \qquad \mathbb{W}_n(t) = n^{-1} \sum_{0 \le i < j < n} (v \circ T^i) \otimes (v \circ T^j).$$

Recall that we have the WIP $W_n \to_\mu W$ in $D([0, 1], \mathbb{R}^m)$ where W is m-dimensional Brownian motion with covariance Σ given by formulas (7) and (8). By [40, Theorem 4.3], we have the *enhanced WIP* (called iterated WIP in [40])

$$(W_n, \mathbb{W}_n) \to_\mu (W, \mathbb{W}) \quad \text{in } \mathscr{D}([0, 1], \mathbb{R}^m \times (\mathbb{R}^m \otimes \mathbb{R}^m)),$$

where $W(t) = \int_0^t W \otimes dW + \Gamma t$. Here $\int W \otimes dW$ is the Itô integral and the area drift $\Gamma \in \mathbb{R}^m \otimes \mathbb{R}^m$ is given by

$$\Gamma = \lim_{n \to \infty} \int_M \mathbb{W}_n(1) \, d\mu, \tag{18}$$

and satisfies

$$\Gamma = \sum_{n=1}^{\infty} \int_M v \otimes (v \circ T^n) \, d\mu. \tag{19}$$

The proof of the enhanced WIP in [40] has two main steps. The first step is to apply [46, Theorem 2.2] (alternatively [37]) to the martingale component m taking into consideration that $\{m \circ T^n, \ n \geq 0\}$ is a reverse martingale difference sequence. This yields an enhanced WIP with zero area drift. The contribution from the coboundary $\chi \circ T - \chi$ is no longer negligible, but a general result [40, Theorem 3.1] for mixing dynamical systems and L^2 coboundaries yields the (typically nonzero) area drift Γ.

Again, strong distributional convergence applies by [78, Theorem 1]. The hypotheses in [78] are verified in the course of the proof of [40, Lemma 6.3]. Hence $(W_n, \mathbb{W}_n) \to_\lambda (W, \mathbb{W})$ in $\mathscr{D}([0, 1], \mathbb{R}^m \times (\mathbb{R}^m \otimes \mathbb{R}^m))$ for all absolutely continuous Borel probability measures λ.

For nonuniformly expanding maps, the enhanced WIP goes through unchanged provided the martingale-coboundary decomposition holds in L^2 (with the usual caveat that formula (19) only holds when μ is mixing). This covers the situation (10) with $\beta > 2$. As before, extra work is required for the case $\beta \in (1, 2]$. By [40, Theorem 10.2], the enhanced WIP holds for all $\beta > 1$ for nonuniformly expanding maps modelled by Young towers, including the intermittent maps (3). For such maps, we obtain optimal results: the enhanced WIP holds precisely when the ordinary CLT holds.

Turning to moments, an immediate consequence of the L^p martingale-coboundary decomposition, $p \geq 2$, and Burkhölder's inequality is that $\|v_n\|_p = O(n^{1/2})$ where the implied constant depends on v and p. As noted in [52, 55, 58], in fact

$$\|v_n\|_{2p} = O(n^{1/2}) \tag{20}$$

and this holds for L^p martingale-coboundary decompositions with $p \geq 1$. This improved result uses the additional information that $v \in L^\infty$ and a maximal inequality of [65].

In the situation (10), we have the martingale-coboundary decomposition for all $1 \leq p < \beta$. As shown in [52, 55], the estimate (20) is sharp; $\|v_n\|_q = O(n^{1/2})$ for $q < 2\beta$ but there are examples where the estimate typically fails for $q > 2\beta$.

We also require estimates for the enhanced (iterated) moment $S_n = \sum_{0 \leq i \leq j < n} (v \circ T^i) \otimes (v \circ T^j)$. Assuming an L^p martingale-coboundary decomposition with $p \geq 3$ and $v \in L^\infty$, it was shown in [40, Proposition 7.1] that $\|S_n\|_{2p/3} = O(n)$. In the Young tower setting, this has been improved in [44] to $\|S_n\|_p = O(n)$ for $p \geq 1$.

The moment estimates discussed above are all in L^p spaces with respect to μ. Clearly if $\lambda \ll \mu$ and $d\lambda/d\mu \in L^\infty$ then the same moment estimates hold also with respect to λ. In particular, we can take $\lambda = \text{Vol}$ for the C^2 uniformly expanding maps. For the intermittent maps (3) it is standard that $d\mu/d\,\text{Leb}$ is bounded below, so we can take $\lambda = \text{Leb}$.

4.1.2 Hyperbolic Diffeomorphism

For Axiom A diffeomorphisms and Young towers with exponential tails, we saw in Sect. 2.3.2 that there is an L^p martingale-coboundary decomposition for all p. Also, for Young towers with exponential contraction and polynomial tails $\mu_Y(\tau > n) = O(n^{-(\beta+1)})$, we have an L^p martingale-coboundary decomposition for $p < \beta$. By [40, 44], we obtain the enhanced WIP provided $\beta > 1$ and optimal moment estimates $\|v_n\|_{2p} = O(n^{1/2})$ and $\|S_n\|_p = O(n)$ for $1 \le p < \beta$. As before, the covariance Σ and drift Γ satisfy (7) and (18), and under additional mixing assumptions we have (8) and (19).

For general Young towers with polynomial tails $\mu_Y(\tau > n) = O(n^{-\beta})$ the enhanced WIP still holds for all $\beta > 1$ by [59] but currently we only have the moment estimates $\|v_n\|_{2p} = O(n^{1/2})$ and $\|S_n\|_{2p/3} = O(n)$ for $3 \le p < \beta$ from [40]. Obtaining optimal moment estimates here is the subject of work in progress.

4.1.3 Hyperbolic Flows

The methods mentioned in Sect. 2.3.3 for passing the WIP from (non)uniformly hyperbolic diffeomorphisms to (non)uniformly hyperbolic flows work just as well for the enhanced WIP [40, Sect. 6]. Define

$$W_n(t) = n^{-1/2} \int_0^{nt} v \circ g_s \, ds, \qquad \mathbb{W}_n(t) = n^{-1} \int_0^{nt} \int_0^s (v \circ g_r) \otimes (v \circ g_s) \, dr \, ds.$$

Then $(W_n, \mathbb{W}_n) \to_\lambda (W, \mathbb{W})$ where W is Brownian motion with covariance Σ and $\mathbb{W}(t) = \int_0^t W \otimes dW + \Gamma_I t$. Here $\Sigma = \lim_{n\to\infty} \mathbb{E}_\lambda(W_n(1) \otimes W_n(1))$ as before, and $\Gamma_I = \lim_{n\to\infty} \mathbb{E}_\lambda \mathbb{W}_n(1)$. Alternatively, $\mathbb{W}(t) = \int_0^t W \otimes \circ dW + \Gamma t$, where $\Gamma = \Gamma_I - \frac{1}{2}\Sigma$ is skew-symmetric. Under extra mixing assumptions,

$$\Sigma = \int_0^\infty \int_\Lambda \{v \otimes (v \circ g_t) + (v \circ g_t) \otimes v\} \, d\mu \, dt, \qquad (21)$$

$$\Gamma = \frac{1}{2} \int_0^\infty \int_\Lambda \{v \otimes (v \circ g_t) - (v \circ g_t) \otimes v\} \, d\mu \, dt. \qquad (22)$$

The situation for moments extends in a straightforward way [40, Sect. 7.2]. Define

$$v_n = \int_0^n v \circ g_s \, ds, \qquad S_n = \int_0^n \int_0^s (v \circ g_r) \otimes (v \circ g_s) \, dr \, ds.$$

Then the estimates described in Sect. 2.3.1 apply equally here.

4.2 Continuous Dynamics

We present now an application of rough path theory to fast-slow systems where the fast variable satisfies a suitable WIP. We consider first continuous dynamics (1) in the case of multiplicative noise, i.e.,

$$\dot{x}_\varepsilon = a(x_\varepsilon) + \varepsilon^{-1} b(x_\varepsilon) v(y_\varepsilon), \qquad \dot{y}_\varepsilon = \varepsilon^{-2} g(y_\varepsilon) . \tag{23}$$

We consider the $\mathbb{R}^m$-valued path

$$W_\varepsilon(t) = \varepsilon \int_0^{t\varepsilon^{-2}} v \circ g_s \, ds$$

and rewrite the slow dynamics in the form of a controlled ODE,

$$dx_\varepsilon = a(x_\varepsilon) dt + b(x_\varepsilon) dW_\varepsilon.$$

Following Kelly–Melbourne [40], this formulation invites an application of finite-dimensional rough path theory; the only modification relative to [40] is our present use of p-variation rough path metrics, which leads to optimal moment assumptions and optimal regularity assumptions on the coefficients. The key is a suitable WIP on the level of rough paths, as discussed in Sect. 3.

To this end, we consider the following two assumptions on the fast dynamics. Following the discussion in Sect. 4.1, we see that a wide range of dynamics satisfy these assumptions. For every $\varepsilon > 0$, we let $\mathbb{W}_\varepsilon$ be the canonical second iterated integral of W_ε, and for $p \in (2, 3)$, we consider the geometric p-rough path $\mathbf{W}_\varepsilon := (W_\varepsilon, \mathbb{W}_\varepsilon)$.

Assumption 4.1 It holds that $(W_\varepsilon, \mathbb{W}_\varepsilon) \to (W, \mathbb{W})$ as $\varepsilon \to 0$ in the sense of finite-dimensional distributions on (M, λ), where W is an m-dimensional Brownian motion and $\mathbb{W}(t) = \int_0^t W \otimes \circ dW + \Gamma t$ for some $\Gamma \in \mathbb{R}^m \otimes \mathbb{R}^m$ deterministic.

Assumption 4.2 There exists $q > 1$ and $K > 0$ such that

$$\left\| \int_s^t v^i \circ g_r \, dr \right\|_{L^{2q}(\lambda)} \leq K |t - s|^{1/2},$$

$$\left\| \int_s^t \int_s^r v^i \circ g_u \, v^j \circ g_r \, du \, dr \right\|_{L^q(\lambda)} \leq K |t - s|,$$

for all $s, t \geq 0$ and $1 \leq i, j \leq m$.

The first assumption identifies the possible limit points of W_ε as a rough path; the second ensures that the WIP holds in a sufficiently strong rough path topology as demonstrated by the following result. As before, all path space norms (p-var, α-Höl, etc.) are relative to the fixed interval $[0, 1]$.

Proposition 4.3 *Under Assumption 4.2, it holds that for all* $p \in (2, 3)$

$$\sup_{\varepsilon \in (0, 1]} \mathbb{E} \| \mathbf{W}_\varepsilon \|_{p\text{-var}}^{2q} < \infty,$$

and for all $\alpha \in (0, \frac{1}{2} - \frac{1}{2q})$

$$\sup_{\varepsilon \in (0, 1]} \mathbb{E} \| \mathbf{W}_\varepsilon \|_{\alpha\text{-Höl}}^{2q} < \infty.$$

Proof Viewing $\mathbf{W}_\varepsilon$ as a path in $G^2(\mathbb{R}^m) \subset \mathbb{R}^m \oplus (\mathbb{R}^m \otimes \mathbb{R}^m)$, the step-2 free nilpotent group equipped with Carnot-Carathéodory metric d, it holds that

$$|d(\mathbf{W}_\varepsilon(s), \mathbf{W}_\varepsilon(t))|_{L^{2q}(\lambda)} \lesssim |W_\varepsilon(s, t)|_{L^{2q}(\lambda)} + |\mathbb{W}_\varepsilon(s, t)|_{L^q(\lambda)}^{1/2} \lesssim |t - s|^{1/2},$$

where the final bound follows from Assumption 4.2. Let $\beta \in [0, 1/2)$. Then $\mathbb{E}[|\mathbf{W}_\varepsilon|_{W^{\beta, 2q}}^{2q}]$ is uniformly bounded in $\varepsilon > 0$, and thus, by the Besov-Hölder and Besov-variation embeddings [28] (see also [29, Corollary A.2, A.3]), so is $\mathbb{E}\|\mathbf{W}_\varepsilon\|_{(\beta - 1/(2q))\text{-Höl}}^q$ and $\mathbb{E}\|\mathbf{W}_\varepsilon\|_{(1/\beta)\text{-var}}^{2q}$.

Theorem 4.4 *Suppose Assumptions 4.1 and 4.2 hold.*

(i) *For every* $p > 2$, *it holds that* $\mathbf{W}_\varepsilon \to_\lambda \mathbf{W}$ *in the* p-*variation rough path topology.*
(ii) *Let* $a \in C^{1+}(\mathbb{R}^d, \mathbb{R}^d)$, $b \in C^{2+}(\mathbb{R}^d, \mathbb{R}^{d \times m})$, *and let* x_ε *be the solution to* (23). *Then* $x_\varepsilon \to_\lambda X$ *in* $C^{p\text{-var}}([0, 1], \mathbb{R}^d)$ *for every* $p > 2$, *where* X *is the solution to the SDE*

$$dX = \left(a(X) + \sum_{i,j=1}^m \Gamma^{i,j} \sum_{k=1}^d b^{i,k} \partial_k b^j(X) \right) dt + b(X) \circ dW, \quad X(0) = \xi.$$

$$(24)$$

Remark 4.5 If follows from Assumptions 4.1 and 4.2 that the covariance matrix Σ and area drift Γ are given by

$$\Sigma = \lim_{\varepsilon \to 0} \mathbb{E}_\lambda(W_\varepsilon(1) \otimes W_\varepsilon(1)), \qquad \Gamma = \lim_{\varepsilon \to 0} \mathbb{E}_\lambda \mathbb{W}_\varepsilon(1) - \frac{1}{2}\Sigma.$$

Under additional mixing assumptions, formulas (21) and (22) hold.

Proof (i) follows from part (i) of Theorem 3.2. For (ii), observe that x_ε solves the ODE

$$dx_\varepsilon = a(x_\varepsilon)dt + b(x_\varepsilon)dW_\varepsilon.$$

We are thus in the framework of part (ii) of Theorem 3.2, from which the conclusion follows.

$\square$

4.3 Discrete Dynamics

We now discuss discrete dynamics (2) in the case of multiplicative noise, i.e.,

$$X_{j+1}^{(n)} = X_j^{(n)} + n^{-1}a(X_j^{(n)}) + n^{-1/2}b(X_j^{(n)})v(Y_j),$$

where, as before, $v : M \to \mathbb{R}^m, b : \mathbb{R}^d \to \mathbb{R}^{d \times m}$, and $a : \mathbb{R}^d \to \mathbb{R}^d$. As usual, $X_0^{(n)} = \xi \in \mathbb{R}^d$ is fixed and Y_0 is drawn randomly from a probability measure λ on M. To consider this system as a controlled ODE, we introduced the càdlàg path

$$x_n : [0, 1] \to \mathbb{R}^d, \quad x_n(t) = X_{\lfloor nt \rfloor}^{(n)} \tag{25}$$

as well as the the càdlàg paths

$$W_n : [0, 1] \to \mathbb{R}^m, \quad W_n(t) = n^{-1/2} \sum_{j=0}^{\lfloor nt \rfloor - 1} v(Y_j),$$

$$z_n : [0, 1] \to \mathbb{R}, \quad z_n(t) = \lfloor tn \rfloor / n.$$

It is easy to verify that x_n defined by (25) is the unique solution of the controlled (discontinuous) ODE

$$dx_n = a(x_n^-)dz_n + b(x_n^-)dW_n, \quad x_n(0) = \xi \in \mathbb{R}^d. \tag{26}$$

Let us denote by $\mathbb{W}_n$ the canonical second iterated integral of W_n

$$\mathbb{W}_{v,n}^{i,j}(s, t) = \int_{(s,t]} (W_{v,n}^{i,-}(r) - W_{v,n}^i(s))dW_{v,n}^j(r), \quad 1 \le i, j \le m.$$

Consider the following analogues of Assumptions 4.1 and 4.2.

Assumption 4.6 It holds that $(W_n, \mathbb{W}_n) \to (W, \mathbb{W})$ as $\varepsilon \to 0$ in the sense of finite-dimensional distributions on (M, λ), where W is a Brownian motion in $\mathbb{R}^m$ and $\mathbb{W}(t) = \int_0^t W \otimes dW + \Gamma t$ for some $\Gamma \in \mathbb{R}^{m \times m}$ deterministic.

Assumption 4.7 There exists $q > 1$ and $K > 0$ such that for all $n \geq 1$ and $0 \leq k, l \leq n$,

$$\left\| W_n(l/n) - W_n(k/n) \right\|_{L^{2q}(\lambda)} \leq Kn^{-1/2}|l - k|^{1/2},$$

$$\left\| \mathbb{W}_n(k/n, l/n) \right\|_{L^q(\lambda)} \leq Kn^{-1}|l - k|.$$

Proposition 4.8 *Under Assumption 4.7, for all $p \in (2, 3)$*

$$\sup_{n \geq 1} \mathbb{E} \left\| (W_n, \mathbb{W}_n) \right\|_{p\text{-}var}^{2q} < \infty.$$

Proof This is a direct application of [30, Proposition 6.17].

$\square$

Remark 4.9 If follows from Assumptions 4.6 and 4.7 that the covariance matrix Σ and the area drift Γ are given by

$$\Sigma = \lim_{n \to \infty} \mathbb{E}_\lambda (W_n(1) \otimes W_n(1)), \qquad \Gamma = \lim_{n \to \infty} \mathbb{E}_\lambda \mathbb{W}_n(1).$$

Under additional mixing assumptions, formulas (8) and (19) hold.

Combining Theorem 3.3 and Proposition 4.8, we arrive at the following convergence result which relaxes the moment conditions required in [40].

Theorem 4.10 *Suppose that Assumptions 4.6 and 4.7 hold.*

(i) *For every $p > 2$, it holds that $\mathbf{W}_n \to_\lambda \mathbf{W}$ in the p-variation rough path topology.*

(ii) *Let $a \in C^{1+}(\mathbb{R}^d, \mathbb{R}^d)$, $b \in C^{2+}(\mathbb{R}^d, \mathbb{R}^{d \times m})$, and let x_n be the solution to (26). Then $x_n \to_\lambda X$ in $C^{p\text{-}var}([0, 1], \mathbb{R}^d)$ for all $p > 2$, where X is the solution to the SDE*

$$dX = \left(a(X) + \sum_{i,j=1}^{m} \Gamma^{i,j} \sum_{k=1}^{d} b^{i,k} \partial_k b^j(X) \right) dt + b(X) \, dW, \quad X(0) = \xi.$$

5 Extension to Families and Non-product Case

Throughout this article, we restricted attention to the case of multiplicative noise given in product form. The general form (1) was addressed in [41], though with suboptimal moment assumptions. By adapting the methods of this article to an infinite-dimensional rough paths setting similar to [41], we are able to handle, with optimal moment assumptions, a generalisation of (1) of the form

$$\frac{d}{dt} x_\varepsilon = a_\varepsilon(x_\varepsilon, y_\varepsilon) + \varepsilon^{-1} b_\varepsilon(x_\varepsilon, y_\varepsilon), \qquad \frac{d}{dt} y_\varepsilon = \varepsilon^{-2} g_\varepsilon(y_\varepsilon), \tag{27}$$

where $a_\varepsilon, b_\varepsilon, g_\varepsilon$ now depend on ε, and so does the probability measure λ_ε from which $y_\varepsilon(0)$ is drawn randomly. We assume we are also given a family μ_ε of ergodic $g_{\varepsilon,t}$-invariant probability measures on M, where $g_{\varepsilon,t}$ is the flow generated by g_ε; we require that $\int_M b_\varepsilon(x, y)\, d\mu_\varepsilon(y) = 0$ for all $\varepsilon \in [0, 1]$ and $x \in \mathbb{R}^d$.

We note that a similar generalisation is also possible for the discrete dynamics (2), which was not addressed in [41] even in the ε-independent setting. Details are found in our forthcoming work [19].

Let $C_\varepsilon^\eta(M, \mathbb{R}^m)$ be the space of C^η functions $v : M \to \mathbb{R}^m$ with $\int_M v\, d\mu_\varepsilon = 0$. Fix $q \in (1, \infty]$, $\kappa, \bar{\kappa} > 0$, $\alpha > 2 + \frac{d}{q}$. Let $a_\varepsilon \in C^{1+\bar{\kappa},0}(\mathbb{R}^d \times M, \mathbb{R}^d)$ and $b_\varepsilon \in C_\varepsilon^{\alpha,\kappa}(\mathbb{R}^d \times M, \mathbb{R}^d)$ satisfying

$$\sup_{\varepsilon \in [0,1]} \|a_\varepsilon\|_{C^{1+\bar{\kappa},0}} < \infty, \quad \sup_{\varepsilon \in [0,1]} \|b_\varepsilon\|_{C^{\alpha,\kappa}} < \infty, \quad \lim_{\varepsilon \to 0} \|b_\varepsilon - b_0\|_{C^{\alpha,\kappa}} = 0.$$

For $v \in C_\varepsilon^\eta(M, \mathbb{R}^m)$, define

$$W_{v,\varepsilon}(t) = \varepsilon \int_0^{\varepsilon^{-2}t} v \circ g_{\varepsilon,s}\, ds, \qquad \mathbb{W}_{v,\varepsilon}(t) = \int_0^t W_{v,\varepsilon} \otimes dW_{v,\varepsilon}.$$

We require the following assumptions.

(1) Moment bounds: there exists $K > 0$ such that for all families $v_\varepsilon,\, w_\varepsilon \in C_\varepsilon^\kappa(M)$, it holds that for all $s, t \geq 0$ and $\varepsilon \in [0, 1]$,

$$\left\| \int_s^t v_\varepsilon \circ g_{\varepsilon,r}\, dr \right\|_{L^{2q}(\lambda_\varepsilon)} \leq K \|v_\varepsilon\|_{C^\kappa} |t - s|^{1/2},$$

$$\left\| \int_s^t \int_s^r v_\varepsilon \circ g_{\varepsilon,u}\, w_\varepsilon \circ g_{\varepsilon,r}\, du\, dr \right\|_{L^q(\lambda_\varepsilon)} \leq K \|v_\varepsilon\|_{C^\kappa} \|w_\varepsilon\|_{C^\kappa} |t - s|.$$

(2) Enhanced WIP: there exists a bilinear operator $\mathfrak{B} : C_0^\eta(M) \times C_0^\eta(M) \to \mathbb{R}$ such that for every family $v_\varepsilon \in C_\varepsilon^\kappa(M, \mathbb{R}^m)$ with $\lim_{\varepsilon \to 0} |v_\varepsilon - v_0|_{C^\kappa} = 0$, there exists an m-dimensional Brownian motion W such that

$$(W_{v_\varepsilon,\varepsilon}, \mathbb{W}_{v_\varepsilon,\varepsilon}) \to_{\lambda_\varepsilon} (W, \mathbb{W}), \quad \text{as } \varepsilon \to 0,$$

in the sense of finite-dimensional distributions, where $\mathbb{W}^{i,j}(t) = \int_0^t W^i\, dW^j + \mathfrak{B}(v_0^i, v_0^j)t$.

(3) Convergence of drift: it holds that

$$\sup_{t \in [0,1]} \|V_\varepsilon(t) - \bar{a}t\|_{C^{1+\bar{\kappa}}} \to_{\lambda_\varepsilon} 0 \quad \text{as } \varepsilon \to 0,$$

where $V_\varepsilon(t) = \int_0^t a_\varepsilon(\cdot, y_\varepsilon(r))dr$ and $\bar{a} = \int_M a_0(\cdot, y)d\mu_0(y)$.

Consider the SDE

$$dX = \tilde{a}(X)\,dt + \sigma(X)\,dB, \quad X(0) = \xi, \tag{28}$$

where B is the standard Brownian motion in $\mathbb{R}^d$ and $\tilde{a}$ and σ are given by

$$\tilde{a}^i(x) = \bar{a}^i(x) + \sum_{k=1}^{d} \mathfrak{B}(b_0^k(x,\cdot), \partial_k b_0^i(x,\cdot)), \quad i = 1,\ldots,d,$$

$$(\sigma(x)\sigma^T(x))^{ij} = \mathfrak{B}(b_0^i(x,\cdot), b_0^j(x,\cdot)) + \mathfrak{B}(b_0^j(x,\cdot), b_0^i(x,\cdot)), \quad i,j = 1,\ldots,d.$$

Under assumptions (1–3) above, the SDE (28) has a unique weak solution X and it holds that $x_\varepsilon \to_{\lambda_\varepsilon} X$.

Acknowledgements I.C. is funded by a Junior Research Fellowship of St John's College, Oxford. P.K.F. acknowledges partial support from the ERC, CoG-683164, the Einstein Foundation Berlin, and DFG research unit FOR2402. A.K. and I.M. acknowledge partial support from the European Advanced Grant StochExtHomog (ERC AdG 320977). H.Z. is supported by the Chinese National Postdoctoral Program for Innovative Talents No: BX20180075. H.Z. thanks the Institute für Mathematik, TU Berlin, for its hospitality.

References

1. Alves, J.F., Freitas, J.M., Luzzatto, S., Vaienti, S.: From rates of mixing to recurrence times via large deviations. Adv. Math. **228**(2), 1203–1236 (2011)
2. Alves, J.F., Luzzatto, S., Pinheiro, V.: Markov structures and decay of correlations for non-uniformly expanding dynamical systems. Ann. Inst. H. Poincaré Anal. Non Linéaire **22**(6), 817–839 (2005)
3. Alves, J.F., Pinheiro, V.: Gibbs-Markov structures and limit laws for partially hyperbolic attractors with mostly expanding central direction. Adv. Math. **223**(5), 1706–1730 (2010)
4. Anosov, D.V.: Geodesic flows on closed Riemannian manifolds of negative curvature. Trudy Mat. Inst. Steklov. **90**, 209 (1967)
5. Araujo, V., Melbourne, I.: Mixing properties and statistical limit theorems for singular hyperbolic flows without a smooth stable foliation. ArXiv e-prints, November 2017
6. Araújo, V., Melbourne, I., Varandas, P.: Rapid mixing for the Lorenz attractor and statistical limit laws for their time-1 maps. Comm. Math. Phys. **340**(3), 901–938 (2015)
7. Bailleul, I., Catellier, R.: Rough flows and homogenization in stochastic turbulence. J. Diff. Equ. **263**(8), 4894–4928 (2017)
8. Benedicks, M., Young, L.-S.: Markov extensions and decay of correlations for certain Hénon maps. Astérisque (261):xi, 13–56, 2000. Géométrie complexe et systèmes dynamiques (Orsay, 1995)
9. Billingsley, P.: The Lindeberg-Lévy theorem for martingales. Proc. Amer. Math. Soc. **12**, 788–792 (1961)
10. Billingsley, P.: Convergence of Probability Measures, Wiley Series in Probability and Statistics: Probability and Statistics. John Wiley & Sons Inc., New York (1999)
11. Birkhoff, G.D.: Proof of the ergodic theorem. Proc. Natl. Acad. Sci. U.S.A. **17**(12), 656–660 (1931)
12. Bowen, R.: Equilibrium states and the ergodic theory of Anosov diffeomorphisms. Lecture Notes in Mathematics, vol. 470. Springer, Berlin-New York (1975)

13. Breuillard, E., Friz, P., Huesmann, M.: From random walks to rough paths. Proc. Amer. Math. Soc. **137**(10), 3487–3496 (2009)
14. Brown, B.M.: Martingale central limit theorems. Ann. Math. Statist. **42**, 59–66 (1971)
15. Buzzi, J., Maume-Deschamps, V.: Decay of correlations for piecewise invertible maps in higher dimensions. Israel J. Math. **131**, 203–220 (2002)
16. Chernov, N., Young. L.S.: Decay of correlations for Lorentz gases and hard balls. In: Hard Ball Systems and the Lorentz Gas, volume 101 of Encyclopaedia Math. Sci., 89–120. Springer, Berlin (2000)
17. Chevyrev, I.: Random walks and Lévy processes as rough paths. Probab. Theor. Relat. Fields **170**(3–4), 891–932 (2018)
18. Chevyrev, I., Friz, P.K.: Canonical RDEs and general semimartingales as rough paths. Ann. Probab. **47**(1), 420–463 (2019)
19. Chevyrev, I., Friz, P.K., Korepanov, A., Melbourne, I., Zhang, H.: Deterministic homogenization for discrete time fast-slow systems under optimal moment assumptions. In preparation
20. Cuny, C., Merlevède, F.: Strong invariance principles with rate for "reverse" martingale differences and applications. J. Theoret. Probab. **28**(1), 137–183 (2015)
21. Davie, A.M.: Differential equations driven by rough paths: an approach via discrete approximation. Appl. Math. Res. Express. AMRX, no. 2:Art. ID abm009, 40 p. (2007)
22. Denker, M., Philipp, W.: Approximation by Brownian motion for Gibbs measures and flows under a function. Ergod. Theor. Dynam. Syst. **4**(4), 541–552 (1984)
23. Donsker,M.D.: An invariance principle for certain probability limit theorems. Mem. Amer. Math. Soc., No. 6:12 (1951)
24. Eagleson, G.K.: Some simple conditions for limit theorems to be mixing. Teor. Verojatnost. i Primenen. **21**(3), 653–660 (1976)
25. Friz, P.K., Gassiat, P., Lyons, T.J.: Physical Brownian motion in a magnetic field as a rough path. Trans. Amer. Math. Soc. **367**(11), 7939–7955 (2015)
26. Friz, P.K., Hairer, M.: A Course on Rough Path Analysis, with an Introduction to Regularity Structures, Springer 2014. Universitext. Springer (2014)
27. Friz, P.K., Shekhar, A.: General rough integration, Lévy rough paths and a Lévy–Kintchine-type formula. Ann. Probab. **45**(4), 2707–2765 (2017)
28. Friz, P.K., Victoir, N.: A variation embedding theorem and applications. J. Funct. Anal. **239**(2), 631–637 (2006)
29. Friz, P.K., Victoir, N.: Multidimensional Stochastic Processes as Rough Paths. Cambridge Studies in Advanced Mathematics, vol. 120. Cambridge University Press, Cambridge (2010)
30. Friz, P.K., Zhang, H.: Differential equations driven by rough paths with jumps. J. Differ. Equ. **264**(10), 6226–6301 (2018)
31. Gordin, M.I.: The central limit theorem for stationary processes. Dokl. Akad. Nauk SSSR **188**, 739–741 (1969)
32. Gottwald, G.A., Melbourne, I.: Homogenization for deterministic maps and multiplicative noise. Proc. R. Soc. Lond. Ser. A Math. Phys. Eng. Sci. **469**(2156), 20130201 (2013)
33. Gottwald, G.A., Melbourne, I.: Central limit theorems and suppression of anomalous diffusion for systems with symmetry. Nonlinearity **29**(10), 2941–2960 (2016)
34. Gouëzel, S.: Central limit theorem and stable laws for intermittent maps. Probab. Theor. Relat. Fields **128**(1), 82–122 (2004)
35. Gouëzel, S.: Statistical properties of a skew product with a curve of neutral points. Ergod. Theor. Dynam. Syst. **27**(1), 123–151 (2007)
36. Hofbauer, F., Keller, G.: Ergodic properties of invariant measures for piecewise monotonic transformations. Math. Z. **180**(1), 119–140 (1982)
37. Jakubowski, A., Mémin, J., Pagès, G.: Convergence en loi des suites d'intégrales stochastiques sur l'espace D^1 de Skorokhod. Probab. Theor. Relat. Fields **81**(1), 111–137 (1989)
38. Keller, G.: Generalized bounded variation and applications to piecewise monotonic transformations. Z. Wahrsch. Verw. Gebiete **69**(3), 461–478 (1985)
39. Kelly, D.: Rough path recursions and diffusion approximations. Ann. Appl. Probab. **26**(1), 424–461 (2016)

40. Kelly, D., Melbourne, I.: Smooth approximation of stochastic differential equations. Ann. Probab. **44**(1), 479–520 (2016)
41. Kelly, D., Melbourne, I.: Deterministic homogenization for fast-slow systems with chaotic noise. J. Funct. Anal. **272**(10), 4063–4102 (2017)
42. Kipnis, C., Varadhan, S.R.S.: Central limit theorem for additive functionals of reversible Markov processes and applications to simple exclusions. Comm. Math. Phys. **104**(1), 1–19 (1986)
43. Korepanov, A., Kosloff, Z., Melbourne, I.: Martingale-coboundary decomposition for families of dynamical systems. Annales l'Institut H Poincare. Anal. Non Lineaire **35**(1), 859–885 (2018)
44. Korepanov, A., Kosloff, Z., Melbourne, I.: Deterministic homogenization for families of fast-slow systems. In preparation
45. Krzyżewski, K., Szlenk, W.: On invariant measures for expanding differentiable mappings. Studia Math. **33**, 83–92 (1969)
46. Kurtz, T.G., Protter, P.: Weak limit theorems for stochastic integrals and stochastic differential equations. Ann. Probab. **19**(3), 1035–1070 (1991)
47. Liverani, C.: Central limit theorem for deterministic systems. In: International Conference on Dynamical Systems (Montevideo, 1995), volume 362 of *Pitman Res. Notes Math. Ser.*, pp. 56–75. Longman, Harlow (1996)
48. Liverani, C., Saussol, B., Vaienti, S.: A probabilistic approach to intermittency. Ergod. Theor. Dynam. Syst. **19**(3), 671–685 (1999)
49. Lyons, T.J.: Differential equations driven by rough signals. Rev. Mat. Iberoam. **14**(2), 215–310 (1998)
50. Maxwell, M., Woodroofe, M.: Central limit theorems for additive functionals of Markov chains. Ann. Probab. **28**(2), 713–724 (2000)
51. McLeish, D.L.: Dependent central limit theorems and invariance principles. Ann. Probab. **2**, 620–628 (1974)
52. Melbourne, I.: Large and moderate deviations for slowly mixing dynamical systems. Proc. Amer. Math. Soc. **137**(5), 1735–1741 (2009)
53. Melbourne, I.: Superpolynomial and polynomial mixing for semiflows and flows. Nonlinearity **31**(10), R268–R316 (2018)
54. Melbourne, I.: Almost sure invariance principle for nonuniformly hyperbolic systems. Comm. Math. Phys. **260**(1), 131–146 (2005)
55. Melbourne, I., Nicol, M.: Large deviations for nonuniformly hyperbolic systems. Trans. Amer. Math. Soc. **360**(12), 6661–6676 (2008)
56. Melbourne, I., Stuart, A.M.: A note on diffusion limits of chaotic skew-product flows. Nonlinearity **24**(4), 1361–1367 (2011)
57. Melbourne, I., Török, A.: Statistical limit theorems for suspension flows. Israel J. Math. **144**, 191–209 (2004)
58. Melbourne, I., Török, A.: Convergence of moments for Axiom A and non-uniformly hyperbolic flows. Ergod. Theor. Dynam. Syst. **32**(3), 1091–1100 (2012)
59. Melbourne, I., Varandas, P.: A note on statistical properties for nonuniformly hyperbolic systems with slow contraction and expansion. Stoch. Dyn. **16**(3), 1660012 (2016)
60. Melbourne, I., Zweimüller, R.: Weak convergence to stable Lévy processes for nonuniformly hyperbolic dynamical systems. Ann. Inst. Henri Poincaré Probab. Stat. **51**(2), 545–556 (2015)
61. Parry, W., Pollicott, M.: Zeta functions and the periodic orbit structure of hyperbolic dynamics. Astérisque, 187–188(268) (1990)
62. Pavliotis, G.A., Stuart, A.M.: Multiscale methods, volume 53 of Texts in Applied Mathematics. Springer, New York, 2008. Averaging and homogenization
63. Pomeau, Y., Manneville, P.: Intermittent transition to turbulence in dissipative dynamical systems. Comm. Math. Phys. **74**(2), 189–197 (1980)
64. Ratner, M.: The central limit theorem for geodesic flows on n-dimensional manifolds of negative curvature. Israel J. Math. **16**, 181–197 (1973)
65. Rio, E.: Théorie asymptotique des processus aléatoires faiblement dépendants. Mathématiques & Applications (Berlin) (Mathematics & Applications), vol. 31. Springer, Berlin (2000)

66. Ruelle, D.: Thermodynamic formalism, volume 5 of Encyclopedia of Mathematics and its Applications. Addison-Wesley Publishing Co., Reading, Mass. (1978). The mathematical structures of classical equilibrium statistical mechanics, With a foreword by Giovanni Gallavotti and Gian-Carlo Rota
67. Ruelle, D.: The thermodynamic formalism for expanding maps. Comm. Math. Phys. **125**(2), 239–262 (1989)
68. Rychlik, M.: Bounded variation and invariant measures. Studia Math. **76**(1), 69–80 (1983)
69. Saussol, B.: Absolutely continuous invariant measures for multidimensional expanding maps. Israel J. Math. **116**, 223–248 (2000)
70. Sinaĭ, Ja.G.: Gibbs measures in ergodic theory. Uspehi Mat. Nauk. **27**(4(166)), 21–64 (1972)
71. Smale, S.: Differentiable dynamical systems. Bull. Amer. Math. Soc. **73**, 747–817 (1967)
72. Tyran-Kamińska, M.: An invariance principle for maps with polynomial decay of correlations. Comm. Math. Phys. **260**(1), 1–15 (2005)
73. Williams, D.R.E.: Path-wise solutions of stochastic differential equations driven by Lévy processes. Rev. Mat. Iberoamericana **17**(2), 295–329 (2001)
74. Young, L.-S.: Statistical properties of dynamical systems with some hyperbolicity. Ann. of Math. (2), **147**(3), 585–650 (1998)
75. Young, L.-S.: Recurrence times and rates of mixing. Israel J. Math. **110**, 153–188 (1999)
76. Young, L.-S.: What are SRB measures, and which dynamical systems have them? J. Statist. Phys. **108**(5–6), 733–754 (2002). Dedicated to David Ruelle and Yasha Sinai on the occasion of their 65th birthdays
77. Young, L.-S.: Generalizations of SRB measures to nonautonomous, random, and infinite dimensional systems. J. Stat. Phys. **166**(3–4), 494–515 (2017)
78. Zweimüller, R.: Mixing limit theorems for ergodic transformations. J. Theoret. Probab. **20**(4), 1059–1071 (2007)

The Deterministic and Stochastic Shallow Lake Problem

G. T. Kossioris, M. Loulakis and P. E. Souganidis

Abstract We study the welfare function of the deterministic and stochastic shallow lake problem. We show that the welfare function is the viscosity solution of the associated Bellman equation, we establish several properties including its asymptotic behaviour at infinity and we present a convergent monotone numerical scheme.

Keywords Shallow Lake · Viscosity solution · Optimal stochastic control

AMS 2010 Mathematics Subject Classification: 93E20 · 60H10 · 49L25

G. T. Kossioris and M. Loulakis: This research has been co-financed by the European Union (European Social Fund ESF) and Greek national funds through the Operational Program "Education and Lifelong Learning" of the National Strategic Reference Framework (NSRF) 2007–2013. Research Funding Program: THALES. Investing in knowledge society through the European Social Fund, Project: Optimal Management of Dynamical Systems of the Economy and the Environment, grant number MIS 377289.
P. E. Souganidis: Partially supported by the National Science Foundation grants DMS-1266383 and DMS-1600129 and the Office for Naval Research Grant N00014-17-1-2095.

G. T. Kossioris
Department of Mathematics and Applied Mathematics, Voutes Campus,
University of Crete, 700 13 Heraklion, Greece
e-mail: kosioris@uoc.gr

M. Loulakis (✉)
School of Applied Mathematical and Physical Sciences,
National Technical University of Athens, 157 80 Athens, Greece
e-mail: loulakis@math.ntua.gr

M. Loulakis
Institute of Applied and Computational Mathematics-Foundation of Research and Technology
Hellas, 700 13 Heraklion, Greece

P. E. Souganidis
Department of Mathematics, University of Chicago 5734 S. University Avenue,
Chicago, IL 60637, USA
e-mail: souganidis@math.uchicago.edu

© Springer Nature Switzerland AG 2019
P. Friz et al. (eds.), *Probability and Analysis in Interacting Physical Systems*, Springer Proceedings in Mathematics & Statistics 283,
https://doi.org/10.1007/978-3-030-15338-0_3

1 Introduction

The scope of this work is the theoretical study of the welfare function describing the economics of shallow lakes. Pollution of shallow lakes is a quite often observed phenomenon because of heavy use of fertilizers on surrounding land and an increased inflow of waste water from human settlements and industries. The shallow lake system provides conflicting services as a resource, due to the provision of ecological services of a clear lake, and as a waste sink, due to agricultural activities.

The economic analysis of the problem requires the study of an optimal control problem or a differential game in case of common property resources by various communities; see, for example [8, 10, 28].

Typically the model is described in terms of the amount, $x(t)$, of phosphorus in algae which is assumed to evolve according to the stochastic differential equation (sde for short)

$$\begin{cases} dx(t) = \left(u(t) - bx(t) + \dfrac{x^2(t)}{x^2(t) + 1} \right) dt + \sigma x(t) dW(t), \\ x(0) = x \geq 0. \end{cases} \tag{1.1}$$

The first term, $u(t)$, in the drift part of the dynamics represents the exterior load of phosphorus imposed by the agricultural community, which is assumed to be positive. The second term is the rate of loss $bx(t)$, which consists of sedimentation, outflow and sequestration in other biomass. The third term is the rate of recycling of phosphorus due to sediment resuspension resulting, for example, from waves, or oxygen depletion. This rate is typically taken to be a sigmoid function $r : \mathbb{R}_+ \to \mathbb{R}_+$; see [10]. For the sake of clarity we make the usual choice in the literature $r(x) = x^2/(1 + x^2)$, $x \in \mathbb{R}_+$. Similar conclusions can be drawn for any non-negative, strictly increasing function r with bounded first and second derivatives and such that $r(x)/x \to 0$, as $x \to 0$ and $(1 - r(x))x \to 0$, as $x \to \infty$. The model also assumes an uncertainty in the recycling rate driven by a linear multiplicative white noise with diffusion strength σ. By suitably adjusting the model parameter b, we may always assume that the stochastic integral in (1.1) is to be interpreted in the Itô sense.

The lake has value as a waste sink for agriculture modeled by $\ln u$ and it provides ecological services that decrease with the amount of phosphorus $-cx^2$. The positive parameter c reflects the relative weight of this welfare component; large c gives more weight to the ecological services of the lake. For the sake of convenience, we assume that $c = 1$.

Assuming an infinite horizon at a discount rate $\rho > 0$, the total benefit is

$$J(x; u) = \mathbb{E}\left[\int_0^\infty e^{-\rho t} \left(\ln u(t) - x^2(t) \right) dt \right], \tag{1.2}$$

where $x(\cdot)$ is the solution to (1.1), for a given $u(\cdot)$, and $x(0) = x$. Optimal management requires to maximize the total benefit, over all exterior loads that act as controls by the social planner. Thus the welfare function is

$$V(x) = \sup_{u \in \mathfrak{U}_x} J(x; u), \tag{1.3}$$

where $\mathfrak{U}_x$ is the set of admissible controls $u : [0, \infty) \to \mathbb{R}^+$ which are specified in the next section.

Dynamic programming arguments lead, under the assumption that the welfare function is decreasing, to the Hamilton-Jacobi-Bellman equation

$$\rho V = \left(\frac{x^2}{x^2 + 1} - bx \right) V_x - \left(\ln(-V_x) + x^2 + 1 \right) + \frac{1}{2} \sigma^2 x^2 V_{xx}, \tag{OHJB}$$

and of the aims of this paper is to provide a rigorous justification of this fact.

In the deterministic case $\sigma = 0$, the optimal dynamics of the problem were fully investigated by analysing the possible equilibria of the dynamics given by Pontryagin maximum principle; see, for example [28, 38]. The possibility to steer the combined economic-ecological system towards the trajectory of optimal management via optimal state-dependent taxes was also considered; see [21].

On the other hand, there is not much in the literature about the stochastic problem. Formal asymptotics expansions of the solution for small σ for Hamilton-Jacobi-Bellman equations like (OHJB) have been presented in [16]. In the same paper, the authors also give a formal phenomenological bifurcation analysis based on a geometric invariant quantity, along with some numerical computations of the stochastic bifurcations based on (formal) asymptotics for small σ.

The connection between stochastic control problems and Hamilton-Jacobi-Bellman equation, which is based on the dynamic programing principle, has been studied extensively. The correct mathematical framework is that of the Crandall-Lions viscosity solutions introduced in [11]; see the review article [12]). The deterministic case leads to the study of a first order Hamilton-Jacobi equation; see, for example [2, 3], and references therein. For the general stochastic optimal control problem we refer to [14, 23, 25–27, 36] and references therein.

The stochastic shallow lake problem has some nonstandard features and, hence, it requires some special analysis. At first, the problem is formulated as a state constraint one on a semi-infinite domain. Viscosity solutions with state constraint boundary conditions were introduced for first order equations by [9, 34]. For second order equations one should consult [1, 19, 24]. Apart from the left boundary condition, the correct asymptotic decay of the solution at infinity is necessary to establish a comparison result; see, for example [13, 18].

The unboundedness of the controls along with the logarithmic term in the cost functional lead to a logarithm of the gradient variable in (OHJB). An a priori knowledge of the solution is required to guarantee that the Hamiltonian is well defined. Moreover, due the presence of the logarithmic term it is necessary to construct an appropriate test function to establish a comparison proof. The commonly used polynomial functions, see, for example [13, 18], are not useful here since they do not yield a supersolution of the equation.

In the present work we first study the stochastic shallow lake problem for a fixed $\sigma > 0$. We first prove the necessary stochastic analysis estimates for the welfare function (1.3). We obtain directly various crucial properties for the welfare function, that is, boundary behaviour, local regularity, monotonicity and asymptotic estimates at infinity.

We prove, including the deterministic case, that (1.3) is the unique decreasing constrained viscosity solution to (OHJB) with quadratic growth at infinity. The comparison theorem is proved by considering a linearized equation and constructing a proper supersolution. Exploiting the well-posedness and stability properties of the problem within the framework of constrained viscosity solutions we investigate the exact asymptotic behavior of the solutions as $\sigma \to 0$ or $x \to \infty$. The latter is used to construct a monotone convergent numerical scheme that along with the optimal dynamics equation can be used to reconstruct numerically the stochastic optimal dynamics.

2   The General Setting and the Main Results

We assume that there exists a filtered probability space $(\Omega, \mathcal{F}, \{\mathcal{F}_t\}_{t \geq 0}, \mathbb{P})$ satisfying the usual conditions, and a Brownian motion $W(\cdot)$ defined on that space. An admissible control $u(\cdot) \in \mathfrak{U}_x$ is an $\mathcal{F}_t$-adapted, $\mathbb{P}$-a.s. locally integrable process with values in $U = (0, \infty)$, satisfying

$$\mathbb{E}\left[\int_0^\infty e^{-\rho t} \ln u(t) dt\right] < \infty, \tag{2.1}$$

such that the problem (1.1) has a unique strong solution $x(\cdot)$.

The shallow lake problem has an infinite horizon. Standard arguments based on the dynamic programming principle (see [14], Sect. 3.7) suggest that the welfare function V given by (1.3) satisfies the HJB equation

$$\rho V = \sup_{u \in U} G(x, u, V_x, V_{xx}), \tag{2.2}$$

with G defined by

$$G(x, u, p, P) = \frac{1}{2}\sigma^2 x^2 P + \left(u - bx + \frac{x^2}{x^2 + 1}\right)p + \ln u - x^2. \tag{2.3}$$

One difficulty in the study of this problem is related to the fact that the control functions u take values in the unbounded set $\mathbb{R}^+$ so that supremum in (2.2) might take infinite values. Indeed, when $U = \mathbb{R}^+$, setting

$$H(x, p, P) = \sup_{u \in \mathbb{R}^+} G(x, u, p, P),$$

we find

$$H(x, p, P) = \begin{cases} \left(\dfrac{x^2}{x^2 + 1} - bx \right) p - \left(\ln(-p) + x^2 + 1 \right) + \dfrac{1}{2} \sigma^2 x^2 P & \text{if } p < 0, \\ +\infty & \text{if } p \geq 0. \end{cases}$$
$$(2.4)$$

It is natural to expect that since shallow lake looses its value with a higher concentration of phosphorus, the welfare function is decreasing as the initial state of phosphorus increases. Assuming that $V_x < 0$, (2.2) becomes (OHJB).

Since the problem is set in $(0, \infty)$, it is necessary to introduce boundary conditions guaranteeing the well-posedness of the corresponding boundary value problem.

Given the possible degeneracies of Hamilton-Jacobi-Bellman equations at $x = 0$, the right framework is that of continuous viscosity solutions in which boundary conditions are considered in the weak sense. Since at the boundary point $x = 0$

$$\inf_{u \in U} \left\{ -u + bx - \frac{x^2}{x^2 + 1} \right\} < 0,$$
$$(2.5)$$

that is, there always exists a control that can drive the system inside $(0, \infty)$, the problem should be considered as a state constraint one on the interval $[0, \infty)$, meaning that V is a subsolution in $[0, \infty)$ and a supersolution in $(0, \infty)$.

Next we present the main results of the paper. The proofs are given in Sect. 4. The first is about the relationship between the welfare function and (OHJB).

Theorem 2.1 *If $\sigma^2 < \rho + 2b$, the welfare function V is a continuous in $[0, \infty)$ constrained viscosity solution of the equation (OHJB) in $[0, \infty)$.*

The second result is the following comparison principle for solutions of (OHJB).

Theorem 2.2 *Assume that $u \in C([0, \infty))$ is a bounded from above strictly decreasing subsolution of (OHJB) in $[0, \infty)$ and $v \in C([0, \infty))$ is a bounded from above strictly decreasing supersolution of (OHJB) in $(0, \infty)$ such that $v \geq -c(1 + x^2)$ and $Du \leq -\frac{1}{c^*}$ in the viscosity sense, for c, c^* positive constants. Then $u \leq v$ in $[0, \infty)$.*

The next theorem describes the exact asymptotic behavior of (1.3) at $+\infty$. Let

$$A = \frac{1}{\rho + 2b - \sigma^2} \quad \text{and} \quad K = \frac{1}{\rho} \left(\frac{2b + \sigma^2}{2\rho} - \frac{A(\rho + 2b)}{(b + \rho)^2} - 1 \right).$$
$$(2.6)$$

Theorem 2.3 *As $x \to \infty$,*

$$V(x) = -A\left(x + \frac{1}{b+\rho}\right)^2 - \frac{1}{\rho}\ln\left[2A(x + \frac{1}{b+\rho})\right] + K + o(1). \qquad (2.7)$$

An important ingredient of the analysis is the following proposition which collects some key properties of V that are used in the proofs of Theorems 2.1 and 2.3 and show that V satisfies the assumptions of Theorem 2.2. The proof is presented in Sect. 3.

Proposition 1 *Suppose $\sigma^2 < \rho + 2b$.*

(i) There exist constants $K_1, K_2 > 0$, such that, for any $x \geq 0$, we have

$$K_1 \leq V(x) + A\left(x + \frac{1}{b+\rho}\right)^2 + \frac{1}{\rho}\ln\left(x + \frac{1}{b+\rho}\right) \leq K_2. \qquad (2.8)$$

(ii) There exist $C > 0$ and $\Phi : [0, +\infty) \to \mathbb{R}$ increasing such that, for any $x_1, x_2 \in [0, +\infty)$ with $x_1 < x_2$,

$$-\Phi(x_2) \leq \frac{V(x_2) - V(x_1)}{x_2 - x_1} \leq -C. \qquad (2.9)$$

(iii) V is differentiable at zero and

$$\ln\left(-V'(0)\right) + \rho V(0) + 1 = 0. \qquad (2.10)$$

It is shown in the next section that the assumption $\sigma^2 < \rho + 2b$ is necessary, otherwise $V(x) = -\infty$.

Relation (2.7) is important for the numerical approximation of (1.3). Given that the computational domain is finite, the correct asymptotic behavior of (1.3) at $+\infty$ is necessary for an accurate computation of V. In this connection, in Sect. 5 we present a monotone numerical scheme approximating (1.3).

3 The Proof of Proposition 1

Properties of the dynamics. The first result states that, if $x \geq 0$, the solution to (1.1) stays nonnegative.

Lemma 3.1 *If $x \geq 0$, $u \in \mathfrak{U}_x$, and $x(\cdot)$ is the solution to (1.1), then $\mathbb{P}\left[x(t) \geq 0, \forall t \geq 0\right] = 1$.*

Proof Elementary stochastic analysis calculations yield that

$$x(t) = xZ_t + \int_0^t \frac{Z_t}{Z_s}\left(u(s) + \frac{x^2(s)}{1 + x^2(s)}\right)ds, \qquad (3.1)$$

where

$$Z_t = e^{\sigma W_t - (b + \frac{\sigma^2}{2})t}, \tag{3.2}$$

and the claim is now obvious, since u takes positive values. $\qquad\square$

The next assertion is that the set of admissible controls is actually independent of the starting point x.

Lemma 3.2 *For all $x, y \geq 0$, $\mathfrak{U}_x = \mathfrak{U}_y$.*

Proof Fix $u \in \mathfrak{U}_x$ and $x \in [0, \infty)$ and let $x(\cdot)$ the unique strong solution to (1.1) with $x(0) = x$, and, for any $y \geq 0$, consider the sde

$$\begin{cases} dw(t) = \left\{ - bw(t) - \frac{(x(t) - w(t))^2}{1 + (x(t) - w(t))^2} + \frac{x(t)^2}{1 + x(t)^2} \right\} dt + \sigma w(t) dW(t), \\ w(0) = x - y, \end{cases} \tag{3.3}$$

and note that the coefficients are Lipschitz and grow at most linearly. It follows that (3.3) has a unique strong solution defined for all $t \geq 0$. It is easy to see now that the process $y(t) = x(t) - w(t)$ satisfies (1.1) with $y(0) = y$. Moreover, the uniqueness of $y(\cdot)$ follows from that of $x(\cdot)$, so $u \in \mathfrak{U}_y$. $\qquad\square$

In view of Lemma 3.2 we will denote the set of admissible controls by $\mathfrak{U}$, regardless of the starting point x in (1.1).

Lemma 3.3 *Suppose $x(\cdot)$, $y(\cdot)$ satisfy (1.1) with controls u_1, $u_2 \in \mathfrak{U}$, respectively, and $x(0) = x$, $y(0) = y$. If $x \leq y$ and $\mathbb{P}\left[u_1(t) \leq u_2(t), \forall t \geq 0\right] = 1$, then*

$$\mathbb{P}\left[y(t) - x(t) \geq (y - x)Z_t, \forall t \geq 0\right] = 1.$$

Proof The proof is an immediate consequence of (3.1) and Gronwall's inequality, since $w(t) = y(t) - x(t)$ satisfies

$$w(t) = (y - x)Z_t + \int_0^t \frac{Z_t}{Z_s}\left(u_2(s) - u_1(s) + w(s)\frac{y(s) + x(s)}{(1 + y^2(s))(1 + x^2(s))}\right) ds.$$

$\qquad\square$

Properties of the welfare function. The results here follow from the properties of (1.1). Throughout this section we refer to quantities depending only on ρ, b and σ^2 as constants.

Remark 3.1 If $\sigma^2 \geq \rho + 2b$, then $V(x) \equiv -\infty$.

Indeed, when $u \in \mathfrak{U}$ and $x(\cdot)$ satisfies (1.1), (3.1) implies that $\mathbb{P}[x(t) \geq M_t(u), \forall t \geq 0] = 1$, with

$$M_t(u) = \int_0^t \frac{Z_t}{Z_s} u(s) \, ds, \tag{3.4}$$

where $\{Z_t\}_{t \geq 0}$ is as in (3.2).

In view of this observation we will hereafter assume that

$$\sigma^2 < \rho + 2b.$$

We will first prove three lemmata before we proceed with the proof of Proposition 1.

Lemma 3.4 *The function* $x \mapsto V(x) + Ax^2$, *where A is defined in (2.6), is decreasing on* $[0, +\infty)$.

Proof Fix $x_1, x_2 \geq 0$ with $x_1 \leq x_2$. It suffices to show that, for J as in (1.2) and for any control $u \in \mathfrak{U}$,

$$J(x_2; u) + Ax_2^2 \leq J(x_1; u) + Ax_1^2,$$

Since this holds trivially if $J(x_2; u) = -\infty$, we may assume that $J(x_2; u) > -\infty$.

Consider now the solutions $x_1(\cdot), x_2(\cdot)$ to (1.1) with initial conditions x_1, x_2 and a common control $u \in \mathfrak{U}$. Lemma 3.3 implies that, $\mathbb{P}$-a.s. and for all $t \geq 0$ and $i = 1, 2$,

$$x_i(t) \geq x_i Z_t, \quad \text{and} \quad x_2(t) - x_1(t) \geq (x_2 - x_1) Z_t.$$

Note that since $u \in \mathfrak{U}$ and $J(x_2; u) > -\infty$,

$$\int_0^\infty e^{-\rho t} x_2^2(t) \, dt < +\infty \Rightarrow \int_0^\infty e^{-\rho t} x_1^2(t) \, dt < +\infty \Rightarrow J(x_1; u) > -\infty.$$

In particular,

$$
\begin{aligned}
J(x_2; u) - J(x_1; u) &= \mathbb{E}\left[\int_0^\infty e^{-\rho t} \left(\ln u(t) - x_2^2(t) \right) dt - \int_0^\infty e^{-\rho t} \left(\ln u(t) - x_1^2(t) \right) dt \right] \\
&= -\mathbb{E}\left[\int_0^\infty e^{-\rho t} \left(x_2(t) - x_1(t) \right) \left(x_2(t) + x_1(t) \right) dt \right] \\
&\leq -(x_2^2 - x_1^2) \int_0^\infty e^{-\rho t} \mathbb{E}[Z_t^2] \, dt = -A(x_2^2 - x_1^2).
\end{aligned}
$$

$\square$

Lemma 3.5 *The welfare function at zero satisfies* $V(0) \leq \frac{1}{\rho} \ln\left(\frac{b+\rho}{\sqrt{2e}} \right)$.

Proof Recall that, for any $u \in \mathfrak{U}$, $x(t) \geq M_t(u)$. Using Jensen's inequality and part (i) of Lemma A.1, we find

$$\mathbb{E}\left[\int_0^\infty e^{-\rho t}\ln u(t)\,dt\right] \le \frac{1}{\rho}\ln\mathbb{E}\left[\int_0^\infty \rho e^{-\rho t}u(t)\,dt\right]$$

$$= \frac{1}{\rho}\ln\mathbb{E}\left[\int_0^\infty \rho(\rho+b)e^{-\rho t}M_t(u)\,dt\right]$$

$$\le \frac{\ln(b+\rho)}{\rho} + \frac{1}{\rho}\ln\mathbb{E}\left[\int_0^\infty \rho e^{-\rho t}x(t)\,dt\right]$$

$$\le \frac{\ln(b+\rho)}{\rho} + \frac{1}{2\rho}\ln\mathbb{E}\left[\int_0^\infty \rho e^{-\rho t}x^2(t)\,dt\right].$$

In view of (2.1), we need only consider $u \in \mathfrak{U}$ such that $D = \mathbb{E}\left[\int_0^\infty e^{-\rho t}x^2(t)\,dt\right]$ $< \infty$. Then

$$\mathbb{E}\left\{\int_0^\infty e^{-\rho t}\left[\ln u(t) - x^2(t)\right]dt\right\} \le \frac{\ln(b+\rho)}{\rho} + \frac{\ln(\rho D)}{2\rho} - D \le \frac{1}{\rho}\ln\left(\frac{b+\rho}{\sqrt{2e}}\right),$$

and the assertion holds.　　□

It follows from Lemmas 3.4 and 3.5 that $V < +\infty$ in $[0, \infty)$. The next result is a special case of the dynamic programming principle.

Lemma 3.6 *Fix $x_1, x_2 \in [0, \infty)$ with $x_1 < x_2$, and, for $u \in \mathfrak{U}$, let $x(\cdot)$ be the solution to (1.1) with control u and $x(0) = x_1$. If τ_u is the hitting time of $x(\cdot)$ on $[x_2, +\infty)$, that is,*

$$\tau_u = \inf\{t \ge 0 : x(t) \ge x_2\},$$

then

$$V(x_1) = \sup_{u \in \mathfrak{U}} \mathbb{E}\left[\int_0^{\tau_u} e^{-\rho t}\left(\ln u(t) - x^2(t)\right)dt + e^{-\rho\tau_u}V(x_2)\right]. \quad (3.5)$$

Proof We have

$$J(x_1; u) = \mathbb{E}\left[\int_0^{\tau_u} e^{-\rho t}\left(\ln u(t) - x^2(t)\right)dt\right]$$

$$+ \mathbb{E}\left[\int_{\tau_u}^\infty e^{-\rho t}\left(\ln u(t) - x^2(t)\right)dt;\ \tau_u < +\infty\right].$$

Conditioning on the σ-field $\mathcal{F}_{\tau_u}$, and applying the strong Markov property, the rightmost term becomes

$$\mathbb{E}\left[e^{-\rho\tau_u}\mathbb{E}\left[\int_{\tau_u}^\infty e^{-\rho(t-\tau_u)}\left(\ln u(t) - x^2(t)\right)dt\,\Big|\mathcal{F}_{\tau_u}\right];\ \tau_u < +\infty\right] \le \mathbb{E}\left[e^{-\rho\tau_u}\right]V(x_2),$$

since on the event $\{\tau_u < +\infty\}, x(\tau_u + \cdot)$ satisfies (1.1) with initial condition $x(\tau_u) = x_2$ and control $u(\tau_u + \cdot)$. Taking the supremum over $u \in \mathfrak{U}$ we see that the left hand side of (3.5) is less than or equal the right hand one.

For the reverse inequality, take any $u \in \mathfrak{U}$ and consider (1.1) driven by the Brownian motion $B(t) = W(\tau_u + t) - W(\tau_u)$, and, for $\varepsilon > 0$, choose a control u_ε such that

$$V(x_2) < J(x_2; u_\varepsilon) + \varepsilon.$$

Define now the new control $u_* \in \mathfrak{U}$ as

$$u_*(t) = \begin{cases} u(t) & \text{for } t \le \tau_u \\ u_\varepsilon(t - \tau_u) & \text{for } t > \tau_u. \end{cases}$$

Just as in the proof of the upper bound we get

$$V(x_1) \ge J(x_1; u_*) = \mathbb{E}\left[\int_0^{\tau_u} e^{-\rho t}\left(\ln u(t) - x^2(t)\right) dt + e^{-\rho \tau_u} J(x_2; u_\varepsilon)\right]$$
$$> \mathbb{E}\left[\int_0^{\tau_u} e^{-\rho t}\left(\ln u(t) - x^2(t)\right) dt + e^{-\rho \tau_u} V(x_2)\right] - \varepsilon,$$

which concludes the proof. $\qquad\qquad\qquad\qquad\qquad\qquad\qquad\qquad\qquad\qquad\quad\square$

We next give the proof of Proposition 1, which is subdivided in several parts.

Proof of Proposition 1: *Proof lower bound, part (i)*: The claim will follow by choosing the control $u(t) = \frac{1+x(t)}{1+x^2(t)}$, which is clearly admissible. Then, (3.1) gives

$$x(t) = xZ_t + \int_0^t \frac{Z_t}{Z_s}\left(1 + \frac{x(s)}{1+x^2(s)}\right) ds = xZ_t + M_t(1) + \int_0^t \frac{Z_t}{Z_s} \frac{x(s)}{1+x^2(s)} ds$$
$$(3.6)$$

and, hence,

$$x^2(t) = x^2 Z_t^2 + 2xZ_t M_t(1) + M_t^2(1) + \left(\int_0^t \frac{Z_t}{Z_s} \frac{x(s)}{1+x^2(s)} ds\right)^2$$
$$+ M_t(1) \int_0^t \frac{Z_t}{Z_s} \frac{2x(s)}{1+x^2(s)} ds + \int_0^t \frac{Z_t^2}{Z_s} \frac{2xx(s)}{1+x^2(s)} ds.$$

To estimate the rightmost term from above, note that, in view of (3.6), $x \le x(s)Z_s^{-1}$, while for the third and fourth terms of the sum we use that $\frac{x(s)}{1+x^2(s)} \le \frac{1}{2}$. It follows that

$$x^2(t) \le x^2 Z_t^2 + 2xZ_t M_t(1) + \frac{9}{4} M_t^2(1) + 2\int_0^t \frac{Z_t^2}{Z_s^2} ds, \qquad (3.7)$$

It is easy to see that

$$\mathbb{E}\left[\int_0^\infty e^{-\rho t} x^2 Z_t^2 dt\right] = x^2 \int_0^\infty e^{-(\rho+2b-\sigma^2)t} dt = Ax^2, \tag{3.8}$$

Lemma A.1 (ii) gives

$$\mathbb{E}\left[\int_0^\infty e^{-\rho t} 2x Z_t M_t(1) dt\right] = 2Ax \int_0^\infty e^{-\rho t}\mathbb{E}[Z_t] dt = \frac{2Ax}{\rho+b}, \tag{3.9}$$

while Lemma A.1 (i) and (iii) yield

$$\int_0^\infty e^{-\rho t}\mathbb{E}\left[M_t^2(1)\right] dt = 2A \int_0^\infty e^{-\rho t}\mathbb{E}\left[M_t(1)\right] dt = \frac{2A}{\rho(\rho+b)}. \tag{3.10}$$

We also have

$$\int_0^\infty e^{-\rho t} \int_0^t \mathbb{E}\left[\frac{Z_t^2}{Z_s^2}\right] ds\, dt = \int_0^\infty e^{-\rho t} \int_0^t e^{(\sigma^2-2b)(t-s)} ds\, dt = \frac{A}{\rho}. \tag{3.11}$$

Using the last four observations in (3.7), we find for some constant B,

$$\int_0^\infty e^{-\rho t}\mathbb{E}\left[x^2(t)\right] dt \le A\left[\left(x + \frac{1}{\rho+b}\right)^2 + B\right]. \tag{3.12}$$

On the other hand, using that, for all $x \ge 0$, $(1+x)^2 \ge (1+x^2)$, and Jensen's inequality, we find

$$\int_0^\infty e^{-\rho t}\mathbb{E}\left[\ln u(t)\right] dt \ge -\int_0^\infty e^{-\rho t}\mathbb{E}\left[\ln\left(1+x(t)\right)\right] dt$$

$$\ge -\frac{1}{\rho}\ln\left(\int_0^\infty \rho e^{-\rho t}\left(1 + \mathbb{E}[x(t)]\right) dt\right)$$

$$= -\frac{1}{\rho}\ln\left(1 + \rho\int_0^\infty e^{-\rho t}\mathbb{E}[x(t)] dt\right).$$

By (3.6) it follows that $\mathbb{E}[x(t)] \le x\mathbb{E}[Z_t] + \frac{3}{2}\mathbb{E}[M_t(1)] = xe^{-bt} + \frac{3}{2}\mathbb{E}[M_t(1)]$. Hence, using Lemma A.1 (i), we obtain

$$\int_0^\infty e^{-\rho t}\mathbb{E}\left[\ln u(t)\right] dt \ge -\frac{1}{\rho}\ln\left(1 + \frac{\rho x}{\rho+b} + \frac{3}{2(\rho+b)}\right).$$

The preceding estimate and (3.12) together imply that, for some suitable constant K_1,

$$V(x) \geq J(x; u) = \mathbb{E}\Big[\int_0^\infty e^{-\rho t}\big(\ln u(t) - x^2(t)\big)\, dt\Big]$$

$$\geq -A\Big(x + \frac{1}{b+\rho}\Big)^2 - \frac{1}{\rho}\ln\Big(x + \frac{1}{b+\rho}\Big) + K_1.$$

Proof of the upper bound, part (i): In view of Lemmas 3.4 and 3.5, it suffices to find $K_2 > 0$, such that the asserted inequality holds for $x \geq 1$.

Fix $u \in \mathfrak{U}$. Then

$$x^2(t) \geq x^2 Z_t^2 + 2x Z_t^2 \int_0^t \frac{1}{Z_s}\Big(u(s) + \frac{x^2(s)}{1 + x^2(s)}\Big) ds$$

$$= x^2 Z_t^2 + 2x Z_t M_t(1 + u) - \int_0^t \frac{Z_t^2}{Z_s^2}\frac{2x Z_s}{1 + x^2(s)}\, ds.$$

Since

$$1 + x^2(t) \geq 1 + x^2 Z_t^2 \geq 2x Z_t,$$

so we can further estimate $x^2(t)$ from below by

$$x^2(t) \geq x^2 Z_t^2 + 2x Z_t M_t(1) + 2x Z_t M_t(u) - \int_0^t \frac{Z_t^2}{Z_s^2}\, ds. \qquad (3.13)$$

Using the elementary inequality that $\ln a \leq ab - \ln b - 1$, which holds for all $a, b > 0$, and Lemma A.1 (ii), we obtain, for some B,

$$\int_0^\infty e^{-\rho t}\mathbb{E}\big[\ln u(t)\big] dt \leq \mathbb{E}\Big[\int_0^\infty e^{-\rho t}\big\{2Ax Z_t u(t) - \ln\big(2Ax Z_t\big)\big\}\, dt\Big]$$

$$= \mathbb{E}\Big[\int_0^\infty e^{-\rho t} 2x Z_t M_t(u)\, dt\Big] - \frac{\ln(2Ax)}{\rho} + \frac{2b + \sigma^2}{2\rho^2}$$

$$\leq \mathbb{E}\Big[\int_0^\infty e^{-\rho t}\Big(x^2(t) - x^2 Z_t^2 - 2x Z_t M_t(1) + \int_0^t \frac{Z_t^2}{Z_s^2}\, ds\Big) dt\Big] - \frac{\ln x + B}{\rho},$$

where in the final step we have used (3.13).

In view of (3.8), (3.9) and (3.11), for every $u \in \mathfrak{U}$ there exists $K_2 > 0$ such that

$$J(x; u) \leq -A\Big(x + \frac{1}{b+\rho}\Big)^2 - \frac{1}{\rho}\ln\Big(x + \frac{1}{\rho+b}\Big) + K_2.$$

The assertion now follows by taking the supremum over $u \in \mathfrak{U}$.

Proof of the lower bound, part (ii): Fix x_1, x_2 as in the statement. It follows from Lemma 3.6 that for any $\epsilon > 0$ there exists a control $u_\epsilon \in \mathfrak{U}$ such that

$$V(x_1) \leq \mathbb{E}\left[\int_0^{\tau_\varepsilon} e^{-\rho t} \ln u_\varepsilon(t)\, dt\right] + \mathbb{E}\left[e^{-\rho \tau_\varepsilon}\right] V(x_2) + \varepsilon(x_2^2 - x_1^2), \qquad (3.14)$$

where τ_ε is the hitting time on $[x_2, +\infty)$ of the solution $x_\varepsilon(\cdot)$ to (1.1) with $x(0) = x_1$ and control u_ε.

Using the elementary inequality

$$\ln u_\varepsilon(t) \leq \ln \kappa + \frac{u_\varepsilon(t)}{\kappa} - 1, \quad \text{with } \kappa = e^{\rho V(x_2)+1},$$

we find

$$V(x_1) - V(x_2) \leq \frac{1}{\kappa} \mathbb{E}\left[\int_0^{\tau_\varepsilon} e^{-\rho t} u_\varepsilon(t)\, dt\right] + \varepsilon(x_2^2 - x_1^2). \qquad (3.15)$$

To conclude it suffices to show that

$$\frac{1}{\kappa} \mathbb{E}\left[\int_0^{\tau_\varepsilon} e^{-\rho t} u_\varepsilon(t)\, dt\right] \leq \Phi(x_2)(x_2 - x_1). \qquad (3.16)$$

To this end, we apply Itô's rule to the semimartingale $Y_t = e^{-\rho t + \gamma x_\varepsilon(t)}$, where $\gamma > 0$ is a constant to be determined, and find

$$Y_t - e^{\gamma x_1} = \int_0^t Y_s \left(-\rho\, ds + \gamma\, dx_\varepsilon(s) + \frac{\gamma^2}{2} d\langle x_\varepsilon \rangle_s\right)$$

$$= \int_0^t Y_s \left(-\rho + \gamma\left(u_\varepsilon(s) - bx_\varepsilon(s) + \frac{x_\varepsilon^2(s)}{1 + x_\varepsilon^2(s)}\right) + \frac{\gamma^2 \sigma^2 x_\varepsilon^2(s)}{2}\right) ds + M_t,$$

where M_t stands for the martingale $\gamma\sigma \int_0^t x_\varepsilon(s)\, dW(s)$.

Next we apply the optional stopping theorem for the bounded stopping time $\tau_N = \min\{\tau_\varepsilon, N\}$, with $N \in \mathbb{N}$, to find

$$\mathbb{E}[Y_{\tau_N}] - e^{\gamma x_1} = \mathbb{E}\left[\int_0^{\tau_N} Y_s\left(-\rho + \gamma\left(u_\varepsilon(s) - bx_\varepsilon(s) + \frac{x_\varepsilon^2(s)}{1 + x_\varepsilon^2(s)}\right) + \frac{\gamma^2 \sigma^2 x_\varepsilon^2(s)}{2}\right) ds\right].$$

Since $0 \leq x_\varepsilon(s) \leq x_2$ in $[0, \tau_\varepsilon]$,

$$e^{\gamma x_2} \mathbb{E}\left[e^{-\rho \tau_N}\right] - e^{\gamma x_1} \geq \mathbb{E}\left[\int_0^{\tau_N} Y_t\left(-\rho + \gamma\left(u_\varepsilon(t) - bx_\varepsilon(t)\right) + \frac{\gamma^2 \sigma^2 x_\varepsilon^2(t)}{2}\right) dt\right],$$

and

$$e^{\gamma x_2} - e^{\gamma x_1} \geq \gamma \mathbb{E}\left[\int_0^{\tau_N} e^{-\rho t} u_\varepsilon(t)\, dt\right] + \mathbb{E}\left[\int_0^{\tau_N} Y_t\left(-b\gamma x_\varepsilon(t) + \frac{\sigma^2 \gamma^2 x_\varepsilon^2(t)}{2}\right) dt\right].$$

Note that the term in the parenthesis above is nonnegative if $\gamma x_\varepsilon(t) \geq 2b/\sigma^2$, and greater than or equal to $-b^2/2\sigma^2$ in any case. Hence,

$$e^{\gamma x_2} - e^{\gamma x_1} \geq \gamma \mathbb{E}\left[\int_0^{T_N} e^{-\rho t} u_\varepsilon(t)\, dt\right] - \frac{b^2 e^{\frac{2b}{\sigma^2}}}{2\sigma^2} \mathbb{E}\left[\int_0^{T_N} e^{-\rho t}\, dt\right].$$

Letting $N \to \infty$ we get

$$e^{\gamma x_2} - e^{\gamma x_1} \geq \gamma \mathbb{E}\left[\int_0^{T_\varepsilon} e^{-\rho t} u_\varepsilon(t)\, dt\right] - \frac{b^2 e^{\frac{2b}{\sigma^2}}}{2\sigma^2} \mathbb{E}\left[\int_0^{T_\varepsilon} e^{-\rho t}\, dt\right]. \tag{3.17}$$

With γ still at our disposal, to show (3.16) it suffices to control the term $\mathbb{E}\left[\int_0^{T_\varepsilon} e^{-\rho t}\, dt\right]$ by $\mathbb{E}\left[\int_0^{T_\varepsilon} e^{-\rho t} u_\varepsilon(t)\, dt\right]$.

Since without loss of generality we may assume that $\varepsilon < A$, Lemma 3.4 and (3.14) give

$$0 \leq V(x_1) - V(x_2) - \varepsilon(x_2^2 - x_1^2) \leq \mathbb{E}\left[\int_0^{T_\varepsilon} e^{-\rho t} \ln u_\varepsilon(t)\, dt\right] - \rho V(x_2)\mathbb{E}\left[\int_0^{T_\varepsilon} e^{-\rho t}\, dt\right].$$

Jensen's inequality then implies that

$$\mathbb{E}\left[\int_0^{T_\varepsilon} e^{-\rho t} u_\varepsilon(t)\, dt\right] \geq e^{\rho V(x_2)} \mathbb{E}\left[\int_0^{T_\varepsilon} e^{-\rho t}\, dt\right],$$

and (3.17) gives, with $C(x_2) = \frac{b^2}{2\sigma^2} e^{\frac{2b}{\sigma^2} - \rho V(x_2)}$,

$$\gamma e^{\gamma x_2}(x_2 - x_1) \geq e^{\gamma x_2} - e^{\gamma x_1} \geq (\gamma - C(x_2)) \mathbb{E}\left[\int_0^{T_\varepsilon} e^{-\rho t} u_\varepsilon(t)\, dt\right]. \tag{3.18}$$

Even though (3.16), and hence the assertion of the Theorem, follows now with a suitable choice of γ, we will optimize the preceding inequality for later use. Choosing $\gamma = q(x_2)/x_2$ in (3.18), where

$$q(x_2) = \frac{x_2 C(x_2)}{2} + \sqrt{\left(\frac{x_2 C(x_2)}{2}\right)^2 + x_2 C(x_2)}, \tag{3.19}$$

we obtain

$$\mathbb{E}\left[\int_0^{T_\varepsilon} e^{-\rho t} u_\varepsilon(t)\, dt\right] \leq (x_2 - x_1)(1 + q(x_2))e^{q(x_2)}.$$

Letting $\varepsilon \to 0$, (3.15) yields

$$\frac{V(x_2) - V(x_1)}{x_2 - x_1} \geq -(1 + q(x_2))e^{q(x_2) - 1 - \rho V(x_2)}, \tag{3.20}$$

the claim now follows. □

Proof of the upper bound, part (ii): In view of Lemma 3.4, it suffices to assume that $x_2 \leq b$, since otherwise we have

$$V(x_2) - V(x_1) \leq -A(x_2^2 - x_1^2) < -Ab(x_2 - x_1).$$

For a positive constant c, choose a $u_c \in \mathfrak{U}$ that is constant an equal to c up to time $\tau_c = \tau_{u_c}$. Then, Lemma 3.6 yields

$$V(x_1) \geq \frac{\ln c - x_2^2}{\rho}(1 - \mathbb{E}[e^{-\rho\tau_c}]) + \mathbb{E}[e^{-\rho\tau_c}]V(x_2),$$

or equivalently,

$$(V(x_2) - V(x_1))\mathbb{E}[e^{-\rho\tau_c}] \leq -(\ln c - \rho V(x_1) - x_2^2)\, \mathbb{E}\left[\int_0^{\tau_c} e^{-\rho t}\, dt\right]. \quad (3.21)$$

Consider now the solution $x_c(\cdot)$ to (1.1) with $x(0) = x_1$ and control u_c. Applying Itô's formula to $e^{-\rho t}x_c(t)$, followed by the optional stopping theorem for the bounded stopping time $\tau_N = \tau_c \wedge N$, we get

$$\mathbb{E}[e^{-\rho\tau_N}x_c(\tau_N)] - x_1 = \mathbb{E}\left[\int_0^{\tau_N} e^{-\rho t}\left(c - (b + \rho)x_c(t) + \frac{x_c^2(t)}{1 + x_c^2(t)}\right) dt\right]. \quad (3.22)$$

The leftmost term of (3.22) is equal to $x_2\mathbb{E}[e^{-\rho\tau_c}; \tau_c \leq N] + e^{-\rho N}\mathbb{E}[x_c(\tau_N); \tau_c > N]$.

On the other hand, since we have assumed that $x_2 \leq b$, we have $x_c(t) \leq b$ up to time τ_c. Thus, the right hand side of (3.22) is bounded by $\mathbb{E}\left[\int_0^{\tau_N} e^{-\rho t}c\, dt\right]$.

Letting $N \to \infty$ in (3.22), by the monotone convergence theorem, we have

$$x_2\mathbb{E}[e^{-\rho\tau_c}] - x_1 \leq c\,\mathbb{E}\left[\int_0^{\tau_c} e^{-\rho t}dt\right] \iff (x_2 - x_1)\mathbb{E}[e^{-\rho\tau_c}] \leq (c + \rho x_1)\,\mathbb{E}\left[\int_0^{\tau_c} e^{-\rho t}dt\right].$$

Substituting this in (3.21) and choosing $\ln c = \rho V(x_1) + 1 + x_2^2$, we find

$$V(x_2) - V(x_1) \leq -(x_2 - x_1)\left(e^{\rho V(x_1)+1+x_2^2} + \rho x_1\right)^{-1}. \quad (3.23)$$

The assertion now follows letting $C = Ab \wedge \left(e^{\rho V(0)+1+b^2} + \rho b\right)^{-1} > 0.$ □

Proof of part (iii): It follows from (3.23) that, for any $\varepsilon \in (0, b]$,

$$\frac{V(\varepsilon) - V(0)}{\varepsilon} \leq -e^{-\rho V(0) - 1 - \varepsilon^2}.$$

Letting $\varepsilon \to 0$ we get

$$\limsup_{\varepsilon \to 0} \frac{V(\varepsilon) - V(0)}{\varepsilon} \leq -e^{-\rho V(0) - 1},$$

while (3.20) gives

$$\frac{V(\varepsilon) - V(0)}{\varepsilon} \geq -\left(1 + q(\varepsilon)\right) e^{q(\varepsilon) - 1 - \rho V(\varepsilon)}.$$

Letting $\varepsilon \to 0$ and noting that $q(\varepsilon) \to 0$, and $V(\varepsilon) \to V(0)$, we have

$$\liminf_{\varepsilon \to 0} \frac{V(\varepsilon) - V(0)}{\varepsilon} \leq -e^{-\rho V(0) - 1},$$

which proves the claim.   □

We conclude observing that since $V \in C([0, \infty))$, the general dynamic programming principle is also satisfied. For a proof we refer to [35].

4   Viscosity Solutions and the Hamilton–Jacobi–Bellman Equation

Since the Hamiltonian (2.4) can take infinite values we have a singular stochastic control problem and the welfare function (1.3) should satisfy the proper variational inequality; see [14], Sect. 8 and [30], Sect. 4. The proof of the next Lemma, except for the treatment of the boundary conditions, follows the lines of Proposition 4.3.2 of [30].

Lemma 4.1 *If $\sigma^2 < \rho + 2b$, the welfare function V defined by (1.3) is a continuous constrained viscosity solution of*

$$\min\left[\rho V - \sup_{u \in \mathbb{R}^+} \left(\frac{1}{2}\sigma^2 x^2 V_{xx} + (u - bx + \frac{x^2}{x^2 + 1})V_x + \ln u - x^2\right), -V_x\right] = 0, \quad in\ [0, \infty). \tag{4.1}$$

Proof That V is a viscosity solution in $(0, \infty)$ follows as in [30], so we omit the details.

Here we briefly discuss the subsolution property at $x = 0$. Let ϕ be a test function such that $V - \phi$ has a maximum at $x = 0$ with $V(0) - \phi(0) = 0$, and, proceeding by contradiction, we assume that

$$\rho\phi(0) - \sup_{u \in \mathbb{R}^+} G\left(0, u, \phi'(0), \phi''(0)\right) > 0 \quad \text{and} \quad -\phi'(0) > 0. \tag{4.2}$$

Since $-\phi'(0) > 0$, the supremum in the above inequality is finite, the Hamiltonian takes the standard form, and the first inequality in (4.2) becomes

$$\rho\phi(0) + 1 + \ln\left(-\phi'(0)\right) > 0.$$

On the other hand, in view of Proposition 1 (iii), we have $\rho V(0) + 1 + \ln(-V'(0)) = 0$, hence $V'(0) > \phi'(0)$, contradicting that $V - \phi$ has a maximum at $x = 0$. We have now obtained all the necessary material for the proof of Theorem 2.1.

Proof The fact that (1.3) is a continuous constrained viscosity solution of the equation (OHJB) is a direct consequence of the above Lemma and the fact that inequality (2.9) implies that $p \leq -C$ for any $p \in D^{\pm}V(x)$, with $x \in (0, \infty)$. The regularity of V in $(0, \infty)$ follows from the classical results for uniformly elliptic operators. $\square$

Due to the extra regularity of the welfare function, the following verification equation holds in $(0, \infty)$, for any optimal pair $(\overline{u}(\cdot)), \overline{x}(\cdot))$,

$$\rho V(\overline{x}(t)) = \sup_{u \in U} G(\overline{x}(t), u, V_x(\overline{x}(t)), V_{xx}(\overline{x}(t)))$$
$$= \left(\frac{\overline{x}^2(t)}{\overline{x}^2(t) + 1} - b\overline{x}(t)\right) V_x(\overline{x}(t)) - \left(\ln(-V_x(\overline{x}(t))) + \overline{x}^2(t) + 1\right)$$
$$+ \frac{1}{2}\sigma^2\overline{x}^2(t) V_{xx}(\overline{x}(t)), \quad t \in [0, \infty) - a.e., \quad \mathbb{P} - a.e.;$$

see [14, 36].

Next we prove the proper comparison principle for (OHJB). The proof is along the lines of the strategy in [17], where given a subsolution u and a supersolution v of (OHJB), $u - v$ is a subsolution of the corresponding linearized equation. Then, one concludes by comparing $u - v$ with the appropriate supersolution of the linearized equation; see also [13, 37]. The difference with the existing results is that, due to the presence of the logarithmic term, the commonly used functions of simple polynomials do not yield a supersolution of the equation.

Having in mind that we are looking for a viscosity solution that is strictly decreasing and satisfies (2.9), we prove the following lemma.

Lemma 4.2 *Suppose u, v satisfy the assumptions of Theorem 2.2. Then $\psi = u - v$ is a subsolution of*

$$\rho\psi + bx D\psi - \left(1 + c^*\right)|D\psi| - \frac{1}{2}\sigma^2 x^2 D^2\psi = 0 \quad \text{in} \quad [0, \infty). \tag{4.3}$$

Proof Let $\bar{x} \geq 0$ a maximum point of $\psi - \phi$ for some smooth function ϕ and set, following [34],

$$\theta(x, y) = \phi(x) + \frac{(x - y + \varepsilon L)^2}{\varepsilon} + \delta(x - \bar{x})^4,$$

where L, δ are positive constants.

The assumptions on u, v imply that the function $(x, y) \mapsto u(x) - v(y) - \theta(x, y)$ is bounded from above and achieves its maximum at, say, $(x_\varepsilon, y_\varepsilon)$. It follows that $x \mapsto u(x) - v(y_\varepsilon) - \theta(x, y_\varepsilon)$ has a local maximum at x_ε and $y \mapsto v(y) - u(x_\varepsilon) + \theta(x_\varepsilon, y)$ has a local minimum at y_ε. Moreover, (see Proposition 3.7 in [12]), as $\varepsilon \to 0$,

$$\frac{|x_\varepsilon - y_\varepsilon|^2}{\varepsilon} \to 0, \ x_\varepsilon \to \bar{x}, \ \text{and} \ u(x_\varepsilon) - v(y_\varepsilon) \to \psi(\bar{x}). \tag{4.4}$$

The inequalities

$$u(x_\varepsilon) - v(y_\varepsilon) - \theta(x_\varepsilon, y_\varepsilon) \leq \psi(\bar{x}) - \phi(\bar{x}) + v(x_\varepsilon) - v(y_\varepsilon) - \frac{|x_\varepsilon - y_\varepsilon + \varepsilon L|^2}{\varepsilon} - \delta(x_\varepsilon - \bar{x})^4$$

and

$$u(x_\varepsilon) - v(y_\varepsilon) - \theta(x_\varepsilon, y_\varepsilon) \geq u(\bar{x}) - v(\bar{x} + \varepsilon L) - \theta(\bar{x}, \bar{x} + \varepsilon L) \geq \psi(\bar{x}) - \phi(\bar{x})$$

together imply that

$$\frac{|x_\varepsilon - y_\varepsilon + \varepsilon L|^2}{\varepsilon} + \delta(x_\varepsilon - \bar{x})^4 \leq v(x_\varepsilon) - v(y_\varepsilon).$$

Since v is decreasing we must have $y_\varepsilon > x_\varepsilon$. In particular, $y_\varepsilon \in (0, \infty)$.

Therefore, setting $p_\varepsilon = 2 \frac{x_\varepsilon - y_\varepsilon + \varepsilon L}{\varepsilon}$ and $q_\varepsilon = \phi_x(x_\varepsilon) + 4\delta(x_\varepsilon - \bar{x})^3$, Theorem 3.2 in [12] implies that, for every $\alpha > 0$, there exist X, $Y \in \mathbb{R}$ such that

$$\rho u(x_\varepsilon) - H(x_\varepsilon, p_\varepsilon + q_\varepsilon, X) \leq 0 \ \text{and} \ \rho v(y_\varepsilon) - H(y_\varepsilon, p_\varepsilon, Y) \geq 0 \tag{4.5}$$

and

$$-\left(\frac{1}{\alpha} + \|M\|\right)I \leq \begin{pmatrix} X & 0 \\ 0 & -Y \end{pmatrix} \leq M + \alpha M^2 \tag{4.6}$$

with $M = D^2\theta(x_\varepsilon, y_\varepsilon)$.

By subtracting the two inequalities in (4.5) we obtain

$$\rho u(x_\varepsilon) - \rho v(y_\varepsilon) + b(x_\varepsilon - y_\varepsilon)p_\varepsilon + \left(\frac{y_\varepsilon^2}{y_\varepsilon^2 + 1} - \frac{x_\varepsilon^2}{x_\varepsilon^2 + 1}\right)p_\varepsilon - \left(\frac{x_\varepsilon^2}{x_\varepsilon^2 + 1} - bx_\varepsilon\right)q_\varepsilon$$
$$\tag{4.7}$$

$$+ \ln(-p_\varepsilon - q_\varepsilon) - \ln(-p_\varepsilon) + x_\varepsilon^2 - y_\varepsilon^2 - \frac{1}{2}\sigma^2 x_\varepsilon^2 X + \frac{1}{2}\sigma^2 y_\varepsilon^2 Y \leq 0.$$

Our assumption on u implies that $p_\varepsilon + q_\varepsilon \le -1/c^*$. Thus, the difference of the logarithmic terms in the above inequality can be estimated from below as

$$\ln\left(\frac{p_\varepsilon + q_\varepsilon}{p_\varepsilon}\right) \ge \frac{q_\varepsilon}{p_\varepsilon + q_\varepsilon} \ge -c^* |q_\varepsilon|,$$

and inequality (4.7) gives

$$\frac{1}{2}\sigma^2 x_\varepsilon^2 X + \frac{1}{2}\sigma^2 y_\varepsilon^2 Y \ge \rho u(x_\varepsilon) - \rho v(y_\varepsilon) + b x_\varepsilon q_\varepsilon - (1 + c^*)|q_\varepsilon| +$$
$$p_\varepsilon(x_\varepsilon - y_\varepsilon)\left(b - \frac{x_\varepsilon + y_\varepsilon}{(1 + x_\varepsilon^2)(1 + y_\varepsilon^2)}\right) + x_\varepsilon^2 - y_\varepsilon^2 \quad (4.8)$$

On the other hand, the right-hand-side in (4.6) yields

$$\frac{1}{2}\sigma^2 x_\varepsilon^2 X - \frac{1}{2}\sigma^2 y_\varepsilon^2 Y \le \frac{1}{2}\sigma^2 x_\varepsilon^2\left(\phi_{xx}(x_\varepsilon) + 12\delta(x_\varepsilon - \bar{x})^2\right) + \frac{\sigma^2}{\varepsilon}(x_\varepsilon - y_\varepsilon)^2 + m\left(\frac{\alpha}{\varepsilon^2}\right),$$
$$(4.9)$$

with m a modulus of continuity independent of α, ε.

By combining (4.8) with (4.9), we conclude the proof taking first $\alpha \to 0$, then $\varepsilon \to 0$ and using (4.4). $\qquad\square$

We continue with the

Proof of Theorem 2.2: The main step is the construction of a solution of the linearized equation. For this, we consider the ode

$$\rho w + \left(bx - (1 + c^*)\right)w' - \frac{1}{2}\sigma^2 x^2 w'' = 0, \quad (4.10)$$

which has a solution of the form

$$w(x) = x^{-k} \mathcal{J}\left(\frac{2 + 2c^*}{\sigma^2 x}\right), \quad (4.11)$$

where k is a root of

$$k^2 + \left(1 + \frac{2b}{\sigma^2}\right)k - \frac{2\rho}{\sigma^2} = 0 \quad (4.12)$$

and $\mathcal{J}$ a solution of the degenerate hypergeometric equation

$$xy'' + (\tilde{b} - x)y' - \tilde{a}y = 0 \quad (4.13)$$

with $\tilde{a} = k$ and $\tilde{b} = 2(k + 1 + b/\sigma^2)$.

Since we are looking for a solution of (4.10) with superquadratic growth at $+\infty$, we choose k to be the negative root of (4.12). The assumption $\sigma^2 < \rho + 2b$ implies $-k > 2$.

We further choose $\mathcal{J}$ to be the Tricomi solution of (4.13) which satisfies

$$\mathcal{J}(0) > 0 \quad \text{and} \quad \mathcal{J}(x) = x^{-k}\Big(1 + \frac{2\rho}{\sigma^2 x} + o(x^{-1})\Big) \quad \text{as } x \to \infty.$$

With this choice, the function w defined in (4.11) for $x > 0$ and by continuity at $x = 0$, satisfies $w(0), w'(0) > 0$ and $w(x) \sim \mathcal{J}(0)x^{-k}$, as $x \to \infty$.

Note that w is increasing in $[0, \infty)$ since it would otherwise have a positive local maximum and this is impossible by (4.10). In particular, w satisfies (4.3).

Set now $\psi = u - v$ and consider $\epsilon > 0$. Since $\psi - \epsilon w < 0$ in a neighborhood of infinity, there exists $x^\epsilon \in [0, \infty)$ such that

$$\max_{x \geq 0} \big(\psi(x) - \epsilon w(x)\big) = \psi(x^\epsilon) - \epsilon w(x^\epsilon).$$

By Lemma 4.2 ψ is a subsolution of (4.3). We now use ϵw as a test function to find that

$$0 \geq \rho\psi(x^\epsilon) + \epsilon b x^\epsilon w(x^\epsilon) - \epsilon\big(1 + c^*\big)|w'(x^\epsilon)| - \frac{1}{2}\sigma^2(x^\epsilon)^2 w''(x^\epsilon) = \psi(x^\epsilon) - \epsilon w(x^\epsilon).$$

Hence, $\psi(x) \leq \epsilon w(x)$ for all $x \in [0, \infty)$. Since ϵ is arbitrary, this proves the claim. $\qquad\square$

The stability property of viscosity solutions yields the following theorem.

Theorem 4.1 *As $\sigma \to 0$, the welfare function V defined by (1.3) converges locally uniformly to the constrained viscosity solution $V^{(d)}$ of the deterministic shallow lake equation in $[0, \infty)$,*

$$\rho V^{(d)} = \Big(\frac{x^2}{x^2 + 1} - bx\Big) V_x^{(d)} - \Big(\ln(-V_x^{(d)}) + x^2 + 1\Big). \qquad \text{(OHJB}_\text{d}\text{)}$$

We next prove Theorem 2.3 that describes the asymptotic behaviour of V as $x \to \infty$. The proof is based on a scaling argument and the stability properties of the viscosity solutions.

Proof of Theorem 2.3: We write V as

$$V(x) = -A\Big(x + \frac{1}{b + \rho}\Big)^2 - \frac{1}{\rho}\ln\Big(2A(x + \frac{1}{b + \rho})\Big) + K + v(x).$$

Straightforwad calculations yield that v is a viscosity solution in $(0, \infty)$ of the equation

$$\rho v + \Big(bx - \frac{x^2}{x^2 + 1}\Big)v' + \ln\left(1 + \frac{1 - \rho\big(x + \frac{1}{b+\rho}\big)v'}{2A\rho\big(x + \frac{1}{b+\rho}\big)^2}\right) - \frac{1}{2}\sigma^2 x^2 v'' + f = 0,$$

$$\text{(4.14)}$$

where

$$f(x) = \frac{b + \frac{\sigma^2}{2} + (b + \rho)\frac{x^2}{1+x^2}}{\rho(1 + x(b + \rho))} + \frac{\sigma^2 x(b + \rho)}{2\rho(1 + x(b + \rho))^2} - 2A\frac{(1 + x(b + \rho))}{1 + x^2}.$$

Note f is smooth on $[0, \infty)$ and vanishes as $x \to \infty$.

Let $v_\lambda(y) = v(\frac{y}{\lambda})$ and observe that, if $v_\lambda(1) \to 0$ as $\lambda \to 0$, then $v(x) \to 0$ as $x \to \infty$. It turns out that v_λ solves

$$\rho v_\lambda + \left(bx - \frac{\lambda x^2}{x^2 + \lambda^2}\right) v'_\lambda + \ln\left(1 + \frac{\lambda^2\left(1 - \rho(x + \frac{\lambda}{b+\rho})v'_\lambda\right)}{2A\rho\left(x + \frac{\lambda}{b+\rho}\right)^2}\right) - \frac{1}{2}\sigma^2 x^2 v''_\lambda + f\left(\frac{x}{\lambda}\right) = 0.$$

Since, by (2.8) v_λ is uniformly bounded, we consider the half-relaxed limits $v^*(y) = \lim\sup_{x \to y, \lambda \to 0} v_\lambda(x)$ and $v_*(y) = \lim\inf_{x \to y, \lambda \to 0} v_\lambda(x)$ in $(0, \infty)$, which are (see [5]) respectively sub- and super-solutions of

$$\rho w + byw' - \frac{1}{2}\sigma^2 y^2 w'' = 0. \tag{4.15}$$

It is easy to check that for any $y > 0$ we have $v^*(y) = \lim\sup_{x \to \infty} v(x)$ and $v_*(y) = \lim\inf_{x \to \infty} v(x)$. The subsolution property of v^* and the supersolution property of v_* give

$$\lim\sup_{x \to \infty} v(x) \leq 0 \leq \lim\inf_{x \to \infty} v(x) \leq \lim\sup_{x \to \infty} v(x).$$

$\square$

5 A Numerical Scheme and Optimal Dynamics

A general argument to prove the convergence of monotone schemes for viscosity solutions of fully nonlinear second-order elliptic or parabolic, possibly degenerate, partial differential equations has been introduced in [6]. Their methodology has been extensively used to approximate solutions to first-order equations, see for example [20, 31–33].

On the other hand, it is not always possible to construct monotone schemes for second-order equations in their full generality. However, various types of nonlinear second-order equations have been approximated via monotone schemes based on [6]; see, for example [4, 7, 15, 29].

Next, following [22] which considered the deterministic problem, we construct a monotone finite difference scheme to approximate numerically the welfare function and recover numerically the stochastic optimal dynamics.

Let Δx denote the step size of a uniform partition $0 = x_0 < x_1 < \ldots < x_{N-1} < x_N = l$ of $[0, l]$ for $l > 0$ sufficiently large. Having in mind (2.9), if V_i is the approximation of V at x_i, we employ a backward finite difference discretization to approximate the first derivative in the linear term of the (OHJB), a forward finite difference discretization for the derivative in the logarithmic term and a central finite difference scheme to approximate the second derivative.

These considerations yield, for $i = 1, \ldots, N - 1$, the approximate equation

$$V_i - \frac{1}{\rho}\left(\frac{x_i^2}{x_i^2 + 1} - bx_i\right)\frac{V_i - V_{i-1}}{\Delta x} + \frac{1}{\rho}\left[x_i^2 + 1 + \ln\left(-\frac{V_{i+1} - V_i}{\Delta x}\right)\right]$$

$$- \frac{\sigma^2}{2\rho}\frac{V_{i+1} + V_{i-1} - 2V_i}{(\Delta x)^2} = 0. \tag{5.1}$$

Setting

$$g(x, w, c, d) = \left[(\Delta x)^2 - \frac{1}{\rho}\left(\frac{x^2}{x^2 + 1} - bx\right)\Delta x + \frac{\sigma^2}{\rho}\right]w + \frac{1}{\rho}(x^2 + 1)(\Delta x)^2$$

$$+ \frac{1}{\rho}(\Delta x)^2\ln\left(-\frac{c - w}{\Delta x}\right) + \frac{1}{\rho}\Delta x\left(\frac{x^2}{x^2 + 1} - bx\right)d - \frac{\sigma^2}{2\rho}(c + d), \tag{5.2}$$

the numerical approximation of V satisfies

$$g(x_i, V_i, V_{i+1}, V_{i-1}) = 0, \quad \text{for } i = 1, \ldots, \text{N-1}. \tag{5.3}$$

Following [6], we write the numerical approximation scheme defined by (5.3) in the form

$$S(r, x, v^r(x), v^r) = 0 \quad \text{in } [0, \infty), \tag{5.4}$$

where $S : \mathbb{R}^+ \times \mathbb{R}^+ \times \mathbb{R} \times M(\mathbb{R}^+) \to \mathbb{R}$, $r = \Delta x$ and v^r is defined by $v^r(y) = V_i$ for $y \in [x_i, x_{i+1})$. Here, $M(\mathbb{R}^+)$ is the space of locally bounded functions defined in $\mathbb{R}^+$. We have the following theorem.

Proposition 2 *A numerical scheme defined by (5.4), with $v^r(x_{i+1}) < v^r(x_i)$, is consistent and monotone. In addition, it is stable and v^r, $r \to 0$, converges locally uniformly to the unique constrained viscosity solution of the equation* (OHJB).

Proof Consistency is obtained by standard finite differences methods. For the monotonicity we observe that for two different approximation grid vectors $(U_0, \ldots, U_N)$ and $(V_0, \ldots, V_N)$ with $U_i \geq V_i$ and $U_i = V_i = w$, we have

$$g(x_i, w, U_{i+1}, U_{i-1}) \leq g(x_i, w, V_{i+1}, V_{i-1}), \tag{5.5}$$

provided $\Delta x(\frac{x^2}{x^2+1} - bx) \leq \sigma^2/2$. Such a condition is satisfied for fixed $\sigma > 0$, if we take Δx small enough. Since the welfare function solves a state constraint problem, the equation is satisfied on the left boundary point and the proof follows that of Theorem 2.1 of [6].

For numerical computations, since the computational domain of the problem is finite, a boundary condition has to be imposed at $x = l$, for l sufficiently large, by exploiting the asymptotic behaviour of the welfare function V as $x \to +\infty$. The boundary condition at the right endpoint $x_N = l$ is provided by the asymptotic estimate (2.7).

The aforementioned numerical scheme suggests a numerical algorithm for the computation of optimal dynamics governing the shallow lake problem. In this direction, the nondegeneracy of the shallow lake equation in $(0, \infty)$ induces extra regularity for the function V in $(0, \infty)$. Hence, the optimal dynamics for the shallow lake problem are described by

$$\begin{cases} d\bar{x}(t) = \left(-\dfrac{1}{V'(\bar{x}(t))} - b\bar{x}(t) + \dfrac{\bar{x}^2(t)}{\bar{x}^2(t) + 1} \right) dt + \sigma\bar{x}(t)dW(t), \\ \bar{x}(0) = x. \end{cases} \tag{5.6}$$

Using the numerical representation of V via (5.3) and properly discretizing the SDE (5.6), we can reconstruct numerically the optimal dynamics. This direct approach to investigate numerically the path properties of the optimal dynamics of the shallow lake problem for the various parameters ρ, b, c, σ of the problem will be presented elsewhere.

Appendix 1

Lemma A.1 *Assume that f is a positive $\mathbb{P}$-a.s. locally integrable $\mathcal{F}_t$ and let $M_t(f)$ be defined as in (3.4). Then,*

(i) $\mathbb{E}\left[\displaystyle\int_0^\infty e^{-\rho t} M_t(f)\, dt \right] = \dfrac{1}{\rho + b}\, \mathbb{E}\left[\displaystyle\int_0^\infty e^{-\rho t} f(t)\, dt \right].$

(ii) $\mathbb{E}\left[\displaystyle\int_0^\infty e^{-\rho t} Z_t M_t(f)\, dt \right] = \begin{cases} A\, \mathbb{E}\left[\int_0^\infty e^{-\rho t} Z_t f(t)\, dt \right] & \text{if } \sigma^2 < \rho + 2b, \\ \infty & \text{if } \sigma^2 \geq \rho + 2b. \end{cases}$

(iii) $\mathbb{E}\left[\displaystyle\int_0^\infty e^{-\rho t} M_t^2(f)\, dt \right] = \begin{cases} 2A\, \mathbb{E}\left[\int_0^\infty e^{-\rho t} f(t) M_t(f)\, dt \right] & \text{if } \sigma^2 < \rho + 2b, \\ \infty & \text{if } \sigma^2 \geq \rho + 2b. \end{cases}$

Proof: (i) Since f is $\mathcal{F}_t$-adapted we have

$$\mathbb{E}\left[M_t(f) \right] = \mathbb{E}\left[\int_0^t \mathbb{E}[Z_t | \mathcal{F}_s] \frac{f(s)}{Z_s}\, ds \right] = \int_0^t e^{-b(t-s)} \mathbb{E}[f(s)]\, ds.$$

Therefore,

$$\mathbb{E}\left[\int_0^\infty e^{-\rho t} M_t(f)\, dt\right] = \int_0^\infty e^{bs}\mathbb{E}[f(s)]\int_s^\infty e^{-(\rho+b)t}dt\, ds$$

$$= \frac{1}{\rho+b}\, \mathbb{E}\left[\int_0^\infty e^{-\rho t} f(t)\, dt\right].$$

(ii) Conditioning first on $\mathcal{F}_s$ we have

$$\mathbb{E}[Z_t M_t(f)] = \mathbb{E}\left[\int_0^t \mathbb{E}[Z_t^2|\mathcal{F}_s]\frac{f(s)}{Z_s}\, ds\right] = \int_0^t e^{(\sigma^2-2b)(t-s)}\mathbb{E}[Z_s f(s)]\, ds,$$

and, hence,

$$\mathbb{E}\left[\int_0^\infty e^{-\rho t} Z_t M_t(f)\, dt\right] = \int_0^\infty e^{(\sigma^2-2b)s}\mathbb{E}[Z_s f(s)]\int_s^\infty e^{-(\rho+2b-\sigma^2)t}\, dt\, ds$$

$$= \begin{cases} A\,\mathbb{E}\left[\int_0^\infty e^{-\rho t} Z_t f(t)\, dt\right] & \text{if } \sigma^2 < \rho + 2b \\ \infty & \text{if } \sigma^2 \geq \rho + 2b. \end{cases}$$

(iii) By Fubini's theorem we have

$$\mathbb{E}[M_t^2(f)] = 2\,\mathbb{E}\left[\int_0^t\int_s^t \frac{Z_t^2}{Z_s Z_q} f(s)f(q)\, dq\, ds\right]$$

$$= 2\,\mathbb{E}\left[\int_0^t\int_s^t \mathbb{E}[Z_t^2|\mathcal{F}_q]\frac{1}{Z_s Z_q} f(s)f(q)\, dq\, ds\right]$$

$$= 2\int_0^t\int_s^t e^{(\sigma^2-2b)(t-q)}\mathbb{E}\left[\frac{Z_q}{Z_s} f(s)f(q)\right] dq\, ds$$

$$= 2\int_0^t e^{(\sigma^2-2b)(t-q)}\mathbb{E}[f(q)M_q(f)]\, dq,$$

and, therefore,

$$\mathbb{E}\left[\int_0^\infty e^{-\rho t} M_t^2(f)\, dt\right] = \int_0^\infty e^{(2b-\sigma^2)q}\mathbb{E}[f(q)M_q(f)]\int_q^\infty e^{-(\rho+2b-\sigma^2)t}dt\, dq$$

$$= \begin{cases} 2A\,\mathbb{E}\left[\int_0^\infty e^{-\rho t} f(t)M_t(f)\, dt\right] & \text{if } \sigma^2 < \rho + 2b, \\ \infty & \text{if } \sigma^2 \geq \rho + 2b. \end{cases}$$

$\square$

References

1. Alvarez, O., Lasry, J.M., Lions, P.L.: Convex viscosity solutions and state constraints. J. Math. Pures. Appl. **76**, 265–288 (1997)
2. Bardi, M., Dolcetta, I.C.: Optimal control and viscosity solutions of Hamilton-Jacobi-Bellman equations, Birkhäuser (1997)
3. Barles, G.: Solutions de viscosité des équations de Hamilton-Jacobi, Mathématiques & Applications, 17. vol. 1995, pp. 129–178. Springer (1994)
4. Barles, G., Jakobsen, E.R.: On the convergence rate of approximation schemes for Hamilton-Jacobi-Bellman equations ESAIM. Math. Model. Numer. Anal. **36**(1), 33–54 (2002)
5. Barles, G., Perthame, B.: Discontinuous solutions of deterministic optimal stopping time problems RAIRO model. Math. Anal. Numer. **21**(4), 557579 (1987)
6. Barles, G., Souganidis, P.E.: Convergence of approximation schemes for fully nonlinear second order equations. Asymptotic Anal. **4**(3), 271283 (1991)
7. Bonnans, J.F., Zidani, H.: Consistency of generalized finite difference schemes for the stochastic HJB equation. SIAM J. Numer. Anal. **41**(3), 1008–1021 (2004)
8. Brock, W.A., Starrett, D.: Managing systems with non-convex positive feedback. Environ. Resour. Econ. **26**, 575–602 (2003)
9. Capuzzo Dolcetta, I., Lions, P.L.: Hamilton-Jacobi equations with state constraints. Trans. Amer. Math. Soc. **318**(2), 643–683 (1990)
10. Carpenter, S.R., Ludwig, D., Brock, W.A.: Management of eutrophication for lakes subject to potentially irreversible change. Ecol. Appl. **9**(3), 751–771 (1999)
11. Crandall, M.G., Lions, P.L.: Viscosity solutions of Hamilton-Jacobi equations. Trans. Amer. Math. Soc. **277**, 1–42 (1983)
12. Crandall, M.G., Ishii, H., Lions, P.L.: User's guide to viscosity solutions of second order partial differential equations. Bull. Amer. Math. Soc. **27**, 1–67 (1992)
13. Da Lio, F., Ley, O.: Uniqueness results for second-order Bellman-Isaacs equations under quadratic growth assumptions and applications. SIAM J. Control Optim. **45**(1), 74–106 (2006)
14. Fleming, W.H., Soner, H.M.: Controlled Markov Processes and Viscosity Solutions. Springer (1993)
15. Froese, B.D., Oberman, A.M.: Convergent finite difference solvers for viscosity solutions of the elliptic Monge-Ampère equation in dimensions two and higher. SIAM J. Numer. Anal. **49**(4), 16921714 (2011)
16. Grass, D., Kiseleva, T., Wagener, F.: Small-noise asymptotics of Hamilton-Jacobi-Bellman equations and bifurcations of stochastic optimal control problems. Commun. Nonlinear Sci. Numer. Simulat. **22**, 38–54 (2015)
17. Ishii, H.: Comparison results for Hamilton-Jacobi equations without growth condition on solutions from above. Appl. Anal. **67**(3–4), 357–372 (1997)
18. Ishii, H., Lions, P.L.: Viscosity solutions of fully nonlinear second-order elliptic partial differential equations. J. Differ. Equations **83**, 26–78 (1990)
19. Katsoulakis, M.A.: Viscosity solutions of second order fully nonlinear elliptic equations with state constraints. Indiana Univ. Math. J. **43**, 493–518 (1994)
20. Kossioris, G., Makridakis, Ch., Souganidis, P.E.: Finite volume schemes for HamiltonJacobi equations. Numer. Math. **83**(3), 427442 (1999)
21. Kossioris, G., Plexousakis, M., Xepapadeas, A., de Zeeuw, A.: On the optimal taxation of common-pool resources. J. Econom. Dynam. Control. **35**(11), 1868–1879 (2011)
22. Kossioris, G., Zohios, Ch.: The value function of the shallow lake problem as a viscosity solution of a HJB equation. Quart. Appl. Math. **70**, 625–657 (2012)
23. Krylov, N.V.: Applications of mathematics. In: Controlled Diffusion Processes, vol. 14. Springer, New York (1980)
24. Lasry, J.M., Lions, P.L.: Nonlinear elliptic equations with singular boundary conditions and stochastic control with state constraints. Math. Ann. **283**, 583–630 (1989)

25. Lions, P.L.: Optimal control of diffusion processes and Hamilton-Jacobi-Bellman equations, part I: the dynamic programming principles and applications. Comm. Partial Differ. Equations **8**(10), 1101–1174 (1983)
26. Lions, P.L.: Optimal control of diffusion processes and Hamilton-Jacobi-Bellman equations, part II: viscosity solutions and uniqueness. Comm. Partial Differ. Equations **8**(11), 1229–1276 (1983)
27. Lions, P.L.: Optimal control of diffusion processes and Hamilton-Jacobi-Bellman equations part III: regularity of the optimal cost function. In: Nonlinear PDE and Applications, College de France Seminar, vol. V. Pitman, Boston (1983)
28. Mäler, K.-G., Xepapadeas, A., de Zeeuw, A.: The economics of shallow lakes. Environ. Resour. Econ. **26**(4), 603–624 (2003)
29. Osher, S., Fedkiw, R.P.: Level Set Methods and Dynamic Implicit Surfaces. Springer, New York (2003)
30. Pham, H.: Continuous-time stochastic control and optimization with financial applications. In: Stochastic Modelling and Applied Probability, 61. Springer (2009)
31. Qian, Q.: Approximations for viscosity solutions of Hamilton-Jacobi equations with locally varying time and space grids. SIAM J. Numer. Anal. **43**(6), 2371–2401 (2006)
32. Rouy, E., Tourin, A.: A viscosity solution approach to shape-from-shading. SIAM J. Num. Anal. **29**, 867884 (1992)
33. Sethian, J.A.: Fast marching methods. SIAM Rev. **41**(2), 199–235 (1999)
34. Soner, H.M.: Optimal control with state-space constraint I and II. SIAM J. Control Optim.**24**, 552–561, 1110–1122 (1986)
35. Touzi, N.: Optimal Stochastic Control, Stochastic Target Problems, and Backward SDE. Fields Institute Monographs, 29, Springer (2013)
36. Yong, J., Zhou, X.Y.: Stochastic Controls: Hamiltonian Systems and HJB Equations. Springer (1999)
37. Zariphopoulou, Th.: Consumption-investement models with constraints. SIAM J. Control Optim. **32**(1), 59–85 (1994)
38. Wagener, F.O.O.: Skiba points and heteroclinic bifurcations, with applications to the shallow lake system. J. Econom. Dynam. Control **27**(9), 1533–1561 (2003)

Independent Particles in a Dynamical Random Environment

Mathew Joseph, Firas Rassoul-Agha and Timo Seppäläinen

Abstract We study the motion of independent particles in a dynamical random environment on the integer lattice. The environment has a product distribution. For the multidimensional case, we characterize the class of spatially ergodic invariant measures. These invariant distributions are mixtures of inhomogeneous Poisson product measures that depend on the past of the environment. We also investigate the correlations in this measure. For dimensions one and two, we prove convergence to equilibrium from spatially ergodic initial distributions. In the one-dimensional situation we study fluctuations of the net current seen by an observer traveling at a deterministic speed. When this current is centered by its quenched mean its limit distributions are the same as for classical independent particles.

F. Rassoul-Agha was partially supported by NSF grant DMS-1407574 and Simons Foundation grant 306576.

M. Joseph and F. Rassoul-Agha were partially supported by NSF grant DMS-0747758.

T. Seppäläinen was partially supported by NSF grants DMS-0701091, DMS-1003651, DMS-1306777 and DMS-1602486, by Simons Foundation grant 338287, and by the Wisconsin Alumni Research Foundation.

M. Joseph
Indian Statistical Institute, Bangalore, 8th Mile Mysore Road, RVCE Post,
Bengaluru 560059, Karnataka, India
e-mail: m.joseph@isibang.ac.in

F. Rassoul-Agha (✉)
Department of Mathematics, University of Utah, 155 South 1400 East,
Salt Lake City, UT 84112, USA
e-mail: firas@math.utah.edu
URL: http://www.math.utah.edu/~firas

T. Seppäläinen
Mathematics Department, University of Wisconsin-Madison,
Van Vleck Hall 480 Lincoln Dr., Madison, WI 53706-1388, USA
e-mail: seppalai@math.wisc.edu
URL: http://www.math.wisc.edu/~seppalai

© Springer Nature Switzerland AG 2019
P. Friz et al. (eds.), *Probability and Analysis in Interacting Physical Systems*, Springer Proceedings in Mathematics & Statistics 283,
https://doi.org/10.1007/978-3-030-15338-0_4

Keywords Random walk in random environment · Particle current · Limit distribution · Fractional brownian motion · EW universality

2000 Mathematics Subject Classification. 60K35 · 60K37

1 Introduction and Results

This paper studies particles that move on the integer lattice $\mathbb{Z}^d$. Particles interact through a common environment that specifies their transition probabilities in space and time. The environment is picked randomly at the outset and fixed for all time. Given the environment, particles evolve independently, governed by the transition probabilities specified by the environment.

We have two types of results. First we characterize those invariant distributions for the particle process that satisfy a spatial translation invariance. These turn out to be mixtures of inhomogeneous Poisson product measures that depend on the past of the environment. Poisson is expected, in view of the classical result that a system of independent random walks has a homogeneous Poisson invariant distribution [5, Sect. 8.5]. For $d = 1, 2$, we use coupling ideas from [7] (as presented in [18]) to prove convergence to this equilibrium from spatially invariant initial distributions.

In the one-dimensional case we study fluctuations of the particle current seen by an observer moving at the characteristic speed. In the present setting the characteristic speed is simply the mean speed v of the particles. More generally, the characteristic speed is the derivative $H'(\rho)$ of the flux H as a function of particle density ρ. The flux $H(\rho)$ is the mean rate of flow across a fixed bond of the lattice when the system is stationary with density ρ. For independent particles $H(\rho) = v\rho$.

It is expected, and supported by known rigorous results, that the current fluctuations are of order $n^{1/4}$ with Gaussian limits if the macroscopic flux H is linear, and of order $n^{1/3}$ with Tracy-Widom type limits if the flux H is strictly convex or concave. In statistical physics terminology, the former is the Edwards-Wilkinson (EW) universality class, and the latter the Kardar-Parisi-Zhang (KPZ) universality class. (See [3] for the physics perspective on these matters, and [4, 20] for mathematical reviews). Our motivation is to investigate the effect of a random environment in the EW class. We find that, when the current is centered by its quenched mean and the environment is averaged out, the fluctuation picture in the dynamical environment is the same as that for classical independent random walks [10, 19]. Consistent with EW universality, the current fluctuations have magnitude $t^{1/4}$ and occur on a spatial scale of $t^{1/2}$ where t denotes the macroscopic time variable.

There is an interesting contrast with the case of static environment investigated in [15]. In the static environment, the quenched mean of the current has fluctuations of magnitude $t^{1/2}$ and converges weakly to a Brownian motion. Our results suggest that under a dynamic environment the quenched mean of the current has fluctuations of magnitude $t^{1/4}$ and that when the particle system is stationary in time these fluctuations are governed by a fractional Brownian motion with Hurst parameter $1/4$.

Other work on the motion of independent particles in a random environment includes articles [9, 14].

We turn to a description of the process and then the results.

1.1 The Particle Process and Its Invariant Distributions

The particles follow independent random walks in a common dynamical random environment (RWRE). More precisely, they move in a space-time environment $\omega = (\omega_{x,s})_{(x,s)\in\mathbb{Z}^d\times\mathbb{Z}}$ indexed by a discrete time variable s and a discrete space variable x. The environment at space-time point $(x, s) \in \mathbb{Z}^d \times \mathbb{Z}$ is a vector $\omega_{x,s} = (\omega_{x,s}(z) : z \in \mathbb{Z}^d, |z| \leqslant R)$ of jump probabilities that satisfy

$$0 \leqslant \omega_{x,s}(z) \leqslant 1 \quad \text{and} \quad \sum_{z\in\mathbb{Z}^d\,:\,|z|\leqslant R} \omega_{x,s}(z) = 1. \tag{1.1}$$

R is a fixed finite constant that specifies the range of jumps. From a space-time point (x, s) admissible jumps are to points $(y, s + 1)$ such that $|y - x| \leqslant R$. In environment ω the transition probabilities governing the motion of a $\mathbb{Z}^d$-valued walk $X_\centerdot = (X_s)_{s\in\mathbb{Z}_+}$ are

$$P^\omega[X_{s+1} = y \mid X_s = x] = \pi^\omega_{s,s+1}(x, y) \equiv \omega_{x,s}(y - x). \tag{1.2}$$

P^ω is the quenched probability measure on the path space of the walk $X_\centerdot$. The environment is "dynamical" because at each time s the particle sees a new environment $\bar{\omega}_s = (\omega_{x,s} : x \in \mathbb{Z}^d)$.

$(\Omega, \mathfrak{S})$ denotes the space of environments ω satisfying the above assumptions, endowed with the product topology and its Borel σ-algebra $\mathfrak{S}$. The environment restricted to levels $s \in \{m, \ldots, n\}$ is denoted by

$$\bar{\omega}_{m,n} = (\bar{\omega}_s)_{m\leqslant s\leqslant n} = (\omega_{x,s} : m \leqslant s \leqslant n, x \in \mathbb{Z}^d).$$

Environments at levels generate σ-algebras $\mathfrak{S}_{m,n} = \sigma\{\bar{\omega}_{m,n}\}$. In these formulations $m = -\infty$ or $n = \infty$ are also possible. $T_{x,s}$ is the shift on Ω, that is $(T_{x,s}\omega)_{y,t} = \omega_{x+y,s+t}$.

Let $\mathbb{P}$ be a probability measure on Ω such that

$$\text{the probability vectors } (\omega_{x,s})_{(x,s)\in\mathbb{Z}^d\times\mathbb{Z}} \text{ are i.i.d. under } \mathbb{P}. \tag{1.3}$$

We make two nondegeneracy assumptions. The first one guarantees that the quenched walk is not degenerate:

$$\mathbb{P}\{\exists z \in \mathbb{Z}^d : 0 < \omega_{0,0}(z) < 1\} > 0. \tag{1.4}$$

Denote the mean transition kernel by $p(u) = \mathbb{E}\pi_{s,s+1}(x, x + u)$. The second key assumption is that

there does not exist $x \in \mathbb{Z}^d$ and an additive subgroup
$\mathbb{G} \subsetneq \mathbb{Z}^d$ such that $\sum_{z \in \mathbb{G}} p(x + z) = 1$.   (1.5)

Another way to state assumption (1.5) is that the averaged walk has span 1, or that it is aperiodic in Spitzer's [22] terminology.

To create a system of particles, let $\{X^{u,j} : u \in \mathbb{Z}^d, j \in \mathbb{N}\}$ denote a collection of random walks on $\mathbb{Z}^d$ such that walk $X^{u,j}_{\cdot}$ starts at site u: $X^{u,j}_0 = u$. When the environment ω is fixed, we use P^ω to denote the joint quenched measure of the walks $\{X^{u,j}_{\cdot}\}$. Under P^ω these walks move independently on $\mathbb{Z}^d$ and each walk obeys transitions (1.2).

Further, assume given an initial configuration $\eta = (\eta(u))_{u \in \mathbb{Z}^d}$ of occupation variables. Variable $\eta(u) \in \mathbb{Z}_+$ specifies the number of particles initially at site u. P^ω_η denotes the quenched distribution of the walks $\{X^{u,j}_{\cdot} : u \in \mathbb{Z}^d, 1 \leqslant j \leqslant \eta(u)\}$. Occupation variables for all times $s \in \mathbb{Z}_+$ are then defined by

$$\eta_s(x) = \sum_{u \in \mathbb{Z}^d} \sum_{j=1}^{\eta(u)} \mathbf{1}\{X^{u,j}_s = x\}, \qquad (x, s) \in \mathbb{Z}^d \times \mathbb{Z}_+.$$

When the initial configuration $\eta = \eta_0$ has probability distribution ν we write $P^\omega_\nu(\cdot) = \int P^\omega_\eta(\cdot) \nu(d\eta)$ for the quenched distribution of the process.

When the environment is averaged over we drop the superscript ω: for any event A that involves the walks and occupation variables, and any event $B \subseteq \Omega$, $P_\nu(A \times B) = \int_B P^\omega_\nu(A) \mathbb{P}(d\omega)$.

It will be convenient to construct initial distributions $\nu = \nu^\omega$ as functions of the environment, so that the quenched process distribution is then $P^\omega_{\nu^\omega}(\cdot) = \int P^\omega_\eta(\cdot) \nu^\omega(d\eta)$. But then it will always be the case that ν^ω depends only on the *past* $\bar{\omega}_{-\infty,-1}$ of the environment. Consequently the initial distribution ν^ω and the quenched distribution of the walks $P^\omega(\{X^{x,j}_{\cdot}\} \in \cdot)$ are independent under the product measure $\mathbb{P}$ on the environment. The averaged process distribution is then

$$\int_\Omega P^\omega_{\nu^\omega}(\cdot) \mathbb{P}(d\omega) = \int_{\mathbb{Z}^{\mathbb{Z}^d}_+} P_\eta(\cdot) \bar{\nu}(d\eta) = P_{\bar{\nu}}(\cdot)$$

where $\bar{\nu}(d\eta) = \int_\Omega \nu^\omega(d\eta) \mathbb{P}(d\omega)$ is the averaged initial distribution. In particular, in both quenched and averaged sense, the initial occupation variables $\{\eta_0(x)\}$ are independent of the walks $\{X^{x,j}_{\cdot}\}$.

The first result describes the invariant distributions of the occupation process $\eta_t = (\eta_t(x))_{x \in \mathbb{Z}^d}$. The starting point is an invariant distribution for the environment process seen by a tagged particle: this is the process $T_{X_n,n}\omega$ where $X_{\cdot}$ denotes a walk that starts at the origin. A familiar martingale argument and Green function bounds (Proposition 3.1 in Sect. 3 below) show the existence of an $\mathfrak{S}_{-\infty,-1}$-measurable density function

f on Ω such that $\mathbb{E}(f) = 1$, $\mathbb{E}(f^2) < \infty$, and the probability measure $\mathbb{P}_\infty(d\omega) = f(\omega)\,\mathbb{P}(d\omega)$ is invariant for the Markov chain $T_{X_n,n}\omega$.

For $0 \leqslant \lambda < \infty$ let Γ^λ denote the mean λ Poisson distribution on $\mathbb{Z}_+$. For $0 \leqslant \rho < \infty$ and $\omega \in \Omega$ define the following inhomogeneous Poisson product probability distribution on particle configurations $\eta = (\eta(x))_{x \in \mathbb{Z}^d}$:

$$\mu^{\rho,\omega}(d\eta) = \bigotimes_{x \in \mathbb{Z}^d} \Gamma^{\rho f(T_{x,0}\omega)}(d\eta(x)). \tag{1.6}$$

(Such a measure is called a Cox process with random intensity $\rho f(T_{x,0}\omega)$). Define the averaged measure by

$$\mu^\rho = \int \mu^{\rho,\omega}\,\mathbb{P}(d\omega). \tag{1.7}$$

Theorem 1.1 *Let the dimension $d \geqslant 1$. Consider independent particles on $\mathbb{Z}^d$ in an i.i.d. space-time environment (as indicated in (1.3)), with bounded jumps, under assumptions (1.4) and (1.5).*

(a) For each $0 \leqslant \rho < \infty$, μ^ρ is the unique invariant distribution for the process η. that is also invariant and ergodic under spatial translations and has mean occupation $\int \eta(x)\,d\mu^\rho = \rho$. Furthermore, the tail σ-field of the state space $\mathbb{Z}_+^{\mathbb{Z}^d}$ is trivial under μ^ρ, and under the path measure P_{μ^ρ} the process η. is ergodic under time shifts.

(b) Suppose $d = 1$ or $d = 2$. Let ν be a probability distribution on $\mathbb{Z}_+^{\mathbb{Z}^d}$ that is stationary and ergodic under spatial translations and has mean occupation $\rho = \int \eta(x)\,d\nu$. Then if ν is the initial distribution for the process η., the process converges in distribution to the invariant distribution with density ρ: $P_\nu\{\eta_t \in \cdot\} \Rightarrow \mu^\rho$ as $t \to \infty$.

Part (b) of the theorem is restricted to $d = 1, 2$ because our proof uses recurrence of random walks (see Proposition 2.2).

Two auxiliary Markov transitions q and $\bar{q}$ on $\mathbb{Z}^d$ play important roles throughout much of the paper:

$$\begin{aligned}
q(x, y) &= \sum_{z \in \mathbb{Z}^d} \mathbb{E}[\omega_{0,0}(z)\omega_{x,0}(z + y)] \\
&= \begin{cases} \sum_{z \in \mathbb{Z}^d} \mathbb{E}[\omega_{0,0}(z)\omega_{0,0}(z + y)] & x = 0,\, y \in \mathbb{Z}^d \\ \sum_{z \in \mathbb{Z}^d} p(z)p(z + y - x) & x \neq 0,\, y \in \mathbb{Z}^d \end{cases}
\end{aligned} \tag{1.8}$$

and

$$\bar{q}(x, y) = \bar{q}(0, y - x) = \sum_{z \in \mathbb{Z}^d} p(z)p(z + y - x). \tag{1.9}$$

Think of q as a symmetric random walk whose transition probability is perturbed at the origin, and of $\bar{q}$ as the corresponding unperturbed homogeneous walk.

For $\theta \in \mathbb{T}^d = (-\pi, \pi]^d$ define characteristic functions

$$\phi^\omega(\theta) = \sum_z \omega_{0,0}(z) e^{i\theta \cdot z}, \tag{1.10}$$

$$\lambda(\theta) = \sum_{z \in \mathbb{Z}^d} q(0, z) e^{i\theta \cdot z} = \mathbb{E}|\phi^\omega(\theta)|^2 \tag{1.11}$$

and

$$\bar{\lambda}(\theta) = \sum_{z \in \mathbb{Z}^d} \bar{q}(0, z) e^{i\theta \cdot z} = |\mathbb{E}\phi^\omega(\theta)|^2. \tag{1.12}$$

(We use the bar notation for quantities associated with the homogeneous walk $\bar{q}$, in addition to a few other particular items such as $\bar{\omega}_s$ for the environment on level s. In the case of $\bar{\lambda}(\theta)$ this must not be confused with complex conjugation). Assumption (1.5) implies that the random walk $\bar{q}$ is not supported on a subgroup smaller than $\mathbb{Z}^d$, hence $\bar{\lambda}(\theta) < 1$ for $\theta \in \mathbb{T}^d \setminus \{0\}$ [22, p. 67, T7.1]. Define a constant β by

$$\beta = \frac{1}{(2\pi)^d} \int_{\mathbb{T}^d} \frac{1 - \lambda(\theta)}{1 - \bar{\lambda}(\theta)} \, d\theta. \tag{1.13}$$

The distribution $q(0, z)$ is not degenerate by assumption (1.4) and hence $\lambda(\theta)$ is not identically 1. Since also $\bar{\lambda}(\theta) \leqslant \lambda(\theta)$, we see that $\beta \in (0, 1]$ is well-defined.

Under the invariant distribution μ^ρ the covariance of the occupation variables is

$$\mathrm{Cov}^{\mu^\rho}[\eta(0), \eta(m)] = \rho^2 \mathrm{Cov}[f(\omega), f(T_{m,0}\omega)] = \rho^2 \mathbb{E}[f(\omega) f(T_{m,0}\omega)] - \rho^2, \quad m \in \mathbb{Z}^d. \tag{1.14}$$

The first equality above comes from the structure of μ^ρ: given ω, the occupation variables are independent with means $E^{\mu^{\rho,\omega}}[\eta(m)] = \rho f(T_{m,0}\omega)$. Our next theorem gives a formula for (1.14).

Theorem 1.2 *Let $d \geqslant 1$. For $m \in \mathbb{Z}^d \setminus \{0\}$*

$$\mathbb{Cov}[f(\omega), f(T_{m,0}\omega)] = -\frac{\beta^{-1}}{(2\pi)^d} \int_{\mathbb{T}^d} \cos(\theta \cdot m) \frac{1 - \lambda(\theta)}{1 - \bar{\lambda}(\theta)} \, d\theta \tag{1.15}$$

and

$$\mathbb{Var}[f(\omega)] = \beta^{-1} - 1. \tag{1.16}$$

The compact analytic formulas (1.13) and (1.15) arise from probabilistic formulas that involve the transitions q and $\bar{q}$ and the potential kernel of $\bar{q}$. The probabilistic arguments are somewhat different in the recurrent ($d \leqslant 2$) and transient ($d \geqslant 3$) cases. The reader can find these in Sect. 4.

By the Riemann-Lebesgue lemma we have that

$$\lim_{m \to \infty} \mathbb{C}\text{ov}[f(\omega), f(T_{m,0}\omega)] = 0.$$

By computing the integral in (1.15) an interesting special case arises:

Corollary 1.3 *For the simplest case where $d = 1$ and $p(x) + p(x + 1) = 1$ for some $x \in \mathbb{Z}$, the fixed time occupation variables in the stationary process are uncorrelated:*

$$\mathbb{C}\text{ov}[f(\omega), f(T_{m,0}\omega)] = 0 \quad \text{for } m \neq 0.$$

1.2 Limit of the Current Process

To study the particle current we restrict to dimension $d = 1$. Define the mean and variance of the averaged walk by

$$v = \sum_{x \in \mathbb{Z}} x p(x) \quad \text{and} \quad \sigma^2 = \sum_{x \in \mathbb{Z}} x^2 p(x) - v^2. \tag{1.17}$$

For $t \in \mathbb{R}_+ = [0, \infty)$ and $r \in \mathbb{R}$, let

$$Y_n(t, r) = \sum_{x > 0} \sum_{j=1}^{\eta_0(x)} \mathbf{1}\{X^{x,j}_{\lfloor nt \rfloor} \leqslant \lfloor nvt \rfloor + \lfloor r\sqrt{n} \rfloor\} - \sum_{x \leqslant 0} \sum_{j=1}^{\eta_0(x)} \mathbf{1}\{X^{x,j}_{\lfloor nt \rfloor} > \lfloor nvt \rfloor + \lfloor r\sqrt{n} \rfloor\}. \tag{1.18}$$

$Y_n(t, r)$ represents the net right-to-left current of particles seen by a moving observer who starts at the origin and travels to $\lfloor nvt \rfloor + \lfloor r\sqrt{n} \rfloor$ in time $\lfloor nt \rfloor$.

We look at the current under the following assumptions. Given ω, initial occupation variables obey a product measure that may depend on the past of the environment, but so that shifts are respected. Precisely,

> given the environment ω, initial occupation variables $(\eta_0(x))_{x \in \mathbb{Z}}$ have distribution $\mu^\omega(d\eta_0) = \bigotimes_{x \in \mathbb{Z}} \mu_x^\omega(d\eta_0(x))$ where μ_x^ω is allowed to depend measurably on $\bar{\omega}_{-\infty, -1}$. Furthermore, $\mu_x^\omega = \mu_0^{T_{x,0}\omega}$. $\tag{1.19}$

Let P^ω denote the quenched distribution $P^\omega_{\mu^\omega}$ of initial occupation variables and walks, and $P = P^\omega(\cdot)\mathbb{P}(d\omega)$ the distribution over everything: particles, walks and environments.

Make this moment assumption:

$$E[\eta_0(0)^2] < \infty. \tag{1.20}$$

Parameters that appear in the results are

$$\rho_0 = E[\eta_0(x)] \quad \text{and} \quad \sigma_0^2 = \mathbb{E}[\text{Var}^\omega(\eta_0(0))]. \tag{1.21}$$

Next we describe the limiting process. Let $\dot{W}$ be space-time white noise corresponding and B a two-sided one-parameter Brownian motion on $\mathbb{R}$, independent of $\dot{W}$. Let W be the two-parameter Brownian motion on $\mathbb{R}_+ \times \mathbb{R}$ given by $W(t, r) = \dot{W}([0, t] \times [0, r])$, if $r > 0$, and $W(t, r) = -\dot{W}([0, t] \times [r, 0])$, if $r < 0$. Define the process $Z(t, r)$ as the unique mild solution of the stochastic heat equation (see [25])

$$Z_t = \frac{\sigma^2}{2} Z_{rr} + \sqrt{\rho_0}\, \dot{W}, \qquad Z(0, r) = \sigma_0 B(r).$$

Process Z is given by

$$\begin{aligned} Z(t, r) = \sqrt{\rho_0} \iint_{[0,t] \times \mathbb{R}} \varphi_{\sigma^2(t-s)}(r - x)\, dW(s, x) \\ + \sigma_0 \int_{\mathbb{R}} \varphi_{\sigma^2 t}(r - x) B(x)\, dx, \end{aligned} \tag{1.22}$$

where $\varphi_{\nu^2}(x) = (2\pi\nu^2)^{-1/2} \exp(-x^2/2\nu^2)$ denotes the centered Gaussian density with variance ν^2, and $\Phi_{\nu^2}(x) = \int_{-\infty}^x \varphi_{\nu^2}(y)dy$ the distribution function.

$\{Z(t, r) : t \in \mathbb{R}_+, r \in \mathbb{R}\}$ is a mean zero Gaussian process. Its covariance can be expressed as follows: with

$$\Psi_{\nu^2}(x) = \nu^2 \varphi_{\nu^2}(x) - x\left(1 - \Phi_{\nu^2}(x)\right) \tag{1.23}$$

define two covariance functions on $(\mathbb{R}_+ \times \mathbb{R}) \times (\mathbb{R}_+ \times \mathbb{R})$ by

$$\Gamma_1\big((s, q), (t, r)\big) = \Psi_{\sigma^2(t+s)}(r - q) - \Psi_{\sigma^2|t-s|}(r - q) \tag{1.24}$$

and

$$\Gamma_2\big((s, q), (t, r)\big) = \Psi_{\sigma^2 s}(-q) + \Psi_{\sigma^2 t}(r) - \Psi_{\sigma^2(t+s)}(r - q). \tag{1.25}$$

Then

$$\mathbf{E}[Z(s, q)Z(t, r)] = \rho_0 \Gamma_1\big((s, q), (t, r)\big) + \sigma_0^2 \Gamma_2\big((s, q), (t, r)\big). \tag{1.26}$$

(Boldface $\mathbf{P}$ and $\mathbf{E}$ denote generic probabilities and expectations not connected with the RWRE model).

The theorem we state is for the finite-dimensional distributions of the current process, scaled and centered by its quenched mean:

$$\overline{Y}_n(t, r) = n^{-1/4}\{Y_n(t, r) - E^\omega[Y_n(t, r)]\}.$$

Fix any $N \in \mathbb{N}$, time points $0 < t_1 < t_2 < \cdots < t_N \in \mathbb{R}_+$, space points $r_1, r_2, \ldots, r_N$ $\in \mathbb{R}$ and an N-vector $\boldsymbol{\theta} = (\theta_1, \ldots, \theta_N) \in \mathbb{R}^N$. Form the linear combinations

$$\overline{Y}_n(\boldsymbol{\theta}) = \sum_{i=1}^{N} \theta_i \overline{Y}_n(t_i, r_i) \quad \text{and} \quad Z(\boldsymbol{\theta}) = \sum_{i=1}^{N} \theta_i Z(t_i, r_i).$$

Theorem 1.4 *Consider independent particles on $\mathbb{Z}$ in an i.i.d. space-time environment with bounded jumps, under assumptions (1.3) and (1.5). Let the ω-dependent initial distribution satisfy (1.19) and (1.20). With definitions as above, quenched characteristic functions converge in $L^1(\mathbb{P})$:*

$$\lim_{n \to \infty} \mathbb{E}\left|E^\omega(e^{i\overline{Y}_n(\boldsymbol{\theta})}) - \mathbf{E}(e^{iZ(\boldsymbol{\theta})})\right| = 0. \tag{1.27}$$

In particular, under the averaged distribution P, convergence in distribution holds for the $\mathbb{R}^N$-valued vectors as $n \to \infty$:

$$\left(\overline{Y}_n(t_1, r_1), \overline{Y}_n(t_2, r_2), \cdots, \overline{Y}_n(t_N, r_N)\right) \Rightarrow \left(Z(t_1, r_1), Z(t_2, r_2), \cdots, Z(t_N, r_N)\right).$$

While we do not have a quenched limit (convergence of distributions under a fixed ω), limit (1.27) does imply that, if a quenched limit exists, it is the same as we have found.

A special case of the above theorem is the stationary situation. The proof of the following corollary comes by a direct computation using (1.26).

Corollary 1.5 *Consider the same setting as in the previous theorem. If furthermore variables η_0 have conditional distribution (1.6), and more generally when $\sigma_0^2 = \rho_0$, process $Z(t, 0)$ has covariance*

$$\mathbf{E}[Z(s, 0)Z(t, 0)] = \frac{\rho_0 \sigma}{\sqrt{2\pi}}(\sqrt{s} + \sqrt{t} - \sqrt{|t - s|}),$$

i.e. $\rho_0^{-1}\sigma^{-1}\sqrt{\pi/2}\, Z(t, 0)$ is a fractional Brownian motion with Hurst parameter $1/4$.

The next two theorems are on fluctuations of the quenched mean process $E^\omega Y_n(t, r)$ in the special case of one-dimensional random walks with admissible steps 0 and 1. Although we expect the result to hold for more general random walks, this is the only case for which we are able to characterize the fluctuations. Let $\sigma_D^2 = \mathbb{V}\mathrm{ar}(\omega_{0,0})$ and $\alpha = \mathbb{E}\omega_{0,0}(1 - \omega_{0,0})$. Note that $v = p(1) = \mathbb{E}\omega_{0,0}$ and $\sigma^2 = v(1 - v)$.

First, we consider the case of an initial configuration η_0 with independent quenched means.

Theorem 1.6 *Let $\{\eta_0(x) : x \in \mathbb{Z}\}$ be such that the quenched means $\{E^\omega \eta_0(x) : x \in \mathbb{Z}\}$ are independent with mean ρ_0 and variance σ_0^2. Assume that there exists $\varepsilon > 0$ such that $\sup_x \mathbb{E}[|E^\omega \eta_0(x)|^{2+\varepsilon}] < \infty$. Assume the η_0-variables are independent of the transition probabilities $\{\omega_{x,t} : (x, t) \in \mathbb{Z} \times \mathbb{Z}_+\}$. Then the finite-dimensional marginals of the process $\{n^{-1/4} E^\omega (Y_n(t, r) - \rho_0 r \sqrt{n}) : t \geq 0, r \in \mathbb{R}\}$ converge weakly as $n \to \infty$ to those of the unique mild solution to the stochastic heat equation*

$$z_t = \frac{\sigma^2}{2} z_{rr} + \frac{\rho_0 \sigma_D}{\sqrt{\alpha}} \dot{W}, \quad z(0, r) = \sigma_0 B(r),$$

where $\dot{W}$ is space-time white noise and B a two-sided Brownian, independent of $\dot{W}$.

The above includes the case when η_0 is independent of ω altogether. In that case, $\sigma_0 = 0$ and thus the initial condition becomes $z(0, r) \equiv 0$.

Next, we look at the stationary case. The reason this is different from the previous theorem is that now the quenched means of the initial occupation variables are not independent.

Theorem 1.7 *Let $\{\eta_0(x) : x \in \mathbb{Z}^d\}$ be distributed according to (1.6) with $\rho = 1$. Assume the averaged probabilities $p_0 = p_1 = 1/2$ so that $v = 1/2$. Then for $t \geq s > 0$ we have*

$$\lim_{n \to \infty} \frac{1}{\sqrt{n}} \mathrm{Cov}\big(E^\omega Y_n(s, 0), E^\omega Y_n(t, 0)\big) = \frac{1}{2\sqrt{2\pi}} (\tfrac{1}{4}\alpha^{-1} - 1)(\sqrt{t} + \sqrt{s} - \sqrt{t - s}).$$

The above limit matches the covariance structure of a constant $((\tfrac{1}{4}\alpha^{-1} - 1)^{1/2}/(2\pi)^{1/4})$ times a fractional Brownian motion with Hurst parameter $1/4$. Theorems 1.6 and 1.7 are proved in Sect. 6. At the end of that section, we explain why we expect the same limiting behavior in the setting of Theorem 1.7 as that of Theorem 1.6 and how this would imply the fractional Brownian motion limit.

Further notational conventions $\mathbb{N} = \{1, 2, 3, \dots\}$ and $\mathbb{Z}_+ = \{0, 1, 2, \dots\}$. Multi-step transition probabilities from time s to time $t > s + 1$ are

$$\pi_{s,t}^\omega(x, y) = \sum_{u_1, \dots, u_{t-s-1} \in \mathbb{Z}^d} \pi_{s,s+1}^\omega(x, u_1) \pi_{s+1,s+2}^\omega(u_1, u_2) \cdots \pi_{t-1,t}^\omega(u_{t-s-1}, y).$$

We omit floor notation from time parameters, and so for the walk, $X_t = X_{\lfloor t \rfloor}$ for real $t \geq 0$. No jumps happen between integer times.

$\mathcal{S}$ denotes the set of all measures μ on $(\mathbb{Z}_+)^{\mathbb{Z}^d}$ that are invariant under spatial translations. $\mathcal{S}_e$ denotes the subset of $\mathcal{S}$ consisting of ergodic measures. $\mathcal{I}$ denotes the set of measures that are invariant for the particle evolution, that is $\mu_t = \mu S(t) = \mu$ for all $t \in \mathbb{Z}_+$ (μ_t and $\mu S(t)$ here denote the measure on configurations at time t when the initial measure on configurations is μ). $\mathbb{E}$, E^ω, E, E_η, $\mathbf{E}$ etc will denote expectations with respect to $\mathbb{P}$, P^ω, P, P_η, $\mathbf{P}$, etc. Variances and covariances are denoted similarly. Constants C can change from term to term.

2 Coupled Process

This section describes the coupling that will be used to prove Theorem 1.1. We couple two processes η_t and ζ_t so that matched particles move together forever, while unmatched particles move independently. To do this precisely, choose for each space-time point (x, t) a collection $\Xi_{x,t} = \{v_{x,t}^{0,j}, v_{x,t}^{+,j}, v_{x,t}^{-,j} : j \in \mathbb{N}\}$ of i.i.d. $\mathbb{Z}^d$-valued jump vectors from distribution $\omega_{x,t}$. Given initial configurations η_0 and ζ_0, perform the following actions. At each site x set

$$\xi_0(x) = \eta_0(x) \wedge \zeta_0(x), \beta_0^+(x) = (\eta_0(x) - \zeta_0(x))^+, \text{ and } \beta_0^-(x) = (\eta_0(x) - \zeta_0(x))^-.$$

$\xi_0(x)$ is the number of matched particles, while $\beta_0^{\pm}(x)$ count the unmatched $(+)$ and $(-)$ particles. Move particles from each site x as follows: the $\xi_0(x)$ matched particles jump to locations $x + v_{x,0}^{0,j}$ for $j = 1, \ldots, \xi_0(x)$, the $\beta_0^+(x)$ $(+)$ particles jump to locations $x + v_{x,0}^{+,j}$ for $j = 1, \ldots, \beta_0^+(x)$, and the $\beta_0^-(x)$ $(-)$ particles jump to locations $x + v_{x,0}^{-,j}$ for $j = 1, \ldots, \beta_0^-(x)$. After all jumps from all sites have been executed, match as many pairs of $(+)$ and $(-)$ particles at the same site as possible. This means that a $(+-)$ pair together at the same site merges to create a single ξ-particle at the same site. (For example, if after the jumps site y contains s ξ-particles, k $(+)$ particles and ℓ $(-)$ particles, then set $\xi_1(y) = s + k \wedge \ell$ and $\beta_1^{\pm}(y) = (k - \ell)^{\pm}$). Since particles are not labeled, it is immaterial which particular $(+)$ particle merges with a particular $(-)$ particle. When this is complete we have defined the state $(\xi_1(x), \beta_1^+(x), \beta_1^-(x))_{x \in \mathbb{Z}^d}$ at time $t = 1$. Then repeat, utilizing the jump variables for time $t = 1$. And so on.

This produces a joint process $(\xi_t, \beta_t^+, \beta_t^-)$ such that

$$\xi_t(x) = \eta_t(x) \wedge \zeta_t(x), \beta_t^+(x) = (\eta_t(x) - \zeta_t(x))^+, \text{ and } \beta_t^-(x) = (\eta_t(x) - \zeta_t(x))^-.$$

The η and ζ processes are recovered from

$$\eta_t(x) = \xi_t(x) + \beta_t^+(x) \quad \text{and} \quad \zeta_t(x) = \xi_t(x) + \beta_t^-(x).$$

The definition has the effect that a matched pair of η and ζ particles stays forever together, while a pair of $(+)$ and $(-)$ particles together at a site annihilate each other and turn into a matched pair. If we are only interested in the evolution of the discrepancies (β_t^+, β_t^-) we can discard all matched pairs as soon as they arise, and simply consider independently evolving $(+)$ and $(-)$ particles that annihilate each other upon meeting.

If we denote by $\Xi = \{\Xi_{x,t} : x \in \mathbb{Z}^d, t \in \mathbb{Z}\}$ the collection of jump variables, and by $G_{0,t}$ the function that constructs the values at the origin at time t:

$$\left(\xi_t(0), \beta_t^+(0), \beta_t^-(0)\right) = G_{0,t}(\eta_0, \zeta_0, \Xi)$$

then it is clear that the values at other sites x are constructed by applying this same function to shifted input:

$$\left(\xi_t(x), \beta_t^+(x), \beta_t^-(x)\right) = G_{0,t}(\theta_x \eta_0, \theta_x \zeta_0, \theta_x \Xi). \tag{2.1}$$

Here θ_x is a spatial shift: $(\theta_x \eta)(y) = \eta(x + y)$ and $(\theta_x \Xi)_{y,t} = \Xi_{x+y,t}$ for $x, y \in \mathbb{Z}^d$. In particular, if the initial distribution $\widetilde{\mu}$ of the pair (η_0, ζ_0) is invariant and ergodic under the shifts θ_x, while $\{\Xi_{x,t} : x \in \mathbb{Z}^d, t \in \mathbb{Z}\}$ are i.i.d. and independent of (η_0, ζ_0), it follows first that the triple (η_0, ζ_0, Ξ) is ergodic, and then from (2.1) that for each fixed t the configuration $(\xi_t, \beta_t^+, \beta_t^-)$ is invariant and ergodic under the shifts θ_x.

Let $\widetilde{\mathcal{S}}$, resp. $\widetilde{\mathcal{S}}_e$, denote the set of spatially invariant, resp. ergodic, probability distributions on pairs (η, ζ) of configurations of occupation variables.

Lemma 2.1 *Let $\widetilde{\mu} \in \widetilde{\mathcal{S}}$. The expectations $E_{\widetilde{\mu}}[\beta_t^+(x)]$ and $E_{\widetilde{\mu}}[\beta_t^-(x)]$ are independent of x and nonincreasing in t.*

Proof The independence of x is due to the shift-invariance from (2.1). That $E_{\widetilde{\mu}}[\beta_t^\pm(x)]$ is nonincreasing in t follows from the fact that discrepancy particles are not created, only annihilated. $\qquad\square$

Proposition 2.2 *Let $d = 1$ or 2. Suppose $\widetilde{\mu} \in \widetilde{\mathcal{S}}_e$. Let $E_{\widetilde{\mu}}[\eta(0)] = \rho_1$ and $E_{\widetilde{\mu}}[\zeta(0)] = \rho_2$. If $\rho_1 \geqslant \rho_2$, we have*

$$E_{\widetilde{\mu}}[\beta_t^-(0)] = E_{\widetilde{\mu}}[(\eta_t(0) - \zeta_t(0))^-] \to 0 \text{ as } t \to \infty.$$

Proof We already know from Lemma 2.1 that $E_{\widetilde{\mu}}[\beta_t^\pm(0)]$ cannot increase. To get a contradiction let us assume that $E_{\widetilde{\mu}}[\beta_t^-(0)] \geqslant \delta$ for all t and some $\delta > 0$. Since $E_{\widetilde{\mu}}[\beta_t^+(0)] - E_{\widetilde{\mu}}[\beta_t^-(0)] = E_{\widetilde{\mu}}[\eta_t(0)] - E_{\widetilde{\mu}}[\zeta_t(0)] = \rho_1 - \rho_2 \geqslant 0$, we also have $E_{\widetilde{\mu}}[\beta_t^+(0)] \geqslant \delta$ for all t.

At time 0, assign labels separately to the $(+)$ and $(-)$ particles from some countable label sets $\mathcal{J}^+$ and $\mathcal{J}^-$ and denote the locations of these particles by $\{w_i^+(t), w_j^-(t) : i \in \mathcal{J}^+, j \in \mathcal{J}^-\}$. Each $(+)$ and $(-)$ particle retains its label throughout its lifetime. The lifetime of $(+)$ particle $j \in \mathcal{J}^+$ is

$$\tau_j^+ = \inf\{t \geqslant 0 : w_j^+ \text{ is annihilated by a } (-) \text{ particle}\}.$$

If $\tau_j^+ = \infty$, then j is *immortal*. Similarly define τ_j^-. Let

$$\beta_{0,t}^\pm(x) = \sum_j \mathbf{1}\{w_j^\pm(0) = x, \tau_j^\pm > t\}$$

denote the number of $(\pm)$ particles initially at site x that live past time t. We would like to claim that for a fixed t the configuration $\{(\beta_{0,t}^+(x), \beta_{0,t}^-(x)) : x \in \mathbb{Z}^d\}$ is invariant and ergodic under the spatial shifts θ_x. This will be true if the evolution is given by a mapping $F_{0,t}$ so that $(\beta_{0,t}^+(x), \beta_{0,t}^-(x)) = F_{0,t}(\theta_x \eta_0, \theta_x \zeta_0, \theta_x \Xi)$ for all $x \in \mathbb{Z}^d$. Such a mapping can be created by specifying precise rules for the movement and annihilation of (+) and (−) particles that are naturally invariant under shifts. For example, we can take $\mathcal{J}^{\pm} \subset \mathbb{Z}$ and give the sites of $\mathbb{Z}^d$ some ordering. Label particles initially in increasing order, so that $i < j$ implies $w_i^{\pm}(0) \leqslant w_j^{\pm}(0)$. Then at each time step particles from a given site are distributed to their subsequent locations in increasing order, and $(+, -)$ pairs are matched beginning with lowest labels. Of course the overall ordering of particles is not preserved, but this mechanism does not depend on the absolute labels, only their ordering, and respects the spatial translations.

Then the ergodic theorem implies that

$$E_{\widetilde{\mu}}[\beta_{0,t}^{\pm}(0)] = \lim_{n \to \infty} \frac{1}{(2n+1)^d} \sum_{|x| \leqslant n} \beta_{0,t}^{\pm}(x) \text{ a.s.}$$

Here $|x|$ is the ℓ^{∞} norm: for a vector $x = (x_1, \ldots, x_d)$, $|x| = \max_{1 \leqslant i \leqslant d} |x_i|$. Since particles take jumps of magnitude at most R,

$$\delta \leqslant E_{\widetilde{\mu}}[\beta_t^{\pm}(0)] = \lim_{n \to \infty} \frac{1}{(2n+1)^d} \sum_{|x| \leqslant n} \beta_t^{\pm}(x)$$

$$\leqslant \lim_{n \to \infty} \frac{1}{(2n+1)^d} \sum_{|x| \leqslant n+Rt} \beta_{0,t}^{\pm}(x) = E_{\widetilde{\mu}}[\beta_{0,t}^{\pm}(0)].$$

The initial occupation numbers of immortal $+/-$ particles are

$$\beta_{0,\infty}^{\pm}(x) = \lim_{t \to \infty} \beta_{0,t}^{\pm}(x).$$

The limit exists by monotonicity. This limit produces again a functional relationship of the type (2.1):

$$\left(\beta_{0,\infty}^{+}(x), \beta_{0,\infty}^{-}(x)\right) = \lim_{t \to \infty} \left(\beta_{0,t}^{+}(x), \beta_{0,t}^{-}(x)\right) = \lim_{t \to \infty} F_{0,t}(\theta_x \eta_0, \theta_x \zeta_0, \theta_x \Xi)$$

$$= F_{0,\infty}(\theta_x \eta_0, \theta_x \zeta_0, \theta_x \Xi).$$

Thereby $\{(\beta_{0,\infty}^{+}(x), \beta_{0,\infty}^{-}(x)) : x \in \mathbb{Z}^d\}$ is spatially invariant and ergodic.

By the ergodic theorem again

$$E_{\widetilde{\mu}}[\beta_{0,\infty}^{\pm}(0)] = \lim_{n \to \infty} \frac{1}{(2n+1)^d} \sum_{|x| \leqslant n} \beta_{0,\infty}^{\pm}(x) \quad \text{a.s.}$$

while by the monotone convergence theorem

$$E_{\widetilde{\mu}}[\beta_{0,\infty}^{\pm}(0)] = \lim_{t \to \infty} E_{\widetilde{\mu}}[\beta_{0,t}^{\pm}(0)] \geqslant \delta.$$

We have shown that the assumption $E_{\tilde{\mu}}[\beta_t^-(0)] \geq \delta$ leads to the existence of positive densities of immortal $(+)$ and $(-)$ particles. However, a situation like this will never arise for $d = 1$ or 2, the reason being that any two particles on the lattice will meet each other infinitely often. More precisely, fix any two particles and let $X_{\cdot}^+$ and $X_{\cdot}^-$ denote the walks undertaken by these two particles. Then $X_{\cdot}^+$ and $X_{\cdot}^-$ are two independent walks in a common environment ω. Let $Y_t = X_t^+ - X_t^-$. If we average out the environment, then Y_t is a Markov chain on $\mathbb{Z}^d$ with transition $q(x, y)$ given by (1.8). Away from the origin this is a symmetric random walk with bounded steps, and hence recurrent when $d = 1$ or 2. Thus $Y_t = 0$ infinitely often. We have arrived at a contradiction and the proposition is proved. $\qquad\square$

3 Invariant Measures

In this section we prove Theorem 1.1. We begin by deriving the well-known invariant density for the environment process seen by a single tagged particle.

Proposition 3.1 *There exists a function $0 \leq f < \infty$ on Ω such that $\mathbb{E}f = 1$, $\mathbb{E}(f^2) < \infty$, $f(\omega)$ is a function of $\bar{\omega}_{-\infty,-1}$, and*

$$f(\omega) = \sum_{x \in \mathbb{Z}^d} f(T_{x,-1}\omega)\pi_{-1,0}^{\omega}(x, 0) \quad \mathbb{P}\text{-almost surely.} \tag{3.1}$$

Proof For $N \in \mathbb{Z}_+$ define

$$f_N(\omega) = \sum_{z \in \mathbb{Z}^d} \pi_{-N,0}^{\omega}(z, 0). \tag{3.2}$$

$f_N(\omega)$ is $\mathfrak{S}_{-N,-1}$-measurable and a martingale with $\mathbb{E}f_N = 1$. By the martingale convergence theorem we can define

$$f(\omega) = \lim_{N \to \infty} f_N(\omega) \quad (\mathbb{P}\text{-almost sure limit}).$$

Property (3.1) follows because all the sums involved are finite:

$$\sum_x f(T_{x,-1}\omega)\pi_{-1,0}^{\omega}(x, 0) = \lim_{N \to \infty} \sum_x f_N(T_{x,-1}\omega)\pi_{-1,0}^{\omega}(x, 0)$$

$$= \lim_{N \to \infty} \sum_{z,x} \pi_{-N-1,-1}^{\omega}(z, x)\pi_{-1,0}^{\omega}(x, 0) = \lim_{N \to \infty} \sum_z \pi_{-N-1,0}^{\omega}(z, 0)$$

$$= \lim_{N \to \infty} f_{N+1}(\omega) = f(\omega).$$

In Lemma 3.4 below we show the L^2 boundedness of the sequence $\{f_N\}$. This implies that $f_N \to f$ also in L^2 and thereby implies the remaining statements $\mathbb{E}f = 1$ and $\mathbb{E}(f^2) < \infty$. $\qquad\square$

The following addresses the positivity of f.

Lemma 3.2 $\mathbb{P}(f > 0) = 1$ *if and only if there exists an x such that* $\mathbb{P}\{\pi_{0,1}(0, x) > 0\} = 1$.

Proof If there does not exist an x as in the claim, then by independence of the environment and the finite step-size assumption we see that

$$\mathbb{P}\{\forall x : \pi_{-1,0}(x, 0) = 0\} > 0.$$

But then (3.1) implies that $\mathbb{P}(f = 0) > 0$. Conversely, if there exists an x as in the claim, then (3.1) implies that if $f(\omega) = 0$ then $f(T_{x,-1}\omega) = 0$. Shift-invariance implies the two events are in fact equal, almost surely. This in turn implies that $\{f = 0\}$ is a trivial event and since $\mathbb{E}[f] = 1$ we have that $f > 0$ a.s. $\qquad\square$

To prove the L^2 estimate for f_N we develop a Green function bound for the Markov chain defined as the difference of two walks. Let X_t^x and $\widetilde{X}_t^y$ be two independent walks in a common environment ω, started at $x, y \in \mathbb{Z}^d$, and $Y_t = X_t^x - \widetilde{X}_t^y$. Under the averaged measure Y_t is a Markov chain on $\mathbb{Z}^d$ with transition probabilities $q(x, y)$ defined by (1.8). Y_t can be thought of as a symmetric random walk on $\mathbb{Z}^d$ whose transition has been perturbed at the origin. The corresponding homogeneous, unperturbed random walk is $\bar{Y}_t$ with transition probability $\bar{q}$ in (1.9). Write P_x and $\bar{P}_x$ for the path probabilities of $Y_.$ and $\bar{Y}_.$. Define hitting times of 0 for both walks Y_t and $\bar{Y}_t$ by

$$\tau = \inf\{n \geqslant 1 : Y_n = 0\} \quad \text{and} \quad \bar{\tau} = \inf\{n \geqslant 1 : \bar{Y}_n = 0\}. \tag{3.3}$$

Denote the k-step transition probabilities by $q^k(x, y)$ and $\bar{q}^k(x, y)$.

Lemma 3.3 *There exists a constant $C < \infty$ such that for all $x \in \mathbb{Z}^d$ and $N \in \mathbb{N}$,*

$$\sum_{k=0}^{N} q^k(x, 0) \leqslant C \sum_{k=0}^{N} \bar{q}^k(x, 0).$$

Proof Suppose we had the bound for $x = 0$. Then it follows for $x \neq 0$:

$$\sum_{k=0}^{N} q^k(x,0) = E_x\Big[\sum_{k=0}^{N} \mathbf{1}_{\{Y_k=0\}}\Big] = E_x\Big[\sum_{i=0}^{N} \mathbf{1}_{\{\tau=i\}} \sum_{k=i}^{N} \mathbf{1}_{\{Y_k=0\}}\Big]$$

$$= \sum_{i=0}^{N} P_x(\tau=i) \sum_{k=0}^{n-i} q^k(0,0) \leqslant C \sum_{i=0}^{N} \bar{P}_x(\bar\tau=i) \sum_{k=0}^{n-i} \bar{q}^k(0,0)$$

$$= C \sum_{k=0}^{N} \bar{q}^k(x,0)$$

It remains to prove the result for $x=0$. Let $\sigma_0=0$ and

$$\sigma_{j+1} = \inf\{n > \sigma_j : Y_n = 0 \text{ and } Y_k \neq 0 \text{ for some } k \in \{\sigma_j+1,\dots,n-1\}\}.$$

These are the successive times of arrivals to 0 following excursions away from 0. Let W_j, $j \geq 0$, be the durations of the sojourns at 0, in other words

$$Y_n = 0 \text{ iff } \sigma_j \leq n < \sigma_j + W_j \text{ for some } j \geq 0.$$

Sojourns are geometric and independent of the past, so on the event $\{\sigma_j < \infty\}$,

$$E_0(W_j \mid \mathcal{F}^Y_{\sigma_j}) = \frac{1}{1-q(0,0)}.$$

Let $J_N = \max\{j \geq 0 : \sigma_j \leq N\}$ mark the last sojourn at 0 that started by time N. Then

$$E_0\Big[\sum_{k=0}^{N} \mathbf{1}\{Y_k=0\}\Big] \leq E_0\Big[\sum_{j=0}^{J_N} W_j\Big] = \sum_{j=0}^{\infty} E_0\big[\mathbf{1}\{\sigma_j \leq N\}W_j\big]$$

$$= \frac{1}{1-q(0,0)} E_0(1+J_N).$$

Assumption (1.4) guarantees that $q(0,0) < 1$.

It remains to bound $E_0(1+J_N)$ in terms of $\sum_{k=0}^{N} \bar{q}^k(0,0)$. The key is that once the Markov chain Y_k has left the origin, it follows the same transitions as the homogeneous walk $\bar{Y}_k$ until the next visit to 0. For $z \neq 0$ let

$$K^z_N = \inf\{k \geq 1 : T^z_1 + T^z_2 + \cdots + T^z_k \geq N\}$$

where the $\{T^z_i\}$ are i.i.d. with common distribution $\bar{P}_z\{\bar\tau \in \cdot\}$. Imagine constructing the path Y_k so that every step away from 0 is followed by an excursion of $\bar{Y}_k$ that ends at 0 (or continues forever if 0 is never reached). The step bound (1.1) implies that $P_0\{|Y_1| \leqslant 2R\} = 1$. Then there is stochastic dominance that gives

$$E_0(J_N) \leqslant \sum_{z \neq 0 : |z| \leqslant 2R} \bar{E}(K_N^z). \tag{3.4}$$

By T32.1 in Spitzer [22, p. 378], for $z \neq 0$

$$\lim_{n \to \infty} \frac{\bar{P}_z(\bar{\tau} > n)}{\bar{P}_0(\bar{\tau} > n)} = \bar{a}(z) \tag{3.5}$$

where the potential kernel $\bar{a}$ is

$$\bar{a}(z) = \lim_{n \to \infty} \left\{ \sum_{k=0}^{n} \bar{q}^k(0,0) - \sum_{k=0}^{n} \bar{q}^k(z,0) \right\}. \tag{3.6}$$

By P30.2 in [22, p. 361] $\bar{a}(z) > 0$ for all $z \neq 0$ for $d = 1, 2$. For $d \geqslant 3$ by transience

$$\bar{a}(z) = \sum_{k=0}^{\infty} \bar{q}^k(0,0) - \sum_{k=0}^{\infty} \bar{q}^k(z,0) = (1 - \bar{F}(z,0)) \sum_{k=0}^{\infty} \bar{q}^k(0,0) > 0$$

where $\bar{F}(z,0) = \bar{P}_z\{\bar{Y}_n = 0 \text{ for some } n \geqslant 1\} < 1$.

From (3.5) and $\bar{a}(z) > 0$, there exist $0 < c(z), C(z) < \infty$ such that for all n,

$$c(z)\bar{P}_0(\bar{\tau} > n) \leqslant \bar{P}(T_1^z > n) \leqslant C(z)\bar{P}_0(\bar{\tau} > n)$$

and hence

$$c(z)\bar{E}_0[\bar{\tau} \wedge N] \leqslant \bar{E}[T_1^z \wedge N] \leqslant C(z)\bar{E}_0[\bar{\tau} \wedge N].$$

By Wald's identity and some simple bounds (see Exercice 4.4.1 in [6, Sect. 4.4])

$$\frac{N}{\bar{E}(T_1^z \wedge N)} \leqslant \bar{E}[K_N^z] \leqslant \frac{2N}{\bar{E}(T_1^z \wedge N)}.$$

Let $\{\bar{\tau}_i\}$ be i.i.d. copies of $\bar{\tau}$ from (3.3) and put

$$M_N = \inf\{k \geqslant 1 : \bar{\tau}_1 + \bar{\tau}_2 + \cdots + \bar{\tau}_k \geqslant N\}.$$

Then we have a similar relation:

$$\frac{N}{\bar{E}_0(\bar{\tau} \wedge N)} \leqslant \bar{E}_0[M_N] \leqslant \frac{2N}{\bar{E}_0(\bar{\tau} \wedge N)}$$

Combining the above lines:

$$\bar{E}[K_N^z] \leqslant \frac{2N}{\bar{E}[T_1^z \wedge N]} \leqslant \frac{2N}{c(z)\bar{E}_0[\bar{\tau} \wedge N]} \leqslant \frac{2}{c(z)}\bar{E}_0[M_N]. \tag{3.7}$$

Considering excursions of the $\bar{Y}$-walk away from 0,

$$\bar{E}_0[M_N] \leqslant 1 + \sum_{k=0}^{N} \bar{q}^k(0,0) \leqslant 2 \sum_{k=0}^{N} \bar{q}^k(0,0). \tag{3.8}$$

The proof is now complete with a combination of (3.4), (3.7) and (3.8). $\qquad\square$

From the previous lemma follows the L^2 estimate for f_N which completes the proof of Proposition 3.1.

Lemma 3.4 *There exists a constant $C < \infty$ such that $\mathbb{E}(f_N^2) \leqslant C$ for all N.*

Proof By translations

$$\mathbb{E}(f_N^2) = \sum_{x,z} \mathbb{E}\pi_{-N,0}^\omega(x,0)\pi_{-N,0}^\omega(z,0)$$

$$= \sum_{y} \mathbb{E}P^\omega\{X_N^y = \tilde{X}_N^0\} = \sum_{y} q^N(y,0).$$

By the submartingale property $\mathbb{E}(f_N^2)$ is nondecreasing in N. Hence it suffices to show the existence of a constant C such that

$$\sum_{k=0}^{N} \mathbb{E}(f_k^2) \leqslant C(N+1) \qquad \text{for all } N. \tag{3.9}$$

From above, by Lemma 3.3 and the spatial homogeneity of the $\bar{Y}$-walk,

$$\sum_{k=0}^{N} \mathbb{E}(f_k^2) = \sum_{k=0}^{N} \sum_{x} q^k(x,0) \leqslant C \sum_{k=0}^{N} \sum_{x} \bar{q}^k(x,0)$$

$$= C \sum_{k=0}^{N} \sum_{x} \bar{q}^k(0,x) = C(N+1). \qquad\square$$

Property (3.1) implies that the probability measure $\mathbb{P}_\infty(d\omega) = f(\omega)\mathbb{P}(d\omega)$ is invariant for the process $T_{X_n,n}\omega$. Recall from (1.6) the product measure

$$\mu^{\rho,\omega}(d\eta) = \bigotimes_{x \in \mathbb{Z}^d} \Gamma^{\rho f(T_{x,0}\omega)}(d\eta(x)) \tag{3.10}$$

where Γ^λ is Poisson(λ) distribution. By the definition of f, $\mu^{\rho,\omega}$ depends on ω only through the levels $\bar{\omega}_{-\infty,-1}$.

Lemma 3.5 *The following holds for $\mathbb{P}$-a.e. ω. Let η_0 be $\mu^{\rho,\omega}$-distributed. Then for all times $t \in \mathbb{Z}_+$, under the evolution in the environment ω, η_t is $\mu^{\rho,T_{0,t}\omega}$-distributed, and in particular independent of the environment $\bar{\omega}_t$ at level t.*

Proof Consider the evolution under a fixed ω. The claim made in the lemma is true at time $t = 0$ by the construction. Suppose it is true up to time $t - 1$. Then over $x \in \mathbb{Z}^d$ the variables $\eta_{t-1}(x)$ are independent Poisson variables with means $\rho f(T_{x,t-1}\omega)$. Each particle at site x chooses its next position y independently with probabilities $\pi_{t-1,t}^\omega(x, y)$. As with marking a Poisson process with independent coin flips, the consequence is that the numbers of particles going from x to y are independent Poisson variables with means $\rho f(T_{x,t-1}\omega)\pi_{t-1,t}^\omega(x, y)$, over all pairs (x, y). Since sums of independent Poissons are Poisson, the variables $(\eta_t(y))_{y\in\mathbb{Z}}$ are again independent Poissons and $\eta_t(y)$ has mean

$$\sum_x \rho f(T_{x,t-1}\omega)\pi_{t-1,t}^\omega(x, y) = \sum_z \rho f(T_{z,-1}T_{y,t}\omega)\pi_{-1,0}^{T_{y,t}\omega}(z, 0)$$

$$= \rho f(T_{y,t}\omega).$$

The last equality is from (3.1).

We have shown that $\eta_t = (\eta_t(y))_{y\in\mathbb{Z}^d}$ has distribution $\mu^{\rho,\,T_{0,t}\omega}$. This measure is a function of $\bar{\omega}_{-\infty,t-1}$, hence independent of $\bar{\omega}_t$ under $\mathbb{P}$. $\qquad\square$

Recall from (1.7) the averaged measure $\mu^\rho = \int \mu^{\rho,\omega}\,\mathbb{P}(d\omega)$.

Lemma 3.6 *The measure μ^ρ is invariant and ergodic under spatial shifts θ_x. The tail σ-field of the state space $\mathbb{Z}_+^{\mathbb{Z}^d}$ is trivial under μ^ρ.*

Proof Invariance under θ_x comes from $\mu^{\rho,\omega} \circ \theta_x^{-1} = \mu^{\rho,T_{x,0}\omega}$ and the invariance of $\mathbb{P}$. Ergodicity will follow from tail triviality.

Let $B \subseteq \mathbb{Z}_+^{\mathbb{Z}^d}$ be a tail event. Then by Kolmogorov's 0-1 law $\mu^{\rho,\omega}(B) \in \{0, 1\}$ for each ω. We need to show that $\mu^{\rho,\omega}(B)$ is $\mathbb{P}$-a.s. constant. For this it suffices to show that $\mu^{\rho,\omega}(B)$ is itself (almost surely) a tail measurable function of ω.

Consider a ball $\Lambda = \{(z, s) : |s| + |z| \leqslant M\}$ in the space-time lattice $\mathbb{Z}^d \times \mathbb{Z}$. Since the step size of the walks is bounded by R, for each $x \in \mathbb{Z}^d$ and $N \geqslant 1$

$$f_N(T_{x,0}\omega) = \sum_{z\in\mathbb{Z}^d} \pi_{-N,0}^\omega(z, x)$$

is a function of the environments $\{\omega_{z,s} : s \leqslant -1, |z - x| \leqslant R|s|\}$. Consequently, if $|x| > (R + 1)M$, the entire sequence $\{f_N(T_{x,0}\omega)\}_{N\in\mathbb{N}}$ is a function of the environments outside Λ, and then so is (almost surely) the limit $f(T_{x,0}\omega)$. Since B is tail measurable, $\mu^{\rho,\omega}(B)$ is a function of $\{f(T_{x,0}\omega) : |x| > (R + 1)M\}$ and thereby a function of environments outside Λ. Since Λ was arbitrary, we conclude that $\mu^{\rho,\omega}(B)$ is (almost surely) a tail measurable function of ω. $\qquad\square$

Proof of the first part of Theorem 1.1 (except for uniqueness) Invariance of μ^ρ for the process follows by averaging out ω in the result of Lemma 3.5. Spatial invariance, ergodicity and tail triviality of μ^ρ are in Lemma 3.6. That $\int \eta(0)\,d\mu^\rho = \rho$ follows from the definition of μ^ρ.

We prove the ergodicity of the process $\eta_\bullet$ under the time-shift-invariant path measure P_{μ^ρ}. We use the notation μ^ρ also for the joint measure $\mu^\rho(d\omega, d\eta) = \mathbb{P}(d\omega)\mu^{\rho,\omega}(d\eta)$ and not only for the marginal on η. Let $\mathcal{J}$ be the σ-algebra of invariant sets on the state space of the particle system:

$$\mathcal{J} = \{B \subseteq \mathbb{Z}_+^{\mathbb{Z}^d} : \mathbf{1}_B(\eta) = P_\eta\{\eta_1 \in B\} \text{ for } \mu^\rho\text{-a.s. } \eta\}$$

By Corollary 5 on p. 97 of [17] it suffices to show that $\mathcal{J}$ is trivial. We establish triviality of $\mathcal{J}$ by showing that $E^{\mu^\rho}[\psi \mid \mathcal{J}]$ is almost surely a constant for an arbitrary bounded cylinder function ψ on $\mathbb{Z}_+^{\mathbb{Z}^d}$.

Let $\eta^{a,b}$ denote the configuration obtained by moving one particle from site a to site b, if possible: $\eta^{a,b} = \eta$ if $\eta(a) = 0$, while if $\eta(a) > 0$,

$$\eta^{a,b}(x) = \begin{cases} \eta(a) - 1 & x = a \\ \eta(b) + 1 & x = b \\ \eta(x) & x \neq a, b. \end{cases}$$

Lemma 3.7 *There exists a version of $E^{\mu^\rho}[\psi \mid \mathcal{J}]$ such that for all $\eta \in \mathbb{Z}_+^{\mathbb{Z}^d}$ and $a, b \in \mathbb{Z}^d$, $E^{\mu^\rho}[\psi \mid \mathcal{J}](\eta) = E^{\mu^\rho}[\psi \mid \mathcal{J}](\eta^{a,b})$.*

Proof By Corollary 2 on p. 93 of [17], we can define a version $\widetilde{\psi}$ of $E^{\mu^\rho}[\psi \mid \mathcal{J}]$ pointwise by

$$\widetilde{\psi}(\eta) = \overline{\lim_{n\to\infty}} \frac{1}{n} \sum_{t=0}^{n-1} E_\eta[\psi(\eta_t)].$$

We show that $\widetilde{\psi}(\eta) = \widetilde{\psi}(\eta^{a,b})$.

Assume $\eta(a) > 0$. Consider the basic coupling $P_{\eta,\eta^{a,b}}$ of two processes (η_t, ζ_t) with initial configurations $(\eta_0, \zeta_0) = (\eta, \eta^{a,b})$, as described in Sect. 2. Let

$$\sigma = \inf\{t : \psi(\eta_s) = \psi(\zeta_s) \text{ for all } s \geq t\}.$$

We observe that $P_{\eta,\eta^{a,b}}\{\sigma < \infty\} = 1$ in all dimensions. In dimensions $d \in \{1, 2\}$ the irreducible $\bar{q}$-random walk is recurrent, hence the two discrepancies of opposite sign that start at a and b annihilate with probability 1. In dimensions $d \geq 3$ the discrepancies are marginally genuinely d-dimensional random walks by assumption (1.5). Thus they are transient, and so either the discrepancies annihilate or eventually they never return to the finite set of sites that support ψ.

The conclusion of the lemma follows:

$$|E_\eta[\psi(\eta_t)] - E_{\eta^{a,b}}[\psi(\eta_t)]| \leq 2\|\psi\|_\infty P_{\eta,\eta^{a,b}}\{\sigma > t\} \longrightarrow 0 \quad \text{as } t \to \infty. \qquad \square$$

Lemma 3.8 *Suppose h is a bounded measurable function on $\mathbb{Z}_+^{\mathbb{Z}^d}$ such that for all $a, b \in \mathbb{Z}^d$, $h(\eta^{a,b}) = h(\eta)$ μ^ρ-a.s. Then there exists a tail measurable function h_1 such that $h = h_1$ μ^ρ-a.s.*

Proof To show approximate tail measurability we approximate by a cylinder function and then move particles far enough one by one. (We learned this trick from [21].) Let η^a denote the configuration obtained by removing one particle from site a if possible:

$$\eta^a(x) = \begin{cases} (\eta(a) - 1)^+ & x = a \\ \eta(x) & x \neq a. \end{cases}$$

Let $\varepsilon > 0$. Pick a bounded cylinder function $\tilde{h}$ such that $E^{\mu^\rho}|h - \tilde{h}|^2 < \varepsilon^2$. For each ω pick $b(\omega) \in \mathbb{Z}^d$ so that $f(T_{b(\omega),0}\omega) \geq 1/4$ and $\tilde{h}$ does not depend on the coordinate $\eta(b(\omega))$. Such $b(\omega)$ exists a.s. by the ergodic theorem since $Ef = 1$. Choose $b(\omega)$ so that it is a measurable function. Since $h(\eta) = h(\eta^{a,b(\omega)}) \, \mu^\rho(d\omega, d\eta)$-a.s. and $\tilde{h}(\eta^a) = \tilde{h}(\eta^{a,b(\omega)})$ by choice of $b(\omega)$,

$$\int |h(\eta) - h(\eta^a)| \, \mu^\rho(d\eta) \leqslant \int \mathbf{1}_{\{\eta(a)>0\}} |h(\eta^{a,b(\omega)}) - \tilde{h}(\eta^{a,b(\omega)})| \, \mu^\rho(d\omega, d\eta)$$

$$+ \int \mathbf{1}_{\{\eta(a)>0\}} |\tilde{h}(\eta^a) - h(\eta^a)| \, \mu^\rho(d\eta).$$

In the next calculation we bound the first integral after the inequality. Write $\eta = (\eta', \eta(a), \eta(b(\omega)))$ to make the coordinates at a and $b(\omega)$ explicit. Change summation indices and apply Cauchy-Schwarz:

$$\int \mathbf{1}_{\{\eta(a)>0\}} |h(\eta^{a,b(\omega)}) - \tilde{h}(\eta^{a,b(\omega)})| \, \mu^\rho(d\omega, d\eta)$$

$$= \mathbb{E} \sum_{\substack{k>0 \\ \ell \geq 0}} \Gamma^{\rho f(T_{a,0}\omega)}(k) \Gamma^{\rho f(T_{b(\omega),0}\omega)}(\ell) \int |h(\eta', k-1, \ell+1)$$

$$- \tilde{h}(\eta', k-1, \ell+1)| \, \mu^{\rho,\omega}(d\eta')$$

$$= \mathbb{E} \sum_{\substack{m \geq 0 \\ n > 0}} \frac{f(T_{a,0}\omega)}{m+1} \cdot \frac{n}{f(T_{b(\omega),0}\omega)} \cdot \Gamma^{\rho f(T_{a,0}\omega)}(m) \Gamma^{\rho f(T_{b(\omega),0}\omega)}(n)$$

$$\times \int |h(\eta', m, n) - \tilde{h}(\eta', m, n)| \, \mu^{\rho,\omega}(d\eta')$$

$$= \int \mathbf{1}_{\{\eta(b(\omega))>0\}} \frac{f(T_{a,0}\omega)}{\eta(a)+1} \cdot \frac{\eta(b(\omega))}{f(T_{b(\omega),0}\omega)} \cdot |h(\eta) - \tilde{h}(\eta)| \, \mu^\rho(d\omega, d\eta)$$

$$\leqslant \left\{ \mathbb{E} \sum_{\substack{m \geq 0 \\ n > 0}} \frac{f(T_{a,0}\omega)^2}{(m+1)^2} \cdot \frac{n^2}{f(T_{b(\omega),0}\omega)^2} \cdot \Gamma^{\rho f(T_{a,0}\omega)}(m) \Gamma^{\rho f(T_{b(\omega),0}\omega)}(n) \right\}^{1/2}$$

$$\times \left\{ E^{\mu^\rho}|h - \tilde{h}|^2 \right\}^{1/2}$$

$$\leqslant \sqrt{5\mathbb{E}[f^2]} \varepsilon.$$

To obtain the second equality we replace $\Gamma^{\rho f(T_{a,0}\omega)}(k)$ by $\Gamma^{\rho f(T_{a,0}\omega)}(k-1) \cdot \frac{f(T_{a,0}\omega)}{k}$ and similarly for $\Gamma^{\rho f(T_{b(\omega),0}\omega)}(l)$.

An analogous argument (but easier since we do not need the $b(\omega)$) gives

$$\int \mathbf{1}_{\{\eta(a)>0\}} |\tilde{h}(\eta^a) - h(\eta^a)| \, \mu^\rho(d\eta) \leqslant C\varepsilon.$$

Since $\varepsilon > 0$ was arbitrary we have $h(\eta) = h(\eta^a) \, \mu^\rho$-a.s.

For given finite $\Lambda \subseteq \mathbb{Z}^d$, applying the mapping $\eta \mapsto \eta^a$ repeatedly to remove all particles from Λ shows that h equals a.s. a function g_Λ that does not depend on $(\eta(x) : x \in \Lambda)$. As $\Lambda \nearrow \mathbb{Z}^d$ along cubes, the limit $h_1 = \lim g_\Lambda$ exists a.s. by martingale convergence and is tail measurable. $\qquad\square$

We can now conclude the proof of (temporal) ergodicity of the process $\eta_.$. Lemmas 3.7 and 3.8 show that $E^{\mu^\rho}[\psi \mid \mathcal{J}]$ is μ^ρ-a.s. tail measurable, and hence a constant by Lemma 3.6. $\qquad\square$

3.1 Proof of Uniqueness

In this subsection, we complete the proof of part (a) of Theorem 1.1 by showing that μ^ρ is the *unique* invariant distribution with the stated properties. We also prove the second part of Theorem 1.1. The proof of uniqueness uses standard techniques of interacting particle systems [11]. We will arrive at the proof of uniqueness through a sequence of lemmas.

For two configurations η, ζ of occupation variables, we say that $\eta \leqslant \zeta$ if $\eta(x) \leqslant \zeta(x)$ for all x. For two probability distributions μ, ν on the configuration space, we say $\mu \leqslant \nu$ if there exists a probability measure $\tilde{\mu}$ on pairs (η, ζ) of configurations of occupation variables such that $\tilde{\mu}(\eta \leqslant \zeta) = 1$ and the marginals of $\tilde{\mu}$ are μ and ν. For a convex set $\mathcal{A}$, $\mathcal{A}_e$ will denote the set of extremal elements.

Recall that $\tilde{S}$, resp. $\tilde{S}_e$, denotes the set of spatially invariant resp. ergodic probability distributions on pairs (η, ζ) of configurations of occupation variables. Let $\tilde{\mathcal{I}}$ denote the set of probability distributions on pairs of configurations of occupation variables, that are invariant under the temporal evolution described at the beginning of Sect. 2.

Lemma 3.9 *If $\rho_1 < \rho_2$ then $\mu^{\rho_1} \leqslant \mu^{\rho_2}$.*

Proof We couple $\mu^{\rho_1,\omega}$ and $\mu^{\rho_2,\omega}$ by letting $\tilde{\mu}^\omega$ be the distribution of (η, ζ) defined by letting occupation variables $\eta(x)$ be independent Poisson with means $\rho_1 f(T_{x,0}\omega)$, $\gamma(x)$ be independent Poisson with means $(\rho_2 - \rho_1) f(T_{x,0}\omega)$, and then setting $\zeta(x) = \eta(x) + \gamma(x)$. Then define the coupling of μ^{ρ_1} and μ^{ρ_2} by $\tilde{\mu}(\cdot) = \mathbb{E}[\tilde{\mu}^\omega(\cdot)]$. $\qquad\square$

We state the next two lemmas without proof. The proofs can be found in Lemmas 4.2 - 4.5 of [1].

Lemma 3.10 *We have*

(a) *If $\mu_1, \mu_2 \in \mathcal{I} \cap \mathcal{S}$, there is a $\widetilde{\mu} \in \widetilde{\mathcal{I}} \cap \widetilde{\mathcal{S}}$ with marginals μ_1 and μ_2.*
(b) *If $\mu_1, \mu_2 \in (\mathcal{I} \cap \mathcal{S})_e$, there is a $\widetilde{\mu} \in (\widetilde{\mathcal{I}} \cap \widetilde{\mathcal{S}})_e$ with marginals μ_1 and μ_2.*

Lemma 3.11 *If $\widetilde{\mu} \in (\widetilde{\mathcal{I}} \cap \widetilde{\mathcal{S}})_e$ and $\widetilde{\mu}\{(\eta, \zeta) : \eta \geqslant \zeta \text{ or } \zeta \geqslant \eta\} = 1$ then*

$$\widetilde{\mu}\{(\eta, \zeta) : \eta \geqslant \zeta\} = 1 \text{ or } \widetilde{\mu}\{(\eta, \zeta) : \zeta \geqslant \eta\} = 1.$$

A crucial lemma needed in the proof of uniqueness is the following.

Lemma 3.12 *Let $\widetilde{\mu} \in \widetilde{\mathcal{S}}_e$ such that $\int [\eta(0) + \zeta(0)]d\widetilde{\mu} < \infty$. Fix $x \neq y \in \mathbb{Z}^d$. Then*

$$\lim_{t \to \infty} \widetilde{\mu}_t\{(\eta, \zeta) : \eta(x) > \zeta(x) \text{ and } \eta(y) < \zeta(y)\} = 0$$

Proof Our proof employs some of the notation developed in Sect. 2. Fix a positive integer m. Let $I = [-m, m]^d$ and let B be the event that I contains both $(+)$ and $(-)$ particles. The theorem will be proved if we can show that $\widetilde{\mu}_t(B) \to 0$ as $t \to \infty$. So let us assume to the contrary that we can find a sequence $t_k \uparrow \infty$ such that

$$\widetilde{\mu}_{t_k}(B) \geqslant \delta > 0. \tag{3.11}$$

By our assumptions on the environment, we can find a positive integer $T = T(m)$ and a positive real number $\rho = \rho(m) > 0$ such that

$$\min_{x, y \in I} P\{X^x_\cdot \text{ and } \widetilde{X}^y_\cdot \text{ meet by time } T\} \geqslant \rho.$$

Let $A(t, y)$ denote the event that a $(+)$ or a $(-)$ particle present in the cube $y + I$ at time t has been annihilated by time $t + T$. It is clear that

$$P\big[A(t, y)\big|\eta_t, \zeta_t\big] \geqslant \rho \cdot \mathbb{1}_B\{\theta_y(\eta_t, \zeta_t)\} \text{ a.s.} \tag{3.12}$$

For what follows, assume that all $t_{k+1} - t_k \geqslant T$. Let $\phi_t(x) = \beta^+_t(x) + \beta^-_t(x)$ be the number of discrepancy particles at x at time t. Let $n = l(2m + 1) + m$ for a positive integer l and divide the cube $[-n, n]^d$ into $(2l + 1)^d$ cubes of side length $2m + 1$. We have

$$\frac{1}{(2n+1)^d} \sum_{y \in [-n,n]^d} \phi_{t_k+T}(y) \leqslant \frac{1}{(2n+1)^d} \sum_{y \in [-n-RT, n+RT]^d} \phi_{t_k}(y)$$

$$- \frac{1}{(2n+1)^d} \sum_{j=1}^{(2l+1)^d} \mathbb{1}_B\{\theta_{u(j)}(\eta_{t_k}, \zeta_{t_k})\} \cdot \mathbb{1}_{A(t_k, u(j))}$$

where $u(j)$ is the center of cube j. Taking expectations and letting $n \to \infty$, we get

$$E_{\widetilde{\mu}}[\phi_{t_k+T}(0)] \leqslant E_{\widetilde{\mu}}[\phi_{t_k}(0)] - \liminf_{n \to \infty} \frac{-1}{(2n+1)^d} \sum_{j=1}^{(2l+1)^d} E_{\widetilde{\mu}}\big[\mathbb{1}_B\{\theta_{u(j)}(\eta_{t_k}, \zeta_{t_k})\} \cdot \mathbb{1}_{A(t_k, u(j))}\big]$$

It follows from (3.12) and (3.11) that

$$E_{\widetilde{\mu}}\big[\mathbb{1}_B\{\theta_{u(j)}(\eta_{t_k}, \zeta_{t_k})\} \cdot \mathbb{1}_{A(t_k, u(j))}\big] \geqslant \rho \widetilde{\mu}_{t_k}(B) \geqslant \rho \delta.$$

We thus have

$$E_{\widetilde{\mu}}[\phi_{t_{k+1}}(0)] \leqslant E_{\widetilde{\mu}}[\phi_{t_k+T}(0)] \leqslant E_{\widetilde{\mu}}[\phi_{t_k}(0)] - \frac{\rho \delta}{(2m+1)^d}.$$

We can conclude from Lemma 2.1 that $E_{\widetilde{\mu}}[\phi_{t_k}(0)] \to -\infty$. But this is a contradiction since $E_{\widetilde{\mu}}[\phi_t(0)] \geqslant 0$. The proof of the lemma is complete. $\qquad\square$

Lemma 3.13 *If $\mu_1, \mu_2 \in (\mathcal{I} \cap \mathcal{S})_e$ and $E_{\mu_i}\eta(0) < \infty$ for $i = 1, 2$, then $\mu_1 \leqslant \mu_2$ or $\mu_2 \leqslant \mu_1$.*

Proof From Lemma 3.10, we can find $\widetilde{\mu} \in (\widetilde{\mathcal{I}} \cap \widetilde{\mathcal{S}})_e$ with marginals μ_1 and μ_2. Using the ergodic decomposition of stationary measures [24, Theorem 6.6],

$$\widetilde{\mu}\big\{(\eta, \zeta) : \eta(x) > \zeta(x) \text{ and } \eta(y) < \zeta(y)\big\}$$
$$= \int_{\widetilde{\mathcal{S}}_e} \widetilde{\nu}\big\{(\eta, \zeta) : \eta(x) > \zeta(x) \text{ and } \eta(y) < \zeta(y)\big\} \Psi(d\widetilde{\nu}),$$

for a probability measure Ψ on $\widetilde{\mathcal{S}}_e$. On applying the operator $S(t)$ to both sides of the above equation, we observe that the right hand side goes to 0. We thus get

$$\widetilde{\mu}\{(\eta, \zeta) : \eta \leqslant \zeta \text{ or } \zeta \leqslant \eta\} = 1$$

An application of Lemma 3.11 completes the proof. $\qquad\square$

Proposition 3.14 *If $\mu \in (\mathcal{I} \cap \mathcal{S})_e$ and $\rho_0 = E_{\mu}\eta(0) < \infty$ then $\mu = \mu^{\rho_0}$.*

Proof Since $\mu^\rho \in \mathcal{S}_e \cap \mathcal{I}$, it follows that $\mu^\rho \in (\mathcal{I} \cap \mathcal{S})_e$. We can then conclude from Lemmas 3.9 and 3.13 that there exists a $\rho_0' \in [0, \infty]$ such that $\mu \leqslant \mu^\rho$ for $\rho > \rho_0'$ and $\mu \geqslant \mu^\rho$ for $\rho < \rho_0'$. In particular, we have $\rho_0 = E_{\mu}\eta(0) \leqslant \rho$ for $\rho > \rho_0'$ and similarly $\rho_0 \geqslant \rho$ for $\rho < \rho_0'$. This says that $\rho_0' = \rho_0$.

Now fix $\rho_1 < \rho_0 < \rho_2$. For all $(x_1, x_2, \cdots, x_n) \in (\mathbb{Z}^d)^n$ and all $(k_1, k_2, \cdots, k_n) \in (\mathbb{Z}_+)^n$, we have

$$\mu^{\rho_1}(\eta(x_i) \geqslant k_i, 1 \leqslant i \leqslant n) \leqslant \mu(\eta(x_i) \geqslant k_i, 1 \leqslant i \leqslant n) \leqslant \mu^{\rho_2}(\eta(x_i) \geqslant k_i, 1 \leqslant i \leqslant n)$$

The first inequality (resp. the second inequality) above can be seen by looking at the coupled measure $\widetilde{\mu}$ corresponding to μ^{ρ_1} (resp. μ^{ρ_2}) and μ so that $\widetilde{\mu}(\eta \leqslant \zeta) = 1$

(resp. $\widetilde{\mu}(\eta \geqslant \zeta) = 1$). Now let $\rho_1 \uparrow \rho_0$ and $\rho_2 \downarrow \rho_0$ to see that μ has the same finite dimensional distributions as μ^{ρ_0}. □

Proof of the remaining parts of Theorem 1.1 We first prove that μ^ρ is the unique measure with the stated properties in part (a) of Theorem 1.1. Indeed, let μ be another measure with those properties. Since $\mu \in \mathcal{S}_e \cap \mathcal{I}$, we can conclude that $\mu \in (\mathcal{I} \cap \mathcal{S})_e$. From Proposition 3.14, we must have that $\mu = \mu^\rho$.

We now turn to part (b) of the theorem. Let ν be a probability measure on $\mathbb{Z}_+^{\mathbb{Z}^d}$ that is stationary and ergodic under spatial translations and has mean occupation $\int \zeta(0) \, d\nu = \rho$. Denote the occupation process with initial distribution ν by ζ_t. Utilizing the ergodic decomposition theorem [24, Theorem 6.6], find $\widetilde{\mu} \in \widetilde{\mathcal{S}}_e$ with marginals μ^ρ and ν. Let $\widetilde{\mu}_t$ be the time t distribution of the joint process (η_t, ζ_t) coupled as described in Sect. 2.

Initial shift invariance implies that mean occupations are constant ρ throughout time and space:

$$E_{\widetilde{\mu}_t}[\zeta(x)] = E_\nu[\zeta_t(x)] = \int \mathbb{E}\left\{\sum_y \zeta(y)\pi_{0,t}^\omega(y,x)\right\} \nu(d\zeta) = \sum_y \mathbb{E}\left(\pi_{0,t}^\omega(y,x)\right) \int \zeta(y) d\nu = \rho.$$

Chebyshev's inequality and Tychonov's theorem (Theorem 37.3 in [12]) can be used to show that the sequence $\{\widetilde{\mu}_t\}_{t \in \mathbb{Z}_+}$ is tight.

Let $\widetilde{\nu}$ be any limit point as $t \to \infty$. Then by Proposition 2.2 $\widetilde{\nu}\{(\eta, \zeta) : \eta = \zeta\} = 1$. This proves that $P_\nu\{\zeta_t \in \cdot\} \Rightarrow \mu^\rho$. This completes the proof of Theorem 1.1. □

4   Covariances of the Invariant Measures

Define the Green's functions for both q and $\bar{q}$ walks by

$$G_N(x, y) = \sum_{k=0}^{N} q^k(x, y) \quad \text{and} \quad \bar{G}_N(x, y) = \sum_{k=0}^{N} \bar{q}^k(x, y).$$

Recall the potential kernel for the $\bar{q}$ walk

$$\bar{a}(x) = \lim_{N \to \infty} \left(\bar{G}_N(0, 0) - \bar{G}_N(x, 0)\right). \tag{4.1}$$

In the transient case $d \geqslant 3$ the limit above exists trivially, since

$$G(x, y) = \sum_{k=0}^{\infty} q^k(x, y) < \infty \quad \text{and} \quad \bar{G}(x, y) = \sum_{k=0}^{\infty} \bar{q}^k(x, y) < \infty.$$

So for $d \geqslant 3$

$$\bar{a}(x) = \bar{G}(0, 0) - \bar{G}(x, 0). \tag{4.2}$$

For the existence of the limit (4.1) in the recurrent case $d \in \{1, 2\}$ see T1 on p. 352 of [22]. In all cases the kernel $\bar{a}(x)$ satisfies these equations:

$$\sum_z \bar{q}(0, z)\bar{a}(z) = 1 \quad \text{and} \quad \sum_z \bar{q}(x, z)\bar{a}(z) = \bar{a}(x) \text{ for } x \neq 0. \tag{4.3}$$

The constant β defined by (1.13) has the alternate representation

$$\beta = \sum_z q(0, z)\bar{a}(z). \tag{4.4}$$

We omit the argument for the equality of the two representations of β. It is a simple version of the one given at the end of this section for (1.15).

To prove Theorem 1.2 we first verify this proposition and then derive the Fourier representation (1.15).

Proposition 4.1 *Let $d \geqslant 1$. For $m \in \mathbb{Z}^d \setminus \{0\}$*

$$\mathbb{C}\mathrm{ov}[f(\omega), f(T_{m,0}\omega)] = \beta^{-1} \sum_z q(0, z)[\bar{a}(-m) - \bar{a}(z - m)] \tag{4.5}$$

and

$$\mathbb{V}\mathrm{ar}[f(\omega)] = \beta^{-1} - 1. \tag{4.6}$$

A few more notations. Recall that Y_n denotes the Markov chain with transition q and $\bar{Y}_n$ the $\bar{q}$ random walk. Successive returns to the origin are marked as follows:

$$\tau_0 = 0 \text{ and for } j > 0, \tau_j = \inf\{n > \tau_{j-1} : Y_n = 0\}. \tag{4.7}$$

Abbreviate $\tau = \tau_1$. The corresponding stopping time for $\bar{Y}_n$ is $\bar{\tau}$. For $m \in \mathbb{Z}^d$ and $N \geqslant 1$ abbreviate

$$C_N(m) = \mathbb{C}\mathrm{ov}[f_N(\omega), f_N(T_{m,0}\omega)] = \sum_{z, w \in \mathbb{Z}^d} \mathbb{C}\mathrm{ov}\big[\pi_{-N,0}(z, 0), \pi_{-N,0}(w, m)\big].$$

Define also the function

$$h(y) = \sum_{z \in \mathbb{Z}^d} \mathbb{C}\mathrm{ov}[\pi_{0,1}(0, y + z), \pi_{0,1}(0, z)] = q(0, y) - \bar{q}(0, y), \quad y \in \mathbb{Z}^d.$$

Symmetry $h(-y) = h(y)$ holds.

Lemma 4.2 *In all dimensions $d \geqslant 1$,*

$$C_N(m) = \sum_{y \in \mathbb{Z}^d} h(y) G_{N-1}(y, m). \tag{4.8}$$

Proof The case $N = 1$ follows from a shift of space and time. To do induction on N use the Markov property and the additivity of covariance. Abbreviate temporarily $\kappa_{x,y} = \pi_{-N,-N+1}(x, y)$ and recall that the mean kernel is $p_{y-x} = \mathbb{E}\kappa_{x,y}$.

$$C_N(m) = \sum_{z,z_1,w,w_1} \text{Cov}\big[\kappa_{z,z_1} \pi_{-N+1,0}(z_1, 0), \, \kappa_{w,w_1} \pi_{-N+1,0}(w_1, m)\big]$$

$$= \sum_{z,z_1,w,w_1} \Big\{ \text{Cov}\big[(\kappa_{z,z_1} - p_{z_1-z})\pi_{-N+1,0}(z_1, 0), \, (\kappa_{w,w_1} - p_{w_1-w})\pi_{-N+1,0}(w_1, m)\big]$$

$$\tag{4.9}$$

$$+ \text{Cov}\big[p_{z_1-z}\pi_{-N+1,0}(z_1, 0), \, (\kappa_{w,w_1} - p_{w_1-w})\pi_{-N+1,0}(w_1, m)\big] \tag{4.10}$$

$$+ \text{Cov}\big[p_{z_1-z}\pi_{-N+1,0}(z_1, 0), \, p_{w_1-w}\pi_{-N+1,0}(w_1, m)\big] \Big\}. \tag{4.11}$$

Working from the bottom up, the terms on line (4.11) add up to $C_{N-1}(m)$. The terms on line (4.10) vanish because $\kappa_{w,w_1} - p_{w_1-w}$ is mean zero and independent of the other random variables inside the covariance. On line (4.9) the covariance vanishes unless $z = w$. Thus by rearranging line (4.9) we get

$$C_N(m) - C_{N-1}(m) = \text{line (4.9)}$$

$$= \sum_{z,z_1,w_1} \text{Cov}(\kappa_{z,z_1}, \kappa_{z,w_1}) \mathbb{E}\big[\pi_{-N+1,0}(z_1, 0)\pi_{-N+1,0}(w_1, m)\big]$$

$$= \sum_{y,x} \text{Cov}(\kappa_{0,x}, \kappa_{0,x+y}) \sum_{\ell} \mathbb{E}\big[\pi_{-N+1,0}(y, m + \ell)\pi_{-N+1,0}(0, \ell)\big]$$

$$= \sum_{y} h(y)q^{N-1}(y, m).$$

$\square$

In the recurrent case we will use Abel summation, hence the next lemma.

Lemma 4.3 *Let $d \in \{1, 2\}$. For $x, m \in \mathbb{Z}^d$, the limit*

$$a(x, m) = \lim_{s \nearrow 1} \sum_{k=0}^{\infty} s^k \big(q^k(0, m) - q^k(x, m)\big) \tag{4.12}$$

exists. For $m = 0$ the limit is

$$a(x, 0) = \frac{\bar{a}(x)}{\beta} \tag{4.13}$$

and for m $\neq 0$

$$a(x, m) = \frac{\bar{a}(x)}{\beta} \sum_z q(0, z)\big[\bar{a}(-m) - \bar{a}(z - m)\big] - \bar{a}(-m) + \bar{a}(x - m). \quad (4.14)$$

Proof Let s vary in $(0, 1)$ and let

$$U(x, m, s) = E_x\Bigg[\sum_{k=0}^{\tau-1} s^k \mathbf{1}\{Y_k = m\}\Bigg] \xrightarrow[s\nearrow1]{} E_x\Bigg[\sum_{k=0}^{\tau-1} \mathbf{1}\{Y_k = m\}\Bigg] = U(x, m).$$

Decompose the summation across intervals $[\tau_j, \tau_{j+1})$ and use the Markov property:

$$\sum_{k=0}^{\infty} s^k q^k(x, m) = E_x\Bigg[\sum_{k=0}^{\tau_1-1} s^k \mathbf{1}\{Y_k = m\}\Bigg] + \sum_{j=1}^{\infty} E_x\Bigg[s^{\tau_j} \sum_{k=\tau_j}^{\tau_{j+1}-1} s^{k-\tau_j} \mathbf{1}\{Y_k = m\}\Bigg]$$

$$= U(x, m, s) + \sum_{j=1}^{\infty} E_x(s^\tau) E_0(s^\tau)^{j-1} U(0, m, s)$$

$$= U(x, m, s) + \frac{E_x(s^\tau)}{1 - E_0(s^\tau)} U(0, m, s).$$

From this,

$$\sum_{k=0}^{\infty} s^k \big(q^k(0, m) - q^k(x, m)\big) = \frac{1 - E_x(s^\tau)}{1 - E_0(s^\tau)} U(0, m, s) - U(x, m, s). \quad (4.15)$$

We analyze the quantities on the right in (4.15).

Suppose first $x \neq 0$. Then $U(x, m)$ is the same for the Markov chain Y_k as for the random walk $\bar{Y}_k$ because these processes agree until the first visit to 0. In the notation of Spitzer [22], with a check added to refer to the random walk $\bar{Y}_k$, $\bar{g}_{\{0\}}(x, m) = U(x, m)$. By P29.4 on p. 355 of [22] and D11.1 on p. 115, for recurrent random walk

$$U(x, m) = \bar{g}_{\{0\}}(x, m) = \bar{a}(x) + \bar{a}(-m) - \bar{a}(x - m).$$

For $x = 0$ we have $U(0, 0) = 1$, and for $m \neq 0$,

$$U(0, m) = \sum_{y\neq 0} q(0, y)U(y, m) = \sum_{y\neq 0} q(0, y)\big[\bar{a}(y) + \bar{a}(-m) - \bar{a}(y - m)\big]$$

$$= \beta + \sum_y q(0, y)\big[\bar{a}(-m) - \bar{a}(y - m)\big].$$

For the asymptotics of the fraction on the right in (4.15) we can assume again $x \neq 0$ for otherwise the value is 1. It will be convenient to look at the reciprocal. A computation gives

$$\frac{1 - E_0(s^\tau)}{1 - E_x(s^\tau)} = \frac{\sum_{k=0}^{\infty} s^k P_0(\tau > k)}{\sum_{k=0}^{\infty} s^k P_x(\tau > k)}$$

$$= \frac{1}{\sum_{k=0}^{\infty} s^k P_x(\tau > k)} + s \sum_{z \neq 0} q(0, z) \frac{\sum_{k=0}^{\infty} s^k P_z(\tau > k)}{\sum_{k=0}^{\infty} s^k P_x(\tau > k)}.$$

Again we can take advantage of known random walk limits because both $x, z \neq 0$ so the probabilities are the same as those for $\bar{Y}_k$. By P32.2 on p. 379 of [22], as $s \nearrow 1$, for recurrent random walk the above converges to (note that $E_x(\tau) = \infty$)

$$\sum_{z \neq 0} q(0, z) \frac{\bar{a}(z)}{\bar{a}(x)} = \frac{\beta}{\bar{a}(x)}.$$

Letting $s \nearrow 1$ in (4.15) gives (4.13) and (4.14).   □

For $m = 0$ we can obtain the convergence as in (4.1) without the Abel summation. But we do not need this for further development.

Proof of Proposition 4.1. Since $f_N \to f$ in $L^2(\mathbb{P})$, the covariance in (4.5) is given by the limit of $C_N(m)$, so by (4.8)

$$\text{Cov}[f(\omega), f(T_{m,0}\omega)] = \lim_{N \to \infty} \left\{ \sum_y q(0, y) G_{N-1}(y, m) - \sum_y \bar{q}(0, y) G_{N-1}(y, m) \right\}.$$

Next,

$$\sum_y q(0, y) G_{N-1}(y, m) = G_N(0, m) - \delta_{0,m} = q^N(0, m) - \delta_{0,m} + G_{N-1}(0, m).$$

Since the Markov chain q follows the random walk $\bar{q}$ away from 0 it is null recurrent for $d = 1, 2$ and transient for $d \geq 3$. So $q^N(0, m) \to 0$ [13, Theorem 1.8.5]. Thus the limiting covariance now has the form

$$-\delta_{0,m} + \lim_{N \to \infty} \sum_y \bar{q}(0, y)[G_{N-1}(0, m) - G_{N-1}(y, m)]. \tag{4.16}$$

At this point the treatment separates into recurrent and transient cases. This is because the Green's function is uniformly bounded only in the transient case.

Case 1. $d \in \{1, 2\}$

Convergence in (4.16) implies Abel convergence (Theorem 12.41 in [26] or Theorem 1.33 in Chap. 3 of [27]), so the limiting covariance equals

$$-\delta_{0,m} + \lim_{s \nearrow 1} \sum_y \bar{q}(0, y) \sum_{k=0}^{\infty} s^k \left(q^k(0, m) - q^k(y, m) \right).$$

By substituting in (4.13) and (4.14) we obtain (4.6) and (4.5).

Case 2. $d \geqslant 3$

In the transient case we can pass directly to the limit in (4.16) and obtain

$$\mathbb{Cov}[f(\omega), f(T_{m,0}\omega)] = -\delta_{0,m} + \sum_y \bar{q}(0, y)[G(0, m) - G(y, m)]. \quad (4.17)$$

The sum above can be restricted to $y \neq 0$. By restarting after the first return to 0,

$$G(y, m) = E_y \left[\sum_{k=0}^{\tau-1} \mathbf{1}\{Y_k = m\} \right] + P_y(\tau < \infty) G(0, m). \quad (4.18)$$

Next,

$$G(0, m) = \sum_{j=0}^{\infty} E_0 \left[\mathbf{1}\{\tau_j < \infty\} \sum_{k=\tau_j}^{\tau_{j+1}-1} \mathbf{1}\{Y_k = m\} \right]$$

$$= \sum_{j=0}^{\infty} P_0(\tau < \infty)^j E_0 \left[\sum_{k=0}^{\tau-1} \mathbf{1}\{Y_k = m\} \right]$$

$$= \frac{1}{P_0(\tau = \infty)} \left(\delta_{0,m} + (1 - \delta_{0,m}) \sum_{z \neq 0} q(0, z) E_z \left[\sum_{k=0}^{\tau-1} \mathbf{1}\{Y_k = m\} \right] \right). \quad (4.19)$$

Now consider first $m \neq 0$. Combining the above,

$$\mathbb{Cov}[f(\omega), f(T_{m,0}\omega)] = \sum_{y \neq 0} \bar{q}(0, y)\{G(0, m) - G(y, m)\}$$

$$= \sum_{y \neq 0} \bar{q}(0, y) \left\{ \frac{P_y(\tau = \infty)}{P_0(\tau = \infty)} \sum_{z \neq 0} q(0, z) E_z \left[\sum_{k=0}^{\tau-1} \mathbf{1}\{Y_k = m\} \right] - E_y \left[\sum_{k=0}^{\tau-1} \mathbf{1}\{Y_k = m\} \right] \right\}$$

using equality of q and $\bar{q}$ away from 0

$$= \sum_{y \neq 0} \bar{q}(0, y) \left\{ \frac{P_y(\bar{\tau} = \infty)}{P_0(\tau = \infty)} \sum_{z \neq 0} q(0, z) E_z \left[\sum_{k=0}^{\bar{\tau}-1} \mathbf{1}\{\bar{Y}_k = m\} \right] - E_y \left[\sum_{k=0}^{\bar{\tau}-1} \mathbf{1}\{\bar{Y}_k = m\} \right] \right\}$$

$$= \frac{P_0(\bar{\tau} = \infty)}{P_0(\tau = \infty)} \sum_{z \neq 0} q(0, z) E_z \left[\sum_{k=0}^{\bar{\tau}-1} \mathbf{1}\{\bar{Y}_k = m\} \right] - \sum_{y \neq 0} \bar{q}(0, y) E_y \left[\sum_{k=0}^{\bar{\tau}-1} \mathbf{1}\{\bar{Y}_k = m\} \right]$$

applying (4.18) and (4.19) to the $\bar{q}$ walk

$$= \frac{P_0(\bar{\tau} = \infty)}{P_0(\tau = \infty)} \sum_{z \neq 0} q(0, z) \left\{ \bar{G}(z, m) - P_z(\bar{\tau} < \infty) \bar{G}(0, m) \right\} - P_0(\bar{\tau} = \infty) \bar{G}(0, m)$$

$$= \frac{P_0(\bar{\tau} = \infty)}{P_0(\tau = \infty)} \sum_{z \neq 0} q(0, z) \left\{ \bar{G}(z, m) - P_z(\bar{\tau} < \infty) \bar{G}(0, m) \right\}$$

$$- \frac{P_0(\bar{\tau} = \infty)}{P_0(\tau = \infty)} \sum_{z \neq 0} q(0, z) P_z(\tau = \infty) \bar{G}(0, m)$$

$$= \frac{P_0(\bar{\tau} = \infty)}{P_0(\tau = \infty)} \sum_{z \neq 0} q(0, z) \left[\bar{G}(z, m) - \bar{G}(0, m) \right].$$

To finish this case, note that

$$\beta = \sum_{z} q(0, z) \bar{a}(z) = \sum_{z \neq 0} q(0, z) (\bar{G}(0, 0) - \bar{G}(z, 0)) = \sum_{z \neq 0} q(0, z) \frac{P_z(\bar{\tau} = \infty)}{P_0(\bar{\tau} = \infty)}$$

$$= \frac{P_0(\tau = \infty)}{P_0(\bar{\tau} = \infty)}.$$

We have arrived at

$$\mathbb{Cov}[f(\omega), f(T_{m,0}\omega)] = \beta^{-1} \sum_{z \neq 0} q(0, z) [\bar{a}(-m) - \bar{a}(z - m)].$$

Return to (4.17)–(4.19) to cover the case $m = 0$:

$$\mathbb{Cov}[f(\omega), f(\omega)] = \sum_{y} \bar{q}(0, y) [G(0, 0) - G(y, 0)] - 1 = \sum_{y \neq 0} \bar{q}(0, y) \frac{P_y(\tau = \infty)}{P_0(\tau = \infty)} - 1$$

$$= \frac{P_0(\bar{\tau} = \infty)}{P_0(\tau = \infty)} - 1 = \beta^{-1} - 1.$$

This completes the proof of Proposition 4.1. □

Completion of the proof of Theorem 1.2 It remains to prove the Fourier representation (1.15) from (4.5). In several stages symmetry of $\bar{a}$ and the transitions is used.

$$\mathbb{Cov}[f(\omega), f(T_{m,0}\omega)] = \beta^{-1}\sum_z q(0, z)[\bar{a}(m) - \bar{a}(m - z)]$$

$$= \lim_{N\to\infty} \beta^{-1}\sum_{k=0}^{N}\sum_z q(0, z)[\bar{q}^k(m - z, 0) - \bar{q}^k(m, 0)]$$

$$= \lim_{N\to\infty} \frac{\beta^{-1}}{(2\pi)^d}\sum_{k=0}^{N}\sum_z q(0, z)\int_{\mathbb{T}^d}[e^{-i\theta\cdot(m-z)} - e^{-i\theta\cdot m}]\bar{\lambda}^k(\theta)\,d\theta$$

$$= \lim_{N\to\infty} \frac{-\beta^{-1}}{(2\pi)^d}\int_{\mathbb{T}^d}\cos(\theta\cdot m)\frac{1 - \lambda(\theta)}{1 - \bar{\lambda}(\theta)}(1 - \bar{\lambda}^{N+1}(\theta))\,d\theta$$

$$= \frac{-\beta^{-1}}{(2\pi)^d}\int_{\mathbb{T}^d}\cos(\theta\cdot m)\frac{1 - \lambda(\theta)}{1 - \bar{\lambda}(\theta)}\,d\theta.$$

The last equality comes from $0 \leqslant \bar{\lambda}(\theta) < 1$ for $\theta \in \mathbb{T}^d \setminus \{0\}$ and dominated convergence. The ratio $(1 - \lambda(\theta))/(1 - \bar{\lambda}(\theta))$ stays bounded as $\theta \to 0$ because both transitions q and $\bar{q}$ have zero mean and $\bar{q}$ has a nonsingular covariance matrix [22, P7 p. 74]. $\qquad\square$

5   Convergence of Centered Current Fluctuations

We prove Theorem 1.4 by proving the following proposition. Recall the definition of the current $Y_n(t, r)$ from (1.18), and let $\{Z(t, r) : (t, r) \in \mathbb{R}_+ \times \mathbb{R}\}$ be the mean zero Gaussian process defined by (1.22) or equivalently through the covariance (1.26). Recall also the definitions

$$\overline{Y}_n(t, r) = n^{-1/4}\{Y_n(t, r) - E^{\omega}[Y_n(t, r)]\},$$

$$\overline{Y}_n(\boldsymbol{\theta}) = \sum_{i=1}^{N}\theta_i\overline{Y}_n(t_i, r_i) \quad\text{and}\quad Z(\boldsymbol{\theta}) = \sum_{i=1}^{N}\theta_i Z(t_i, r_i).$$

Proposition 5.1

$$E^{\omega}\big[\exp\{i\overline{Y}_n(\boldsymbol{\theta})\}\big] \to E\big[\exp\{iZ(\boldsymbol{\theta})\}\big] \text{ in } \mathbb{P}\text{-probability.} \tag{5.1}$$

The remainder of the section proves this proposition and thereby Theorem 1.4. We write $\overline{Y}_n(\boldsymbol{\theta})$ as a sum of independent mean zero random variables (under P^{ω}) so that we can apply Lindeberg-Feller [6]:

$$\overline{Y}_n(\boldsymbol{\theta}) = n^{-1/4}\sum_{i=1}^{N}\theta_i\{Y_n(t_i, r_i) - E^{\omega}Y_n(t_i, r_i)\} = W_n = \sum_{m=-\infty}^{\infty}\bar{U}_m \tag{5.2}$$

with

$$\bar{U}_m = \sum_{i=1}^{N} \theta_i \Big(U_m(t_i, r_i) \, \mathbf{1}\{m > 0\} - V_m(t_i, r_i) \, \mathbf{1}\{m \leqslant 0\} \Big), \tag{5.3}$$

and

$$U_m(t, r) = n^{-1/4} \sum_{j=1}^{\eta_0(m)} \mathbf{1}\{X^{m,j}_{\lfloor nt \rfloor} \leqslant \lfloor nvt \rfloor + \lfloor r\sqrt{n} \rfloor\}$$

$$- n^{-1/4} E^{\omega}(\eta_0(m)) P^{\omega}(X^m_{\lfloor nt \rfloor} \leqslant \lfloor nvt \rfloor + \lfloor r\sqrt{n} \rfloor), \tag{5.4}$$

$$V_m(t, r) = n^{-1/4} \sum_{j=1}^{\eta_0(m)} \mathbf{1}\{X^{m,j}_{\lfloor nt \rfloor} > \lfloor nvt \rfloor + \lfloor r\sqrt{n} \rfloor\}$$

$$- n^{-1/4} E^{\omega}(\eta_0(m)) P^{\omega}(X^m_{\lfloor nt \rfloor} > \lfloor nvt \rfloor + \lfloor r\sqrt{n} \rfloor).$$

The n-dependence is suppressed from the notations $\bar{U}_m$, $U_m(t, r)$ and $V_m(t, r)$. The variables $\{\bar{U}_m\}_{m \in \mathbb{Z}}$ are independent under P^{ω} because initial occupation variables and walks are independent. We will also use repeatedly this formula, a consequence of the independence of η_0 and the walks under P^{ω}:

$$E^{\omega}[U_m(t, r)^2] = n^{-1/2} \operatorname{Var}^{\omega} \left(\sum_{j=1}^{\eta_0(m)} \mathbf{1}\{X^{m,j}_{\lfloor nt \rfloor} \leqslant \lfloor nvt \rfloor + \lfloor r\sqrt{n} \rfloor\} \right)$$

$$= n^{-1/2} E^{\omega}(\eta_0(m)) P^{\omega}(X^m_{\lfloor nt \rfloor} \leqslant \lfloor nvt \rfloor + \lfloor r\sqrt{n} \rfloor) P^{\omega}(X^m_{\lfloor nt \rfloor} > \lfloor nvt \rfloor + \lfloor r\sqrt{n} \rfloor)$$

$$+ n^{-1/2} \operatorname{Var}^{\omega}(\eta_0(m)) P^{\omega}(X^m_{\lfloor nt \rfloor} \leqslant \lfloor nvt \rfloor + \lfloor r\sqrt{n} \rfloor)^2 \tag{5.5}$$

and the corresponding formula for $V_m(t, r)$.

Let $a(n) \nearrow \infty$ be a sequence that will be determined precisely in the proof. Define the finite sum

$$W_n^* = \sum_{|m| \leqslant a(n)\sqrt{n}} \bar{U}_m. \tag{5.6}$$

We observe that the terms $|m| > a(n)\sqrt{n}$ can be discarded from (5.2).

Lemma 5.2 $E|W_n - W_n^*|^2 \to 0$ as $n \to \infty$.

Proof By the mutual independence of occupation variables and walks under P^{ω}, and as eventually $a(n) > |r_i|$, the task boils down to showing that sums of this type vanish:

$$E\left[\left(\sum_{m>a(n)\sqrt{n}} U_m(t,r)\right)^2\right] = \mathbb{E}\sum_{m>a(n)\sqrt{n}} E^\omega[U_m(t,r)^2]$$

$$\leqslant n^{-1/2}\,\mathbb{E}\sum_{m>a(n)\sqrt{n}}\left[E^\omega(\eta_0(m)) + \mathrm{Var}^\omega(\eta_0(m))\right]P^\omega\{X_{\lfloor nt\rfloor}^{m,j} \leqslant \lfloor nvt\rfloor + \lfloor r\sqrt{n}\rfloor\}$$

$$\leqslant Cn^{-1/2}\sum_{m>a(n)\sqrt{n}} P\{X_{\lfloor nt\rfloor} \leqslant \lfloor nvt\rfloor + \lfloor r\sqrt{n}\rfloor - m\}$$

$$= CE\left[\left(\frac{X_{\lfloor nt\rfloor} - \lfloor nvt\rfloor}{\sqrt{n}} - r + a(n)\right)^{-}\right].$$

Under the averaged measure P the walk X_s is a sum of bounded i.i.d. random variables, hence by uniform integrability the last line vanishes as $a(n) \nearrow \infty$. There is also a term for $m < a(n)\sqrt{n}$ involving $V_m(t,r)$ that is handled in the same way.

$\square$

The limit $\boldsymbol{\theta} \cdot \mathbf{Z}$ in our goal (5.1) has variance

$$\sigma_\theta^2 = \sum_{1\leqslant i,j\leqslant N} \theta_i\theta_j\left[\rho_0\Gamma_1\big((t_i,r_i),(t_j,r_j)\big) + \sigma_0^2\Gamma_2\big((t_i,r_i),(t_j,r_j)\big)\right] \qquad (5.7)$$

and the two Γ-terms, defined earlier in (1.24) and (1.25), have the following expressions in terms of a standard 1-dimensional Brownian motion B_t:

$$\Gamma_1\big((s,q),(t,r)\big) = \int_{-\infty}^{\infty}\Big(\mathbf{P}[B_{\sigma^2 s} \leqslant q - x]\mathbf{P}[B_{\sigma^2 t} > r - x]$$

$$- \mathbf{P}[B_{\sigma^2 s} \leqslant q - x, B_{\sigma^2 t} > r - x]\Big)\,dx \qquad (5.8)$$

and

$$\Gamma_2\big((s,q),(t,r)\big) = \int_0^{\infty} \mathbf{P}[B_{\sigma^2 s} \leqslant q - x]\mathbf{P}[B_{\sigma^2 t} \leqslant r - x]\,dx$$

$$+ \int_{-\infty}^{0} \mathbf{P}[B_{\sigma^2 s} > q - x]\mathbf{P}[B_{\sigma^2 t} > r - x]\,dx. \qquad (5.9)$$

By Lemma 5.2, the desired limit (5.1) follows from showing

$$E^\omega(e^{iW_n^*}) \to e^{-\sigma_\theta^2/2} \quad \text{in } \mathbb{P}\text{-probability as } n \to \infty. \qquad (5.10)$$

This limit will be achieved by showing that the usual conditions of the Lindeberg-Feller theorem hold in $\mathbb{P}$-probability:

$$\sum_{|m|\leqslant a(n)\sqrt{n}} E^\omega(\bar{U}_m^2) \to \sigma_\theta^2 \qquad (5.11)$$

and

$$\sum_{|m| \leqslant a(n)\sqrt{n}} E^\omega \left(|\bar{U}_m|^2 \mathbf{1}\{|\bar{U}_m| \geqslant \varepsilon\} \right) \to 0. \tag{5.12}$$

The standard Lindeberg-Feller theorem can then be applied to subsequences. The limits (5.11)–(5.12) in $\mathbb{P}$-probability imply that every subsequence has a further subsequence along which these limits hold for $\mathbb{P}$-almost every ω. Thus along this further subsequence W_n^* converges weakly to $\mathcal{N}(0, \sigma_\theta^2)$ under P^ω for $\mathbb{P}$-almost every ω. So, every subsequence has a further subsequence along which the limit (5.10) holds for $\mathbb{P}$-almost every ω. This implies the limit (5.10) in $\mathbb{P}$-probability.

We check the negligibility condition (5.12) in the L^1 sense.

Lemma 5.3 *Under assumption* (1.20),

$$\lim_{n \to \infty} \sum_{|m| \leqslant a(n)\sqrt{n}} E\left[|\bar{U}_m|^2 \mathbf{1}\{|\bar{U}_m| \geqslant \varepsilon\} \right] = 0. \tag{5.13}$$

Proof First

$$\bar{U}_m^2 = \left(\sum_{i=1}^{N} \theta_i \left[U_m(t_i, r_i) \mathbf{1}\{m \geqslant 0\} - V_m(t_i, r_i) \mathbf{1}\{m < 0\} \right] \right)^2$$

$$\leqslant C \sum_{i=1}^{N} U_m(t_i, r_i)^2 \mathbf{1}\{m \geqslant 0\} + C \sum_{i=1}^{N} V_m(t_i, r_i)^2 \mathbf{1}\{m < 0\}.$$

The arguments for the terms above are the same. So take a term from the first sum, let $(t, r) = (t_i, r_i)$, and the task is now

$$\lim_{n \to \infty} \sum_{m=0}^{a(n)\sqrt{n}} E\left[U_m(t, r)^2 \mathbf{1}\{|\bar{U}_m| \geqslant \varepsilon\} \right] = 0. \tag{5.14}$$

Since

$$|\bar{U}_m| \leqslant C n^{-1/4} \left[\eta_0(m) + E^\omega(\eta(m)) \right]$$

and by adjusting ε, limit (5.14) follows if we can show the limit for these sums:

$$\sum_{m=0}^{a(n)\sqrt{n}} E\left[U_m(t, r)^2 \mathbf{1}\{\eta_0(m) > n^{1/4}\varepsilon\} \right]$$

$$+ \sum_{m=0}^{a(n)\sqrt{n}} E\left[U_m(t, r)^2 \mathbf{1}\{E^\omega(\eta_0(m)) > n^{1/4}\varepsilon\} \right]. \tag{5.15}$$

Abbreviate

$$A_m = \{X_{nt}^m \leq \lfloor nvt \rfloor + \lfloor r\sqrt{n} \rfloor \}.$$

The terms of the second sum in (5.15) develop as follows, using (5.5), the independence of $\bar{\omega}_{-\infty,-1}$ and $\bar{\omega}_{0,\infty}$, and the shift invariance:

$$\mathbb{E}\big[E^\omega(U_m(t,r)^2)\mathbf{1}\{E^\omega(\eta_0(m)) > n^{1/4}\varepsilon\}\big]$$
$$\leq n^{-1/2}\mathbb{E}\big[\big(E^\omega(\eta_0(m)) + \mathrm{Var}^\omega(\eta_0(m))\big)\mathbf{1}\{E^\omega(\eta_0(m)) > n^{1/4}\varepsilon\}\big]P(A_m)$$
$$= n^{-1/2}\mathbb{E}\big[\big(E^\omega(\eta_0(0)) + \mathrm{Var}^\omega(\eta_0(0))\big)\mathbf{1}\{E^\omega(\eta_0(0)) > n^{1/4}\varepsilon\}\big]P(A_m).$$

Since the averaged walk is a walk with bounded i.i.d. steps,

$$\sum_{m=0}^{a(n)\sqrt{n}} P(A_m) \leq E\big[(X_{\lfloor nt \rfloor} - \lfloor nvt \rfloor - \lfloor r\sqrt{n} \rfloor)^-\big] \leq C(n^{1/2} + 1). \qquad (5.16)$$

Thus

$$\sum_{m=0}^{a(n)\sqrt{n}} E\big[U_m(t,r)^2\mathbf{1}\{E^\omega(\eta_0(m)) > n^{1/4}\varepsilon\}\big]$$
$$\leq C\mathbb{E}\big[\big(E^\omega(\eta_0(0)) + \mathrm{Var}^\omega(\eta_0(0))\big)\mathbf{1}\{E^\omega(\eta_0(0)) > n^{1/4}\varepsilon\}\big].$$

The last line vanishes as $n \to \infty$ by dominated convergence, by assumption (1.20).

For the first sum in (5.15) first take quenched expectation of the walks while conditioning on η_0, to get the bound

$$E_{\eta_0}^\omega[U_m(t,r)^2] \leq 2n^{-1/2}P^\omega(A_m)\big[\eta_0(m)^2 + E^\omega(\eta_0(m))^2\big].$$

Using again the independence of $\bar{\omega}_{-\infty,-1}$ and $\bar{\omega}_{0,\infty}$, shift-invariance, and (5.16),

$$\sum_{m=0}^{a(n)\sqrt{n}} E\big[U_m(t,r)^2\mathbf{1}\{\eta_0(m) > n^{1/4}\varepsilon\}\big]$$
$$\leq Cn^{-1/2}\sum_{m=0}^{a(n)\sqrt{n}} P(A_m) \cdot E\big[\big(\eta_0(0)^2 + E^\omega(\eta_0(0))^2\big)\mathbf{1}\{\eta_0(0) > n^{1/4}\varepsilon\}\big]$$
$$\leq CE\big[\big(\eta_0(0)^2 + E^\omega(\eta_0(0))^2\big)\mathbf{1}\{\eta_0(0) > n^{1/4}\varepsilon\}\big]$$

The last line vanishes as $n \to \infty$ by dominated convergence, by assumption (1.20).

$\square$

We turn to checking (5.11).

$$\sum_{|m|\leqslant a(n)\sqrt{n}} E^\omega\big[\bar{U}_m^2\big] = \sum_{1\leqslant i,j\leqslant N}\theta_i\theta_j \sum_{|m|\leqslant a(n)\sqrt{n}} \Big[\mathbf{1}_{\{m>0\}} E^\omega\big(U_m(t_i,r_i)U_m(t_j,r_j)\big)$$

$$+\ \mathbf{1}_{\{m\leqslant 0\}} E^\omega\big(V_m(t_i,r_i)V_m(t_j,r_j)\big)\Big].$$

Each quenched expectation of a product of two mean zero random variables is handled in the manner of (5.5) that we demonstrate with the second expectation:

$$E^\omega\big(V_m(t_i,r_i)V_m(t_j,r_j)\big)$$

$$= n^{-1/2}\,\mathrm{Cov}^\omega\bigg(\sum_{k=1}^{\eta_0(m)}\mathbf{1}\{X^{m,k}_{\lfloor nt_i\rfloor} > \lfloor nvt_i\rfloor + r_i\sqrt{n}\},\ \sum_{\ell=1}^{\eta_0(m)}\mathbf{1}\{X^{m,\ell}_{\lfloor nt_j\rfloor} > \lfloor nvt_j\rfloor + r_j\sqrt{n}\}\bigg)$$

$$= n^{-1/2} E^\omega(\eta_0(m))\Big[P^\omega(X^m_{\lfloor nt_i\rfloor} > \lfloor nvt_i\rfloor + r_i\sqrt{n},\ X^m_{\lfloor nt_j\rfloor} > \lfloor nvt_j\rfloor + r_j\sqrt{n})$$

$$-\ P^\omega(X^m_{\lfloor nt_i\rfloor} > \lfloor nvt_i\rfloor + r_i\sqrt{n})P^\omega(X^m_{\lfloor nt_j\rfloor} > \lfloor nvt_j\rfloor + r_j\sqrt{n})\Big]$$

$$+ n^{-1/2}\,\mathrm{Var}^\omega(\eta_0(m))P^\omega(X^m_{\lfloor nt_i\rfloor} > \lfloor nvt_i\rfloor + r_i\sqrt{n})P^\omega(X^m_{\lfloor nt_j\rfloor} > \lfloor nvt_j\rfloor + r_j\sqrt{n}).$$

After some rearranging of the resulting probabilities, we arrive at

$$\sum_{|m|\leqslant a(n)\sqrt{n}} E^\omega\big[\bar{U}_m^2\big]$$

$$= n^{-1/2}\sum_{1\leqslant i,j\leqslant N}\theta_i\theta_j\bigg[\sum_{|m|\leqslant a(n)\sqrt{n}} E^\omega(\eta_0(m))$$

$$\times\Big\{P^\omega(X^m_{\lfloor nt_i\rfloor} \leqslant \lfloor nvt_i\rfloor + \lfloor r_i\sqrt{n}\rfloor)P^\omega(X^m_{\lfloor nt_j\rfloor} > \lfloor nvt_j\rfloor + \lfloor r_j\sqrt{n}\rfloor)$$

$$-\ P^\omega(X^m_{\lfloor nt_i\rfloor} \leqslant \lfloor nvt_i\rfloor + \lfloor r_i\sqrt{n}\rfloor,\ X^m_{\lfloor nt_j\rfloor} > \lfloor nvt_j\rfloor + \lfloor r_j\sqrt{n}\rfloor)\Big\}$$

$$+\sum_{|m|\leqslant a(n)\sqrt{n}} \mathrm{Var}^\omega(\eta_0(m))$$

$$\times\Big\{\mathbf{1}_{\{m>0\}}P^\omega(X^m_{\lfloor nt_i\rfloor} \leqslant \lfloor nvt_i\rfloor + \lfloor r_i\sqrt{n}\rfloor)P^\omega(X^m_{\lfloor nt_j\rfloor} \leqslant \lfloor nvt_j\rfloor + \lfloor r_j\sqrt{n}\rfloor)$$

$$+\mathbf{1}_{\{m\leqslant 0\}}P^\omega(X^m_{\lfloor nt_i\rfloor} > \lfloor nvt_i\rfloor + \lfloor r_i\sqrt{n}\rfloor)P^\omega(X^m_{\lfloor nt_j\rfloor} > \lfloor nvt_j\rfloor + \lfloor r_j\sqrt{n}\rfloor)\Big\}\bigg].$$

$$(5.17)$$

The terms above have been arranged so that the sums match up with the integrals in (5.7)–(5.9). Limit (5.11) is now proved by showing that, term by term, the sums above converge to the integrals. In each case the argument is the same. We illustrate the case of the sum of the first term with the factor $\mathrm{Var}^\omega(\eta_0(m))$ in front. To simplify notation we let $((s,q),(t,r)) = ((t_i,r_i),(t_j,r_j))$. In other words, we show this convergence in $\mathbb{P}$-probability:

$$S_0(n) \equiv n^{-1/2} \sum_{0 < m \leqslant a(n)\sqrt{n}} \mathrm{Var}^\omega(\eta_0(m)) P^\omega(X_{\lfloor ns \rfloor}^m \leqslant \lfloor nvs \rfloor + \lfloor q\sqrt{n} \rfloor)$$

$$\times P^\omega(X_{\lfloor nt \rfloor}^m \leqslant \lfloor nvt \rfloor + \lfloor r\sqrt{n} \rfloor) \tag{5.18}$$

$$\xrightarrow[n \to \infty]{} \sigma_0^2 \int_0^\infty \mathbf{P}[B_{\sigma^2 s} \leqslant q - x]\mathbf{P}[B_{\sigma^2 t} \leqslant r - x]\, dx \equiv I.$$

The proof of $S_0(n) \xrightarrow{\mathbb{P}} I$ is divided into two lemmas. Let

$$S_1(n) = n^{-1/2} \sum_{0 < m \leqslant a(n)\sqrt{n}} \mathrm{Var}^\omega(\eta_0(m))$$

$$\times \mathbf{P}\left(B_{\sigma^2 s} \leqslant q - \frac{m}{\sqrt{n}}\right)\mathbf{P}\left(B_{\sigma^2 t} \leqslant r - \frac{m}{\sqrt{n}}\right). \tag{5.19}$$

Lemma 5.4 $\lim_{n \to \infty} \mathbb{E}|S_0(n) - S_1(n)| = 0.$

Proof By the quenched central limit theorem for space-time RWRE [16], for each $x \in \mathbb{R}$ the limit

$$P^\omega(X_{\lfloor ns \rfloor} \leqslant \lfloor nvs \rfloor + \lfloor x\sqrt{n} \rfloor) \to \mathbf{P}(B_{\sigma^2 s} \leqslant x)$$

holds for $\mathbb{P}$-a.e. ω. Since these are distribution functions (monotone and between 0 and 1) with a continuous limit the convergence is uniform in x. Set

$$D_n(\omega) = \sup_{x,y \in \mathbb{R}} \left| P^\omega(X_{\lfloor ns \rfloor} \leqslant \lfloor nvs \rfloor + \lfloor x\sqrt{n} \rfloor) P^\omega(X_{\lfloor nt \rfloor} \leqslant \lfloor nvt \rfloor + \lfloor y\sqrt{n} \rfloor) \right.$$

$$\left. - \mathbf{P}(B_{\sigma^2 s} \leqslant x)\mathbf{P}(B_{\sigma^2 t} \leqslant y) \right|$$

and then $D_n(\omega) \to 0$ $\mathbb{P}$-a.s. By shift-invariance

$$\mathbb{E}|S_0(n) - S_1(n)| \leqslant n^{-1/2} \sum_{0 < m \leqslant a(n)\sqrt{n}} \mathbb{E}\,\mathrm{Var}^{T_{m,0}\omega}(\eta_0(0))$$

$$\times \left| P^{T_{m,0}\omega}(X_{\lfloor ns \rfloor} \leqslant \lfloor nvs \rfloor + \lfloor q\sqrt{n} \rfloor - m) P^{T_{m,0}\omega}(X_{\lfloor nt \rfloor} \leqslant \lfloor nvt \rfloor + \lfloor r\sqrt{n} \rfloor - m) \right.$$

$$\left. - \mathbf{P}\left(B_{\sigma^2 s} \leqslant q - \frac{m}{\sqrt{n}}\right)\mathbf{P}\left(B_{\sigma^2 t} \leqslant r - \frac{m}{\sqrt{n}}\right) \right|$$

$$\leqslant n^{-1/2} \sum_{0 < m \leqslant a(n)\sqrt{n}} \mathbb{E}\left[\mathrm{Var}^{T_{m,0}\omega}(\eta_0(0)) D_n(T_{m,0}\omega)\right]$$

$$\leqslant 2a(n)\mathbb{E}\left[\mathrm{Var}^\omega(\eta_0(0)) D_n(\omega)\right]. \tag{5.20}$$

Moment assumption (1.20) and dominated convergence guarantee that

$$\mathbb{E}\left[\mathrm{Var}^\omega(\eta_0(0)) D_n(\omega)\right] \longrightarrow 0.$$

Thus we can take

$$a(n) = \left(\sup_{k:k\geqslant n} \mathbb{E}\big[\mathrm{Var}^\omega(\eta_0(0))D_k(\omega)\big]\right)^{-1/2} \tag{5.21}$$

to have $a(n) \nearrow \infty$ while still line (5.20) vanishes as $n \to \infty$. □

The choice of $a(n)$ made above depends on s, t but that is not problematic since we have only finitely many time points t_i to handle.

Lemma 5.5 $\lim_{n\to\infty} \mathbb{E}|S_1(n) - I| = 0.$

Proof First we discard tails of the sum and integral. Given $\varepsilon > 0$, we can choose a large enough $c < \infty$ such that

$$S_1^*(n) = n^{-1/2} \sum_{0<m\leqslant c\sqrt{n}} \mathrm{Var}^\omega(\eta_0(m))$$

$$\times \mathbf{P}\left(B_{\sigma^2 s} \leqslant q - \frac{m}{\sqrt{n}}\right)\mathbf{P}\left(B_{\sigma^2 t} \leqslant r - \frac{m}{\sqrt{n}}\right)$$

satisfies $\mathbb{E}|S_1(n) - S_1^*(n)| \leqslant \varepsilon$, and so that

$$I^* = \sigma_0^2 \int_0^c \mathbf{P}[B_{\sigma^2 s} \leqslant q - x]\mathbf{P}[B_{\sigma^2 t} \leqslant r - x]\,dx$$

satisfies $I - I^* \leqslant \varepsilon$. Thus it suffices to prove $S_1^*(n) \to I^*$.

Next, since the Gaussian distribution functions are Lipschitz continuous,

$$S_1^*(n) - I^* = n^{-1/2} \sum_{0<m\leqslant c\sqrt{n}} \big[\mathrm{Var}^\omega(\eta_0(m)) - \sigma_0^2\big]$$

$$\times \mathbf{P}\left(B_{\sigma^2 s} \leqslant q - \frac{m}{\sqrt{n}}\right)\mathbf{P}\left(B_{\sigma^2 t} \leqslant r - \frac{m}{\sqrt{n}}\right) + O(n^{-1/2}).$$

Introduce an intermediate scale $1 << L << \sqrt{n}$ and use again Lipschitz continuity of the probabilities:

$$S_1^*(n) - I^* = \frac{L}{n^{1/2}} \sum_{0\leqslant j<\frac{c\sqrt{n}}{L}-1} \left(\frac{1}{L}\sum_{m=jL+1}^{(j+1)L} \mathrm{Var}^\omega(\eta_0(m)) - \sigma_0^2\right)$$

$$\times \left\{\mathbf{P}\left(B_{\sigma^2 s} \leqslant q - \frac{jL}{\sqrt{n}}\right)\mathbf{P}\left(B_{\sigma^2 t} \leqslant r - \frac{jL}{\sqrt{n}}\right) + O\left(\frac{L}{\sqrt{n}}\right)\right\} + \frac{R_n}{\sqrt{n}} + O(n^{-1/2}).$$

The error term R_n consists of order L terms bounded by $|\mathrm{Var}^\omega(\eta_0(m)) - \sigma_0^2|$ that appear because the collection of summation intervals $(jL, (j+1)L]$ may not exactly cover the original summation interval $0 < m \leqslant c\sqrt{n}$. It satisfies $\mathbb{E}R_n \leqslant CL$. Finally, bounding the probabilities crudely by 1 and by shift-invariance,

$$\mathbb{E}|S_1^*(n) - I^*| \leqslant C\mathbb{E}\left|\frac{1}{L}\sum_{m=1}^{L} \text{Var}^{\omega}(\eta_0(m)) - \sigma_0^2\right| + O(Ln^{-1/2}).$$

This vanishes as we let first $n \to \infty$ and then $L \to \infty$ and apply the L^1 ergodic theorem. $\qquad\square$

Limit (5.18) has now been verified. All terms in (5.17) are treated the same way to show that they converge, in $L^1(\mathbb{P})$ and therefore in $\mathbb{P}$-probability, to the corresponding integrals in (5.7)–(5.9). This verifies limit (5.11). Since both (5.11) and (5.12) have been checked, the Gaussian limit in (5.10) has been proved, as explained in the paragraph following (5.12). The proof of Proposition 5.1 and thereby also the proof of Theorem 1.4 are complete. $\qquad\square$

6 The Quenched Mean Process

We now prove Theorems 1.6 and 1.7. We will use a simplified notation for the quenched jump probabilities: $\omega_{x,n} = \omega_{x,n}(1)$ and $\omega'_{x,n} = \omega_{x,n}(0) = 1 - \omega_{x,n}(1)$. Note that when the steps are 0 and 1 we have $v = p(1) = \mathbb{E}\omega_{0,0}$. Potential kernel $\bar{a}$ can be easily computed from Eq. (4.3) and seen to equal $\bar{a}(x) = \frac{|x|}{2v(1-v)}$. Recall that $\alpha = \mathbb{E}\omega_{0,0}\omega'_{0,0}$. Then formula (4.4) gives

$$\beta = \frac{\alpha}{v(1-v)}.$$

Proof of Theorem 1.6 Define

$$H_n(x) = E^{\omega}\left[\sum_{y>0}\sum_{j=1}^{\eta_0(y)} \mathbf{1}\{X_n^{y,j} \leqslant x\} - \sum_{y\leqslant 0}\sum_{j=1}^{\eta_0(y)} \mathbf{1}\{X_n^{y,j} > x\}\right].$$

Then $Y_n(t,r) = H_{\lfloor nt \rfloor}(\lfloor nvt \rfloor + \lfloor r\sqrt{n}\rfloor)$. Compute

$$H_{n+1}(x) = E^{\omega}\left[\sum_{y>0}\sum_{j=1}^{\eta_0(y)} \mathbf{1}\{X_n^{y,j} \leqslant x-1\}\right] + E^{\omega}\left[\sum_{y>0}\sum_{j=1}^{\eta_0(y)} \mathbf{1}\{X_n^{y,j} = x\}\right]\omega'_{x,n}$$

$$- \sum_{y\leqslant 0}\sum_{j=1}^{\eta_0(y)} \mathbf{1}\{X_n^{y,j} > x\}\right] - \sum_{y\leqslant 0}\sum_{j=1}^{\eta_0(y)} \mathbf{1}\{X_n^{y,j} = x\}\right]\omega_{x,n}.$$

Also,

$$\omega_{x,n} H_n(x-1) + \omega'_{x,n} H_n(x)$$

$$= E^\omega\left[\sum_{y>0}\sum_{j=1}^{\eta_0(y)} \mathbf{1}\{X_n^{y,j} \leqslant x-1\}\right]\omega_{x,n} - \sum_{y\leqslant 0}\sum_{j=1}^{\eta_0(y)} \mathbf{1}\{X_n^{y,j} > x-1\}\Big]\omega_{x,n}$$

$$+ E^\omega\left[\sum_{y>0}\sum_{j=1}^{\eta_0(y)} \mathbf{1}\{X_n^{y,j} \leqslant x\}\right]\omega'_{x,n} - \sum_{y\leqslant 0}\sum_{j=1}^{\eta_0(y)} \mathbf{1}\{X_n^{y,j} > x\}\Big]\omega'_{x,n}.$$

Taking the difference of the two expressions one finds that

$$H_{n+1}(x) = \omega_{x,n} H_n(x-1) + \omega'_{x,n} H_n(x).$$

In other words, H is the random average process introduced by Ferrari and Fontes [8]. The initial conditions are given by

$$H_0(x) = \begin{cases} 0 & \text{if } x = 0, \\[2mm] \displaystyle\sum_{y=1}^{x} E^\omega \eta_0(y) & \text{if } x > 0, \quad \text{and} \\[4mm] \displaystyle\sum_{y=x+1}^{0} E^\omega \eta_0(y) & \text{if } x < 0. \end{cases}$$

The claim now follows by applying [20, Theorem 4.1] and the characterization on p. 13 of [20]. ([20, Theorem 4.1] as reproduced from [2, Theorem 2.1] where the limiting stochastic heat equation is slightly altered because the process studied was $H_{\lfloor nt\rfloor}(\lfloor nvt\rfloor + \lfloor r\sqrt{n}\rfloor) - H_0(\lfloor r\sqrt{n}\rfloor)$). $\qquad\square$

Proof of Theorem 1.7 Now, we have $\beta = \alpha/(v(1-v)) = 4\alpha$. We will write $p^k_{x,y}$ for the k-step averaged transition. For $t \geqslant 0$ define

$$Y(t) = \sum_{x>0}\sum_{j=1}^{\eta_0(x)} \mathbf{1}\{X_{\lfloor t\rfloor}^{x,j} \leqslant \lfloor vt\rfloor\} - \sum_{x\leqslant 0}\sum_{j=1}^{\eta_0(x)} \mathbf{1}\{X_{\lfloor t\rfloor}^{x,j} > \lfloor vt\rfloor\}.$$

By stationarity

$$E^\omega Y_n(t,0) - E^\omega Y_n(s,0) = E^\omega Y(nt) - E^\omega Y(ns)$$

has the same distribution as the E^ω-mean of

$$Y' = \sum_{x>0}\sum_{j=1}^{\eta_0(x)} \mathbf{1}\{X_{\lfloor nt\rfloor-\lfloor ns\rfloor}^{x,j} \leqslant \lfloor nvt\rfloor - \lfloor nvs\rfloor\} - \sum_{x\leqslant 0}\sum_{j=1}^{\eta_0(x)} \mathbf{1}\{X_{\lfloor nt\rfloor-\lfloor ns\rfloor}^{x,j} > \lfloor nvt\rfloor - \lfloor nvs\rfloor\}.$$

The difference $|Y' - Y(\lfloor nt \rfloor - \lfloor ns \rfloor)|$ is bounded by the number of particles that are at time $\lfloor nt \rfloor - \lfloor ns \rfloor$ between $\lfloor nvt \rfloor - \lfloor nvs \rfloor$ and $\lfloor (\lfloor nt \rfloor - \lfloor ns \rfloor)v \rfloor$. Since $|\lfloor nvt \rfloor - \lfloor nvs \rfloor - \lfloor (\lfloor nt \rfloor - \lfloor ns \rfloor)v \rfloor| \leqslant 2$ we are talking about at most 5 sites and, consequently, $\mathbb{E}[|E^\omega Y' - E^\omega Y(\lfloor nt \rfloor - \lfloor ns \rfloor)|] \leqslant 5\mathbb{E}[f] = 5$. A similar reasoning gives a bound on $\mathbb{E}[|Y(nt) - Y(\lfloor nt \rfloor)|]$ and $\mathbb{E}[|Y(ns) - Y(\lfloor ns \rfloor)|]$. Therefore,

$$\lim_{n\to\infty} \frac{1}{\sqrt{n}}\mathbb{V}\mathrm{ar}\big(E^\omega Y(\lfloor nt \rfloor - \lfloor ns \rfloor)\big) = \lim_{n\to\infty} \frac{1}{\sqrt{n}}\mathbb{V}\mathrm{ar}\big(E^\omega Y_n(t, 0) - E^\omega Y_n(s, 0)\big)$$

$$= \lim_{n\to\infty} \frac{1}{\sqrt{n}}\Big[\mathbb{V}\mathrm{ar}\big(E^\omega Y(\lfloor nt \rfloor)\big) + \mathbb{V}\mathrm{ar}\big(E^\omega Y(\lfloor ns \rfloor)\big) - 2\mathbb{C}\mathrm{ov}\big(E^\omega Y_n(s, 0), E^\omega Y_n(t, 0)\big)\Big].$$

Hence, it is enough to prove that

$$\lim_{n\to\infty} \frac{1}{\sqrt{n}}\mathbb{V}\mathrm{ar}\big(E^\omega Y(n)\big) = \frac{1}{\sqrt{2\pi}}\big(\tfrac{1}{4}\alpha^{-1} - 1\big).$$

Since

$$E^\omega Y(2n + 1) - E^\omega Y(2n) = -f(T_{n,2n}\omega)\omega_{n,2n}$$

we see that it is enough to prove the above limit along the subsequence of even integers.

Let

$$h(\omega) = f(T_{1,0}\omega)\omega'_{1,0}\omega'_{1,1} - f(\omega)\omega_{0,0}\omega_{1,1}.$$

Then $\mathbb{E}(h) = 0$, $\mathbb{E}(h^2) = \frac{1}{8\alpha} - \frac{1}{2}$ (here we use Corollary 1.3 and $p_0 = p_1 = 1/2$), and

$$E^\omega Y(2n + 2) - E^\omega Y(2n) = h(T_{n,2n}\omega).$$

Let $c_0 = \mathbb{V}\mathrm{ar}(f) = \beta^{-1} - 1$. To compute $\mathbb{E}h(\omega)h(T_{n,2n}\omega)$ write

$$h(\omega) = (f(T_{1,0}\omega) - 1)\omega'_{1,0}\omega'_{1,1} - (f(\omega) - 1)\omega_{0,0}\omega_{1,1} + \omega'_{1,0}\omega'_{1,1} - \omega_{0,0}\omega_{1,1}$$

and

$$h(T_{n,2n}\omega) = \sum_{-n+1\leqslant x\leqslant n+1} (f(T_{x,0}\omega) - 1)\pi_{0,2n}(x, n + 1)\omega'_{n+1,2n}\omega'_{n+1,2n+1}$$

$$- \sum_{-n\leqslant y\leqslant n} (f(T_{y,0}\omega) - 1)\pi_{0,2n}(y, n)\omega_{n,2n}\omega_{n+1,2n+1}$$

$$+ \sum_{-n+1\leqslant x\leqslant n+1} \pi_{0,2n}(x, n + 1)\omega'_{n+1,2n}\omega'_{n+1,2n+1}$$

$$- \sum_{-n\leqslant y\leqslant n} \pi_{0,2n}(y, n)\omega_{n,2n}\omega_{n+1,2n+1}.$$

Due to $\mathfrak{S}_{-\infty,-1}$-measurability the f-terms are independent of the ω's. Also, distinct shifts are uncorrelated by Corollary 1.3. Multiplying these terms together and separating the expectations of the factors on levels $2n$ and $2n + 1$ leads to

$$\mathbb{E}h(\omega)h(T_{n,2n}\omega) = \tfrac{1}{4}c_0\mathbb{E}\pi_{0,2n}(1, n + 1)\omega'_{1,0}\omega'_{1,1} \tag{6.1}$$

$$- \tfrac{1}{4}c_0\mathbb{E}\pi_{0,2n}(0, n + 1)\omega_{0,0}\omega_{1,1} \tag{6.2}$$

$$- \tfrac{1}{4}c_0\mathbb{E}\pi_{0,2n}(1, n)\omega'_{1,0}\omega'_{1,1} \tag{6.3}$$

$$+ \tfrac{1}{4}c_0\mathbb{E}\pi_{0,2n}(0, n)\omega_{0,0}\omega_{1,1} \tag{6.4}$$

$$+ \tfrac{1}{4} \sum_{-n+1\leqslant x\leqslant n+1} \mathbb{E}\pi_{0,2n}(x, n + 1)\big(\omega'_{1,0}\omega'_{1,1} - \omega_{0,0}\omega_{1,1}\big) \tag{6.5}$$

$$- \tfrac{1}{4} \sum_{-n\leqslant y\leqslant n} \mathbb{E}\pi_{0,2n}(y, n)\big(\omega'_{1,0}\omega'_{1,1} - \omega_{0,0}\omega_{1,1}\big). \tag{6.6}$$

Thinking through the possible jumps shows that the terms in (6.5) and (6.6) survive only for $x, y \in \{0, 1\}$. And some of these terms can be combined with the ones above. This gives

$$\mathbb{E}h(\omega)h(T_{n,2n}\omega) = \tfrac{1}{4}(c_0 + 1)\mathbb{E}\pi_{0,2n}(1, n + 1)\omega'_{1,0}\omega'_{1,1} \tag{6.7}$$

$$- \tfrac{1}{4}(c_0 + 1)\mathbb{E}\pi_{0,2n}(0, n + 1)\omega_{0,0}\omega_{1,1} \tag{6.8}$$

$$- \tfrac{1}{4}(c_0 + 1)\mathbb{E}\pi_{0,2n}(1, n)\omega'_{1,0}\omega'_{1,1} \tag{6.9}$$

$$+ \tfrac{1}{4}(c_0 + 1)\mathbb{E}\pi_{0,2n}(0, n)\omega_{0,0}\omega_{1,1} \tag{6.10}$$

$$+ \tfrac{1}{4}\mathbb{E}\big[\pi_{0,2n}(0, n + 1)\omega'_{1,0}\omega'_{1,1} - \pi_{0,2n}(1, n + 1)\omega_{0,0}\omega_{1,1}\big] \tag{6.11}$$

$$- \tfrac{1}{4}\mathbb{E}\big[\pi_{0,2n}(0, n)\omega'_{1,0}\omega'_{1,1} - \pi_{0,2n}(1, n)\omega_{0,0}\omega_{1,1}\big]. \tag{6.12}$$

Now transform each term. For example, term (6.7) becomes

(6.7)

$$= \tfrac{1}{4}(c_0 + 1)\mathbb{E}\big[\big(\omega'_{1,0}\omega'_{1,1}p^{2n-2}_{1,n+1} + \omega'_{1,0}\omega_{1,1}p^{2n-2}_{2,n+1} + \omega_{1,0}\omega'_{2,1}p^{2n-2}_{2,n+1} + \omega_{1,0}\omega_{1,1}p^{2n-2}_{3,n+1}\big)\omega'_{1,0}\omega'_{1,1}\big]$$

$$= \tfrac{1}{4}(c_0 + 1)\big\{(\tfrac{1}{2} - \alpha)^2 p^{2n-2}_{1,n+1} + (\tfrac{3}{4}\alpha - \alpha^2)p^{2n-2}_{2,n+1} + \tfrac{1}{4}\alpha p^{2n-2}_{3,n+1}\big\}.$$

After these steps we get

$$\mathbb{E}h(\omega)h(T_{n,2n}\omega)$$

$$= \tfrac{1}{4}(c_0 + 1)\Big[-\tfrac{1}{4}\alpha(p^{2n-2}_{0,n+1} + p^{2n-2}_{0,n-3})$$

$$+ (2\alpha^2 - \tfrac{3}{2}\alpha + \tfrac{1}{4})(p^{2n-2}_{0,n} + p^{2n-2}_{0,n-2}) + (-4\alpha^2 + \tfrac{7}{2}\alpha - \tfrac{1}{2})p^{2n-2}_{0,n-1}\Big]$$

$$+ \tfrac{1}{4}\Big[\tfrac{1}{16}(p^{2n-2}_{0,n+1} + p^{2n-2}_{0,n-3}) + (\tfrac{1}{8} - \tfrac{1}{2}\alpha)(p^{2n-2}_{0,n} + p^{2n-2}_{0,n-2}) + (\alpha - \tfrac{3}{8})p^{2n-2}_{0,n-1}\Big].$$

Letting X_k denote the (averaged) Markov chain with transition $p_{x,y}$, introduce $Z_k = X_{2k} - k$ with transition $r_{0,0} = 1/2$, $r_{0,\pm 1} = 1/4$. For higher order transitions $p_{0,n+i}^{2n-2} = r_{i+1}^{n-1}$. Replace the p's with r's and combine them using symmetry: $r_2^{n-1} = r_{-2}^{n-1}$, etc. Then

$$
\begin{aligned}
\mathbb{E}h(\omega)h(T_{n,2n}\omega) &= \tfrac{1}{4}(c_0 + 1)\Big[\tfrac{1}{2}\alpha(r_0^{n-1} - r_2^{n-1}) - (4\alpha^2 - 3\alpha + \tfrac{1}{2})(r_0^{n-1} - r_1^{n-1})\Big] \\
&\quad + \tfrac{1}{4}\Big[-\tfrac{1}{8}(r_0^{n-1} - r_2^{n-1}) + (\alpha - \tfrac{1}{4})(r_0^{n-1} - r_1^{n-1})\Big] \\
&= \tfrac{1}{4}(\tfrac{1}{2} - \tfrac{1}{8}\alpha^{-1})(r_0^{n-1} - r_1^{n-1})
\end{aligned}
\tag{6.13}
$$

where in the last step we used $c_0 + 1 = (4\alpha)^{-1}$.

Use the potential kernel a^Z of the r-walk: the variance is $1/2$ so $a^Z(x) = 2|x|$. From [22] and symmetry, $a^Z(x) = \lim_{m\to\infty} a_m^Z(x)$ with

$$
a_m^Z(x) = \sum_{k=0}^{m}(r_{0,0}^k - r_{x,0}^k) = \sum_{k=0}^{m}(r_0^k - r_x^k).
\tag{6.14}
$$

Then

$$
\begin{aligned}
\mathrm{Var}\big(E^\omega Y(2n)\big) &= \mathrm{Var}\Bigg[\sum_{k=0}^{n-1} h(T_{k,2k}\omega)\Bigg] \\
&= n\mathbb{E}(h^2) + 2\sum_{k=1}^{n-1}(n-k)\mathbb{E}h(\omega)h(T_{k,2k}\omega) \\
&= \big(\tfrac{1}{16}\alpha^{-1} - \tfrac{1}{4}\big)\Bigg[2n - \sum_{k=1}^{n-1}(n-k)(r_0^{k-1} - r_1^{k-1})\Bigg] \\
&= \big(\tfrac{1}{16}\alpha^{-1} - \tfrac{1}{4}\big)\Bigg[2n - \sum_{j=1}^{n-1}\sum_{k=1}^{j}(r_0^{k-1} - r_1^{k-1})\Bigg] \\
&= \big(\tfrac{1}{16}\alpha^{-1} - \tfrac{1}{4}\big)\Bigg[na^Z(1) - \sum_{j=1}^{n-1}a_{j-1}^Z(1)\Bigg]\Bigg] \\
&= \big(\tfrac{1}{16}\alpha^{-1} - \tfrac{1}{4}\big)\Bigg[a^Z(1) + \sum_{j=1}^{n-1}(a^Z(1) - a_{j-1}^Z(1))\Bigg].
\end{aligned}
\tag{6.15}
$$

Let us look at $a_k^Z(x)$. The characteristic function is

$$
\zeta(\theta) = \sum_x r_x e^{ix\theta} = \tfrac{1}{2}(1 + \cos\theta).
$$

By symmetry:

$$a_m^Z(x) = \sum_{k=0}^{m}(r_0^k - r_x^k) = \sum_{k=0}^{m}\frac{1}{2\pi}\int_{-\pi}^{\pi}\left(\zeta^k(\theta) - e^{-ix\theta}\zeta^k(\theta)\right)d\theta$$

$$= \frac{1}{2\pi}\int_{-\pi}^{\pi}\frac{1-\cos x\theta}{1-\zeta(\theta)}\left(1 - \zeta^{m+1}(\theta)\right)d\theta$$

$$= \frac{2}{2\pi}\int_{-\pi}^{\pi}\frac{1-\cos x\theta}{1-\cos\theta}\left(1 - \zeta^{m+1}(\theta)\right)d\theta$$

$$= 2|x| - \frac{1}{\pi}\int_{-\pi}^{\pi}\frac{1-\cos x\theta}{1-\cos\theta}\left(\frac{1+\cos\theta}{2}\right)^{m+1}d\theta.$$

The value of the first integral above is on p. 61 in [22]. But actually we only need $x = 1$:

$$a_m^Z(1) = a^Z(1) - \frac{1}{\pi}\int_{-\pi}^{\pi}\left(\frac{1+\cos\theta}{2}\right)^{m+1}d\theta = a^Z(1) - \frac{2}{\pi}\int_{-\pi/2}^{\pi/2}\cos^{2m+2}x\,dx$$

$$= a^Z(1) - 2\prod_{\ell=1}^{m+1}\left(1 - \frac{1}{2\ell}\right) = a^Z(1) - \frac{2}{(m+1)!}\prod_{\ell=1}^{m+1}(\ell - \tfrac{1}{2}).$$

Put this back into (6.15):

$$\mathbb{Var}\left(E^\omega Y(2n)\right) = \left(\tfrac{1}{16}\alpha^{-1} - \tfrac{1}{4}\right)\left[a^Z(1) + \sum_{j=1}^{n-1}\frac{2}{j!}\prod_{\ell=1}^{j}(\ell - \tfrac{1}{2})\right]$$

$$= \left(\tfrac{1}{8}\alpha^{-1} - \tfrac{1}{2}\right)\left[1 + \sum_{j=1}^{n-1}\frac{j^{-1/2}}{\Gamma_j(-1/2)\cdot(-1/2)}\right].$$

Above we used the definition

$$\Gamma_m(x) = \frac{m!\,m^x}{x(x+1)\cdots(x+m)}.$$

According to p. 461 of [23], $\Gamma_m(x) \to \Gamma(x)$ for $x \notin \mathbb{Z}_-$. Plugging back into the above:

$$\mathbb{Var}\left(E^\omega Y(2n)\right) = \left(\tfrac{1}{8}\alpha^{-1} - \tfrac{1}{2}\right)4\sqrt{n}\left[\frac{1}{4\sqrt{n}} + \frac{1}{2\sqrt{n}}\sum_{j=1}^{n-1}\frac{j^{-1/2}}{-\Gamma_j(-1/2)}\right]$$

$$\sim \sqrt{2n}\cdot\sqrt{2}(\tfrac{1}{4}\alpha^{-1} - 1)\cdot\frac{1}{-\Gamma(-1/2)}$$

$$= \sqrt{2n}\cdot\sqrt{2}(\tfrac{1}{4}\alpha^{-1} - 1)\cdot\frac{1}{2\sqrt{\pi}}.$$

The theorem is proved. □

We close this section with a remark regarding the expected limit of the quenched mean process in the stationary case.

In the setting of Theorem 1.7 the random average process H from the proof of Theorem 1.6 has the initial profile

$$H_0(x) = \begin{cases} 0 & \text{if } x = 0, \\ \displaystyle\sum_{y=1}^{x} f(T_{y,0}\omega) & \text{if } x > 0, \quad \text{and} \\ \displaystyle\sum_{y=x+1}^{0} E^{\omega} f(T_{y,0}\omega) & \text{if } x < 0. \end{cases}$$

Thus, to extend the convergence result of Theorem 1.6 to include the stationary case we need to prove a functional central limit theorem for the partial sums $\sum_{y=1}^{x} f(T_{y,0}\omega)$.

Finally, note that if indeed the claim of Theorem 1.6 holds in the stationary setting of Theorem 1.7, then we would have

$$\sigma_0^2 = \mathbb{V}\text{ar}(f) = \frac{1}{\beta} - 1 = \frac{1}{4\alpha} - 1 = \frac{1/4 - \alpha}{\alpha} = \frac{\rho_0^2 \sigma_D^2}{\alpha}.$$

(Recall that we assumed the mean $\rho_0 = 1$ in Theorem 1.7). Thus one would have

$$\begin{aligned} \mathbf{E}[z(s,0)z(t,0)] &= \frac{\rho_0^2 \sigma_D^2}{\alpha} \frac{\sigma}{\sqrt{2\pi}} (\sqrt{t} + \sqrt{s} - \sqrt{|t-s|}) \\ &= \frac{1}{2\sqrt{2\pi}} (\tfrac{1}{4}\alpha^{-1} - 1)(\sqrt{t} + \sqrt{s} - \sqrt{|t-s|}), \end{aligned}$$

as stated in Theorem 1.7.

References

1. Andjel, E.D.: Invariant measures for the zero range processes. Ann. Probab. **10**(3), 525–547 (1982)
2. Balázs, M., Rassoul-Agha, F., Seppäläinen, T.: The random average process and random walk in a space-time random environment in one dimension. Comm. Math. Phys. **266**, 499–545 (2006)
3. Barabási, A.L., Stanley, H.E.: Fractal Concepts in Surface Growth. Cambridge University Press (1995)
4. Corwin, I.: The Kardar-Parisi-Zhang equation and universality class. Random Matrices Theor. Appl. **1**(1), 1130001 (2012)
5. Doob, J.L.: Stochastic Processes. Wiley Classics Library. John Wiley & Sons Inc., New York. Reprint of the 1953 original, A Wiley-Interscience Publication (1990)
6. Durrett, R.: Duxbury advanced series. In: Probability: Theory and Examples, 3rd edn. Brooks/Cole-Thomson, Belmont, CA (2004)

7. Ekhaus, M., Gray, L.: Convergence to equilibrium and a strong law for the motion of restricted interfaces. Unpublished manuscript (1994)
8. Ferrari, P.A., Fontes, L.R.G.: Fluctuations of a surface submitted to a random average process. Electron. J. Probab. **3**(6), 34 (electronic) (1998)
9. Jara, M., Peterson, J.: Hydrodynamic limit for a system of independent, sub-ballistic random walks in a common random environment. Ann. Inst. Henri Poincaré Probab. Stat. **53**(4), 1747–1792 (2017)
10. Kumar, R.: Space-time current process for independent random walks in one dimension. ALEA Lat. Am. J. Probab. Math. Stat. **4**, 307–336 (2008)
11. Liggett, Thomas M.: Interacting particle systems. In: Grundlehren der Mathematischen Wissenschaften [Fundamental Principles of Mathematical Sciences], vol. 276. Springer, New York (1985)
12. James, R.: Munkres, 2nd edn. Topology. Prentice-Hall Inc, Upper Saddle River, N.J. (2000)
13. Norris, J.R.: Markov chains. In: Cambridge Series in Statistical and Probabilistic Mathematics, vol. 2. Cambridge University Press, Cambridge. Reprint of 1997 original (1998)
14. Peterson, J.: Systems of one-dimensional random walks in a common random environment. Electron. J. Probab. **15**(32), 1024–1040 (2010)
15. Peterson, J., Seppäläinen, T.: Current fluctuations of a system of one-dimensional random walks in random environment. Ann. Probab. **38**(6), 2258–2294 (2010)
16. Rassoul-Agha, F., Seppäläinen, T.: An almost sure invariance principle for random walks in a space-time random environment. Probab. Theory Related Fields **133**(3), 299–314 (2005)
17. Rosenblatt, M.: Markov Processes, Structure and Asymptotic Behavior. Springer, New York. Die Grundlehren der mathematischen Wissenschaften, Band 184 (1971)
18. Seppäläinen, T.: Translation invariant exclusion processes. Lecture notes available at: http://www.math.wisc.edu/~seppalai/excl-book/etusivu.html
19. Seppäläinen, T.: Second-order fluctuations and current across characteristic for a one-dimensional growth model of independent random walks. Ann. Probab. **33**(2), 759–797 (2005)
20. Seppäläinen, T.: Current fluctuations for stochastic particle systems with drift in one spatial dimension. Ensaios Matemáticos [Mathematical Surveys], vol. 18. Sociedade Brasileira de Matemática, Rio de Janeiro (2010)
21. Sethuraman, S.: On extremal measures for conservative particle systems. Ann. Inst. H. Poincaré Probab. Statist. **37**(2), 139–154 (2001)
22. Spitzer, F.: Graduate texts in mathematics. In: Principles of Random Walks, vol. 34, 2nd edn. Springer, New York (1976)
23. Stromberg, K.R.: Introduction to Classical Real Analysis. Wadsworth International, Belmont, Calif. Wadsworth International Mathematics Series (1981)
24. Varadhan, S.R.S.: Probability Theory. Courant Lecture Notes in Mathematics, vol. 7. New York University Courant Institute of Mathematical Sciences, New York (2001)
25. Walsh, J.B.: An introduction to stochastic partial differential equations. In: École d'été de probabilités de Saint-Flour, XIV—1984, vol. 1180 of Lecture Notes in Math., pp. 265–439. Springer, Berlin (1986)
26. Wheeden, R.L., Zygmund, A.: Pure and applied mathematics. In: Measure and Integral: An Introduction to Real Analysis, vol. 43. Marcel Dekker Inc., New York (1977)
27. Zygmund, A.: Trigonometric Series. vol. I, II. Cambridge Mathematical Library. Cambridge University Press, Cambridge. Reprint of the 1979 edition (1988)

Stable Limit Laws for Reaction-Diffusion in Random Environment

Gérard Ben Arous, Stanislav Molchanov and Alejandro F. Ramírez

Abstract We prove the emergence of stable fluctuations for reaction-diffusion in random environment with Weibull tails. This completes our work around the quenched to annealed transition phenomenon in this context of reaction diffusion. In Ben Arous et al (Transition asymptotics for reaction-diffusion in random media. Probability and mathematical physics, American Mathematical Society, Providence, RI, pp 1–40, 2007, [8]), we had already considered the model treated here and had studied fully the regimes where the law of large numbers is satisfied and where the fluctuations are Gaussian, but we had left open the regime of stable fluctuations. Our work is based on a spectral approach centered on the classical theory of rank-one perturbations. It illustrates the gradual emergence of the role of the higher peaks of the environments. This approach also allows us to give the delicate exact asymptotics of the normalizing constants needed in the stable limit law.

Gérard Ben Arous: Partially supported by NSF DMS1209165 and BSF 2014019.
Alejandro F. Ramírez: Partially supported by Fondo Nacional de Desarrollo Científico y Tecnológico 1141094 and 1180259 and Iniciativa Científica Milenio.

G. Ben Arous
Courant Institute of Mathematical Sciences, New York University,
New York, NY 10012, USA
e-mail: benarous@cims.nyu.edu

S. Molchanov
Department of Mathematics, University of North Carolina-Charlotte,
376 Fretwell Bldg. 9201 University City Blvd., Charlotte, NC 28223-0001, USA
e-mail: smolchan@math.uncc.edu

S. Molchanov
National Research University Higher School of Economics, Myasnitskaya ul., 20, Moscow,
Russia 101000

A. F. Ramírez (✉)
Facultad de Matemáticas, Pontificia Universidad Católica de Chile,
Santiago 7820436, Chile
e-mail: aramirez@mat.puc.cl
URL: http://www.mat.puc.cl/~aramirez

© Springer Nature Switzerland AG 2019
P. Friz et al. (eds.), *Probability and Analysis in Interacting Physical Systems*, Springer Proceedings in Mathematics & Statistics 283,
https://doi.org/10.1007/978-3-030-15338-0_5

Keywords Rank one perturbation theory · Random walk · Principal eigenvalue · Stable distributions

AMS 2010 Subject Classifications Primary 82B41 · 82B44 · Secondary 60J80 · 82C22

1 Introduction

We establish the emergence of stable fluctuations for branching random walks in a random environment. More precisely, we consider branching random walks on the lattice $\mathbb{Z}^d$, and denote by $v(x)$ the rate of branching at site $x \in \mathbb{Z}^d$. We assume that the branching is binary (each particle gives birth to two offsprings) and that the rates $(v(x))_{x \in \mathbb{Z}^d}$ are i.i.d random variables, which we call here the random environment.

We are interested in the spatial fluctuations of the number $N_x(t)$ of particles at time $t > 0$, whose ancestor at time 0 was at site $x \in \mathbb{Z}^d$, or rather in the behavior of its mean $m(t, x)$, as a function of the "quenched" random environment.

This question can be formulated as an equivalent problem about reaction-diffusion in random environment, since the random function $m(x, t)$ is the solution of the reaction-diffusion equation

$$\frac{\partial m(x, t)}{\partial t} = \kappa \Delta m(x, t) + v(x)m(x, t), \qquad t \geq 0, x \in \mathbb{Z}^d, \tag{1}$$

with initial condition,

$$m(x, 0) = 1, \qquad x \in \mathbb{Z}^d, \tag{2}$$

where Δ is the discrete Laplacian on $\mathbb{Z}^d$, defined for every function $f \in l^1(\mathbb{Z}^d)$ as

$$\Delta f(x) := \sum_{e \in E} (f(x + e) - f(x)),$$

where

$$E := \{e \in \mathbb{Z}^d : |e|_1 = 1\} \tag{3}$$

and $|\cdot|_1$ is the $l^1(\mathbb{Z}^d)$ norm. Equation (1) is also known as the parabolic Anderson model (a detailed survey of this equation can be found in König [13]).

Following [8], we will study the mean number of particles at time t, if one starts with a particle at time 0, whose position is picked uniformly at random in the box Λ_L of size L. More precisely we will consider the spatial average

$$m_L(t) = \frac{1}{|\Lambda_L|} \sum_{x \in \Lambda_L} m(x, t) \tag{4}$$

This quantity exhibits a very rich dynamical transition, when t increases, as a function of L. This transition was introduced a decade ago, as a mechanism for the "transition from annealed to quenched asymptotics" for Markovian dynamics, in [7] in the context of random walks on random obstacles, and in [8] in the current context of reaction-diffusion in random environment. The basic intuition behind this rich picture can in fact be understood in the much simpler context of sums of i.i.d random exponentials, as was done in [6]. In this simple context of i.i.d random variables, the transition boils down to the graduate emergence of the role of extreme values, which gradually impose a breakdown of the CLT first and eventually of the LLN, and induce stable fluctuations.

In the present context of branching random walks in random environment, we have proved in [8], that the extreme values of the random environment play a similar role, and established the precise breakdown of the Central Limit Theorem and of the Law of Large Numbers. But we had left open the much more delicate question of stable fluctuations. This is the purpose of this work.

Our main result in this work establishes that, when the tails of the branching rate are Weibull distributed (as in [8]) and in the regime where the Central Limit Theorem fails, $m_L(t)$, once properly centered and scaled, converges to a stable distribution. Before establishing these stable limit laws, one major difficulty is to understand the needed exact asymptotics of the centering and scaling constants, as functions of t. This asymptotic behavior is unusually delicate, as we will see below. The validity of these stable limits was until now only proved in two much simpler situations: either the simple case treated in [6] of sums of i.i.d random variables, or in the one dimensional case of random walks among random obstacles treated in [7]. In a future work, we plan to extend our results to branching rates with double exponentially decaying tails.

Our result was also claimed in the recent work of Gartner and Schnitzler in [11]. Nevertheless we believe that the proof given in [11] is incomplete. In fact the statement of Theorem 1 in [11], giving the stable limit law, is ambiguous. The problem lays with the understanding of the normalizing function. In fact, this is one of the crucial places where in this article we use decisively the rank-one perturbation theory. We believe that the estimate of the normalizing function in [11] (called here $e^{th(t)}$ and there $B_\alpha(t)$) is flawed (for more details see Appendix B).

The main tool here is naturally based on spectral theory, and more precisely on the classical theory of rank-one perturbations for self-adjoint operators, which we recall briefly in Appendix A. Using rank-one perturbation theory in this context of random media is quite natural, as for instance in [9] (Biskup-Konig), where the case of faster tails (doubly exponential) is studied in great depth for the parabolic Anderson model. More broadly, spectral tools have been pushed quite far for the understanding of the extreme values of the spectrum of the Anderson operator in the beautiful series of work [1–5] (Astrauskas). In particular, in [5], a general survey of asymptotic geometric properties of extreme order statistics of an i.i.d. random field as the volume of the set of indexes increases is given, and several applications to the parabolic Anderson model are given. Nevertheless, the implications of this analysis to the convergence to stable distributions within the quenched-annealed transition mechanism studied here is not explicitly discussed in [5].

Let us now describe more precisely the content of this article. We state precisely our results in Sect. 2. We begin by giving our notations, and then state, our results for the exact behavior of the centering constant in Theorem 1. We then state our main result Theorem 3, stating the convergence to a stable distribution for $m_L(t)$, once properly centered and scaled, and then in Theorem 2 we display the structure of the scaling function with the gradual emergence of a transition mechanism as the parameter of the Weibull distribution of the potential increases. In Sect. 3, we establish the needed spectral results about our random Schrödinger operator, using the rank-one perturbation theory recalled in Appendix A. In Sect. 4, we recall the basic facts of extreme value theory for i.i.d. Weibull distributions. It should be stressed that most the analysis of Sect. 3 is a well-known consequence of rank-one perturbation theory, while also Sect. 4 is standard extreme value analysis of random variable, but in order to be self-contained we have decided to include it here. In Sect. 5, we prove the existence of the scaling function of Theorems 3 and 2. In Sect. 6, we prove Theorem 3. In Sect. 7 we prove Theorem 3, giving the precise asymptotics of the annealed moments. In Appendix B, we describe the main passages of the article [11] where there is some mathematical issue.

2 Notations and Results

Let $v := (v(x))_{x \in \mathbb{Z}^d}$ where $v(x) \in [0, \infty)$. Consider the space $W := [0, \infty)^{\mathbb{Z}^d}$, endowed with its natural σ-algebra. Let μ be the probability measure on W such that the coordinates of v are i.i.d. Let us now consider a simple symmetric random walk of rate $\kappa > 0$, which branches at a site x at rate $v(x)$ giving birth to two particles. Let us call $m(x, t, v)$ the expectation of the total number of random walks at time t given that initially there was only one random walk at site x, in the environment v. We will frequently write $m(x, t)$ instead of $m(x, t, v)$. In Proposition 2.2 of [8] (see also Theorem 2.1 of [10]) it is shown that whenever

$$\left\langle \left(\frac{v_+(0)}{\log_+ v(0)} \right)^d \right\rangle < \infty, \tag{5}$$

then μ-a.s. for all $x \in \mathbb{Z}^d$ and $t \geq 0$, $m(x, t)$ is finite and it satisfies the parabolic Anderson equations (1) and (2).

Throughout, we will call v the *potential* of the Eq. (1) and we will assume that it has a Weibull law of parameter $\rho > 1$ so that

$$\mu(v(0) > y) = \exp\left\{ -\frac{y^\rho}{\rho} \right\} \qquad \text{for } y \geq 0. \tag{6}$$

For any function F of the potential, we will use the notations $\langle F \rangle := \int F d\mu$, and $Var_\mu(F) := \langle (F - \langle F \rangle)^2 \rangle$ whenever they are well defined. Let us also introduce the *conjugate exponent* ρ' of $\rho > 1$, defined by the equation

$$\frac{1}{\rho'} + \frac{1}{\rho} = 1.$$

Our first result gives the precise asymptotic behavior of the average of the expectation of the total number of random walks.

Theorem 1 *Consider the solution of (1) and (2). Then,*

$$\langle m(0, t) \rangle \sim \left(\frac{\pi}{\rho - 1} \right)^{1/2} t^{1 - \frac{\rho'}{2}} e^{\frac{t^{\rho'}}{\rho'} - 2d(\kappa t - t^{2 - \rho'})}. \tag{7}$$

The average $\langle m(0, t) \rangle$ of Theorem 1 will turn out to be the adequate centering of the empirical average of the field $(m(x, t)_{x \in \mathbb{Z}^d}$ in certain regimes, leading to the appearance of stable laws. To state the corresponding results we still need to introduce additional notation.

Consider on $\mathbb{Z}^d$ the norm $||x|| := \sup\{|x_i| : 1 \leq i \leq d\}$. For each $r \geq 0$ and $x \in \mathbb{Z}^d$ consider the subset $\Lambda(x, r) := \{y \in \mathbb{Z}^d : ||y|| \leq r\}$. For $L > 0$, we will use the notation Λ_L instead of $\Lambda(0, L)$. We now define the averaged first moment at scale L as

$$m_L(t) := \frac{1}{|\Lambda_L|} \sum_{x \in \Lambda_L} m(x, t).$$

We will say that a sequence $\gamma = (z_1, z_2, \ldots, z_p)$ in $\mathbb{Z}^d$ is a *path* if for each $1 \leq j \leq p - 1$, the sites z_j and z_{j+1} are nearest neighbors. The *length* of a path γ, denoted by $|\gamma|$, is equal to the number p of sites defining it. Furthermore, we will say that γ connects sites x and y if $z_1 = x$ and $z_p = y$. Denote the set of paths contained in a set $U \subset \mathbb{Z}^d$ and connecting x to y, by $\mathbb{P}_U(x, y)$; the set of paths starting from x and contained in U by $\mathbb{P}_U(x)$. For each $n \geq 1$, define $\mathbb{P}_U^n(x, y) := \{\gamma \in \mathbb{P}_U(x, y) : |\gamma| = n\}$. Furthermore, to each path $\gamma \in \mathbb{P}_U^n(0)$ we can associate a set $\{y_1, \ldots, y_k\}$, which represents the different sites visited by the path, and for $j \in \mathbb{N}$ a set $\{n_1, \ldots, n_j\}$ such that $\sum_{i=1}^k n_i = n$, and n_i represents the number of times the site y_i was visited by γ. Finally, whenever $U = \mathbb{Z}^d$ we will use the notation $\mathbb{P}(x)$, $\mathbb{P}(x, y)$, $\mathbb{P}^n(0)$, instead of $\mathbb{P}_U(x)$, $\mathbb{P}_U(x, y)$, and $\mathbb{P}^n(0)$, respectively. For each $v \in W$ and natural N, consider the function defined for $s \geq 0$ as

$$B_N(s, v) := \frac{s + 2d\kappa}{1 + \sum_{j=1}^N \sum_{\gamma \in \mathbb{P}^{2j+1}(0,0)} \prod_{z \in \gamma, z \neq z_1} \frac{\kappa}{2d\kappa + (s - v_0(z))_+}}, \tag{8}$$

where $v_0(z) := v(z)$ for $z \neq 0$ while $v_0(0) = 0$. Also, define the constants

$$M := \min\{j \geq 1 : 2j\rho' > 2(j+1)\}, \tag{9}$$

$$B_N := \frac{2d\kappa}{1 + \sum_{j=1}^{N} \sum_{\gamma \in \mathbb{P}^{2j+1}(0,0)} \left(\frac{1}{2d}\right)^{2j}}, \tag{10}$$

$$\mathcal{A}_0 = \mathcal{A}_0(\gamma, \rho) := \left(\frac{\gamma\rho}{\rho'}\right)^{1/\rho},$$

$$\alpha = \alpha(\gamma, \rho) := \left(\frac{\gamma\rho}{\rho'}\right)^{1/\rho'} \tag{11}$$

and

$$\gamma_1 := \frac{\rho'}{\rho} \quad \text{and} \quad \gamma_2 := \frac{\rho'}{\rho} 2^{1/\rho'}. \tag{12}$$

We also will need to define for each natural N the function $\zeta_N : [0, \infty) \to \left(0, e^{-\frac{1}{\rho} B_N^\rho}\right)$ by

$$\zeta_N(s) := E_\mu \left[e^{-\frac{1}{\rho} B_N(s, v)^\rho}\right], \tag{13}$$

where E_μ is the expectation with respect to μ.

Throughout, for $\alpha \in (0, 2)$ we will call S_α the distribution of the totally asymmetric stable law of exponent α (and skewness parameter 1) with characteristic function

$$\phi_\alpha(u) := \begin{cases} \exp\left\{-\Gamma(1-\alpha)|u|^\alpha e^{-\frac{i\pi\alpha}{2} \text{sgn}(u)}\right\} & \text{if } 0 < \alpha < 1 \\ \exp\left\{iu(1-\bar{\gamma}) - \frac{\pi}{2}|u|(1 + i \text{ sgn}(u) \cdot \frac{2}{\pi} \log|u|)\right\} & \text{if } \alpha = 1 \\ \exp\left\{\frac{\Gamma(2-\alpha)}{\alpha-1}|u|^\alpha e^{-\frac{i\pi\alpha}{2} \text{sgn}(u)}\right\} & \text{if } 1 < \alpha < 2, \end{cases} \tag{14}$$

where $\Gamma(s) := \int_0^\infty x^{s-1}e^{-x}dx$ is the gamma function, $\text{sgn}(u) := \frac{u}{|u|}$ for $u \neq 0$ and $\text{sgn}(u) := 0$ for $u = 0$ and $\gamma = 0.5772...$ is the Euler constant.

Our second result establishes the existence of a function h, which we will call the *scaling function* and which provides the adequate scaling factor $e^{th(t)}$ which gives the limiting stable laws. To state it, given an increasing function $L : [0, \infty) \to [1, \infty)$, we define the function $\tau : [0, \infty) \to \mathbb{R}$ by

$$\tau(t) := \frac{d\rho'}{\gamma} \log L(t) \quad \text{for} \quad t \geq 0. \tag{15}$$

Theorem 2 *Consider a potential v having a Weibull distribution of parameter $\rho > 1$. Consider an increasing function $L(t)$ with $\tau(t)$ defined in (15) and such that*

$$\tau(t) \sim t. \tag{16}$$

Then, there exists a unique function $h(t) : [t_0, \infty) \to [0, \infty)$ where t_0 is defined by $\frac{1}{L(t_0)^d} = e^{-\frac{1}{\rho}B_M^\rho}$ [c.f. (10)] and M in (9), which satisfies the equation

$$\zeta_M(h(t)) = \frac{1}{L(t)^d} \tag{17}$$

and such that

$$\lim_{t \to \infty} \frac{m_L(t) - A(t)}{e^{th(t)}|\Lambda_L|^{-1}} = S_\alpha,$$

where

$$A(t) := \begin{cases} 0 & \text{if } 0 < \gamma < \gamma_1 \\ \left\langle m(0, t) \mathbb{1}_{\left\{\sum_{x \in \Lambda_t} m(x,t) \leq e^{th(t)}\right\}}\right\rangle & \text{if } \gamma = \gamma_1 \\ \langle m(0, t)\rangle & \text{if } \gamma_1 < \gamma < \gamma_2, \end{cases} \tag{18}$$

and where the convergence is in distribution and $\alpha = \alpha(\gamma, \rho)$ is given by (11).

Our third result gives precise information about the structure of the scaling function h which satisfies (17) of Theorem 2.

Theorem 3 *Let $0 < \gamma < \gamma_2$, L an increasing function with $\tau(t)$ defined in (15) and such that $\lim_{t \to \infty} L(t) = \infty$. Then, Eq. (17) has a unique solution h which which admits the expansion*

$$h(t) = A_0 \tau^{\rho'-1}(t) - 2d\kappa + \sum_{1 \leq j \leq M} h_j(t) + O\left(\frac{1}{t^{(2M+1)\frac{\rho'}{\rho}}}\right), \tag{19}$$

where $h_j(t)$, $1 \leq j \leq M$ are functions recursively defined as

$$h_0(t) := A_0 \tau^{\rho'-1}(t) - 2d\kappa, \tag{20}$$

and for $1 \leq j \leq M$

$$h_j(t) := \frac{1}{A_0^{\rho-1}}\left(\frac{\gamma}{\rho'}\tau^{\rho'-1}(t) + \frac{1}{\tau(t)}\log E_\mu\left[e^{-\frac{1}{\rho}B_j(h_0(t)+\cdots+h_{j-1}(t),v)^\rho}\right]\right) \tag{21}$$

and

$$h_j(t) = O\left(\frac{1}{t^{(2j-1)\frac{\rho'}{\rho}}}\right). \tag{22}$$

An immediate consequence of Theorem 3, is the appearance of a transition mechanism in the asymptotic behavior of the quenched-annealed transition, where as the value of ρ increases in $(1, \infty)$, the number of terms in the asymptotic expansion of the scaling function h which are relevant also increase. for even values of ρ. Indeed, we have the following corollary of Theorem 3, which also shows that the first terms in the expansion of the scaling function can be computed explicitly.

Corollary 1 *Under the assumptions of Theorem 2, for $0 < \gamma < \gamma_2$, the scaling function $h(t)$ defined in (17) admits the expansion:*

(i) for $1 < \rho < 2$ $h(t) = \mathcal{A}_0 \tau^{\rho'-1}(t) - 2d\kappa$,
(ii) for $2 \leq \rho < 3$, $h(t) = \mathcal{A}_0 \tau^{\rho'-1}(t) - 2d\kappa + 2d\kappa^2 \mathcal{A}_0^3 \tau^{1-\rho'}(t)$ and
(iii) for $3 \leq \rho < \frac{3+\sqrt{17}}{2}$,

$$h(t) = \mathcal{A}_0 \tau^{\rho'-1}(t) - 2d\kappa + K_2 \tau^{1-\rho'}(t) + K_3 \tau^{\frac{\rho}{\rho-1}(3-2\rho')-1}(t) + K_4 \frac{1}{\tau(t)}$$

$$+ K_5 \log \tau(t) + o\left(\frac{1}{t}\right),$$

where

$$K_2 := \mathcal{A}_0^{2-\rho} K_1, \quad K_3 := \frac{K_1 (\mathcal{A}_0 K_1)^{\frac{1}{\rho-1}}}{\mathcal{A}_0^{\rho-1}} \left(1 - \frac{1}{\rho} \mathcal{A}_0\right),$$

$$K_4 := (3 - 2\rho') \log \mathcal{A}_0 + \frac{1}{2} \log \pi(\rho - 1) - \mathcal{A}_0^{3-\rho} K_1^2 \quad \text{and} \quad K_5 := \frac{\rho}{2(\rho - 1)}.$$

For ρ larger than $(3 + \sqrt{17})/2$ extra terms in the expansion of h have to be computed in Corollary 1. In order to keep the length of the computations limited, we have decided to stop there. There seems to be no straightforward interpretation on the appearance of this number.

The proof of Theorem 3 is based on rank-one perturbation methods to obtain asymptotic expansions of the largest eigenvalues of the Laplacian operator in a potential having high peaks.

Throughout this article, a constant C will always denote a non-random number, independent of time. We will use the letters C or $C_1, C_2, \ldots$ to denote them. We will use the notation $O(s) : [0, \infty) \to [0, \infty)$ to denote a function (possibly random) which satisfies for all $s \geq 0$,

$$|O(s)| \leq Cs$$

for some constant C.

3 Rank-One Perturbation for Schrödinger Operators

Here we will apply perturbation theory to study the asymptotic behavior of the principal Dirichlet eigenvalue and eigenfunction of the Laplacian operator plus a potential with a rank-one perturbation. The results we will present are deterministic, in the sense that we do not assume that the potential is random, and are not particularly original, since they correspond to a standard application of rank-one perturbation theory (see for example [15]). With the aim of giving a self-contained presentation, the basic tools of rank-one perturbation theory that will be used are presented in Appendix A. In Sect. 3.1, we give a precise statement about the principal Dirichlet eigenvalue and eigenfunction of the perturbed operator in Theorem 4. In Sect. 3.2 we derive some estimates and formulas about the spectrum of the unperturbed Schrödinger operator and its Green function. In Sect. 3.3 we relate the results of Sect. 3.2 to the spectrum of the perturbed Schrödinger operator through the rank-one perturbation theory presented in Appendix A, to prove Theorem 3. Throughout we assume that U is a finite connected subset of $\mathbb{Z}^d$.

3.1 Principal Eigenvalue and Eigenfunction

Let U be a finite connected subset of $\mathbb{Z}^d$. Consider the *Schrödinger operator*

$$H^0_{U,w} := \kappa \Delta_U + w,$$

on U, with Dirichlet boundary conditions, where $\kappa > 0$ and w is a *non-negative potential* on U: a set $w = (w(x))_{x \in U}$, where $w(x) \geq 0$. In other words, $H^0_{U,w}$ is the operator defined on $l^2(U)$ acting as,

$$H^0_{U,w} f(x) = \kappa \sum_{j=1}^{2d} (f(x + e_j) - f(x)) + w(x) f(x), \qquad x \in U,$$

with the convention that $f(y) = 0$ if $y \notin U$, and where $\{e_j : 1 \leq j \leq 2d\}$ are the canonical generators of $\mathbb{Z}^d$ and their corresponding inverses. Note that $H^0_{U,w}$ is a bounded symmetric operator on $l^2(U)$. Define

$$\bar{w}_U := \max\{w(x) : x \in U\}.$$

Theorem 4 *Let $U \subset \mathbb{Z}^d$ and w a potential on U and $x_0 \in U$, and assume that $w(x_0) = 0$. Consider the Schrödinger operator $H^0_{U,w}$. Then, whenever $h > \bar{w}_U$,*

$$H_{U,w,h} := H^0_{U,w} + h\delta_{x_0}$$

has a simple principal Dirichlet eigenvalue λ_0 and a principal Dirichlet eigenfunction ψ_0. Furthermore, the following are satisfied.

(i) The principal Dirichlet eigenvalue has the expansion

$$\lambda_0 = h - 2d\kappa + h \sum_{j=1}^{\infty} \sum_{\gamma \in \mathbb{P}_U^{2j+1}(x_0, x_0)} \prod_{z \in \gamma, z \neq z_1} \frac{\kappa}{\lambda_0 + 2d\kappa - w(z)}. \tag{23}$$

(ii) There is a constant K such that for all $x \in U$ one has that

$$\psi_0(x) = K \sum_{\gamma \in \mathbb{P}_U(x, x_0)} \prod_{z \in \gamma} \frac{\kappa}{\lambda_0 + 2d\kappa - w(z)}.$$

(iii) Whenever $h \geq 4d\kappa$ we have that

$$\psi_0(x) = \mathbb{1}_{x_0}(x) + \varepsilon(x),$$

where $\varepsilon(x)$ satisfies for all $x \in U$

$$|\varepsilon(x)| \leq C \frac{1}{(h - \bar{w}_U)^{|x - x_0|_1 + 1}},$$

for some constant C that does not depend on h nor w.

(iv) We have that

$$\sup\{\lambda \in \sigma(H_{U,\omega,h}) : \lambda < \lambda_0\} \leq \bar{w}_U.$$

Theorem 4 will be proved in Sect. 3.3.

3.2 Green Function of the Unperturbed Operator

Throughout, given any self-adjoint operator A defined on $l^2(U)$, we will denote by $res(A)$ and $\sigma(A)$ the resolvent set and the spectrum of A respectively. For $\lambda \in res(H^0_{U,w})$, let us introduce on $U \times U$ the function

$$g_\lambda^{U,w}(x, y) := (\delta_x, (\lambda I - H^0_{U,w})^{-1}\delta_y).$$

Note that $g_\lambda^{U,w}$ is the *Green function* of the operator $H^0_{U,w} - \lambda I$.

Note that if $\gamma \in \mathbb{P}_U(x, y)$, then

$$|\gamma| \geq |x - y|_1 + 1. \tag{24}$$

Furthermore

$$|\mathbb{P}_U^k(x, y)| \le (2d)^{k-1}. \tag{25}$$

Lemma 1 *Let U be a bounded connected subset of $\mathbb{Z}^d$, w a non-negative potential on U and $\kappa > 0$. Let $x, y \in U$. Then, the Green function $g_\lambda^{U,w}(x, y)$ is analytic if $\lambda \in res(H_{U,w}^0)$ and*

$$g_\lambda^{U,w}(x, y) = \frac{1}{\kappa} \sum_{\gamma \in \mathbb{P}_U(x,y)} \prod_{z \in \gamma} \frac{\kappa}{\lambda + 2d\kappa - w(z)}. \tag{26}$$

Proof Now, let us remark that for $x, y \in \mathbb{Z}^d$,

$$g_\lambda^{U,w}(x, y) = \int_0^\infty E_x \left[e^{\int_0^t (w(X_s) - \lambda) ds} \delta_y(X_t) \right] dt$$
$$= \sum_{\gamma \in \mathbb{P}_U(x,y)} \int_0^\infty E_x \left[e^{\int_0^t (w(X_s) - \lambda) ds} C_\gamma \right] dt, \tag{27}$$

where $(X_t)_{t \ge 0}$ is a simple symmetric random walk of total jump rate $2d\kappa$ starting from x and E_x the corresponding expectation, while C_γ is the event that the random walk follows the path γ in the time interval $[0, t]$. For each $\gamma = \{x_1 = x, \ldots, x_n = y\} \in \mathbb{P}_U(x, y)$ define $B_\gamma := E_x \left[e^{\int_0^t (w(X_s) - \lambda) ds} C_\gamma \right]$. Then, noting that the probability density on the path γ is given by $e^{-2d\kappa s_1 - \cdots - 2d\kappa s_{n-1} - 2d\kappa(t - s_1 - \cdots - s_{n-1})}$, where s_i is the time spent on x_i, we see that

$$B_\gamma = \frac{(2d\kappa)^{n-1}}{(2d)^{n-1}} e^{-2d\kappa t} \int_{S_{n-1}} e^{\bar{w}(x_1)s_1 + \cdots + \bar{w}(x_{n-1})s_{n-1} + \bar{w}(x_n)(t - s_1 - \cdots - s_{n-1})} ds_1 \cdots ds_{n-1}$$
$$= \kappa^{n-1} e^{-2d\kappa t + \bar{w}_n t} \int_{S_{n-1}} e^{(\bar{w}_1 - \bar{w}_n)s_1 + \cdots + (\bar{w}_{n-1} - \bar{w}_n)s_{n-1}} ds_1 \cdots ds_{n-1},$$

where $\bar{w}_i = w(x_i) - \lambda$ and $S_{n-1} = \{s_1 + \cdots + s_{n-1} < t\}$. Using induction on n we can compute the above integral to obtain,

$$B_\gamma = \frac{1}{\kappa} \prod_{z \in \gamma} \frac{\kappa}{\lambda + 2d\kappa - w(z)}.$$

Substituting this expression back in (27) finishes the proof. ∎

Note that the largest eigenvalue of $H_{U,w}$ can be expressed as $\lambda_+^{U,w} := \sup\{\lambda \in \sigma(H_{U,w})\}$.

Lemma 2 *For every finite connected set U and non-negative potential w on U.*

$$\bar{w}_U - 2d\kappa \le \lambda_+^{U,w} \le \bar{w}_U. \tag{28}$$

Proof Let x_m be some site where $\bar{w}_U = w(x_m)$. The first inequality of (28) follows from the computation $(f, H_{U,w}f) \geq w(x_m) - 2d\kappa$ for $f = \delta_{x_m}$. The second from the estimate $(f, H_{U,w}f) \leq w(x_m)$, for arbitrary $f \in l^2(U)$ with unit norm. □

Corollary 2 *Let* $x, y \in U$. *If* $\lambda - \lambda_+^{U,w} > 2d\kappa$, *then*

$$
\frac{1}{\kappa}\left(\frac{\kappa}{\lambda + 2d\kappa}\right)^{|x-y|_1+1} \leq g_\lambda^{U,w}(x, y) \leq \frac{(2d\kappa)^{|x-y|_1}}{(\lambda - \lambda_+^{U,w})^{|x-y|_1}} \frac{1}{\lambda - \lambda_+^{U,w} - 2d\kappa}. \tag{29}
$$

Proof For every $z \in U$, $\lambda + 2d\kappa \geq \lambda + 2d\kappa - w(z)$. By Lemma 1 and the fact that the shortest path between x and y has length $|x - y|_1 + 1$, we prove the first inequality of display (29). Also, by Lemma 1 and the inequality (24) we see that

$$
g_\lambda^{U,w}(x, y) \leq \frac{1}{\kappa}\sum_{k=|x-y|_1+1}^{\infty} \sum_{\gamma \in \mathbb{P}_U^k(x,y)} \left(\frac{\kappa}{\lambda - \lambda_+^{U,w}}\right)^k.
$$

Now, from (25), we conclude the proof. □

We can now derive the following lemma.

Lemma 3 *Assume that* $h > \bar{w}_U + 2d\kappa$. *Then, there exists a unique* $\lambda_0 > \lambda_+^{U,w}$, *which satisfies the equation,*

$$
h g_{\lambda_0}^{U,w}(x, x) = 1. \tag{30}
$$

Proof By Lemma 12, note that Eq. (30) has a unique solution $\lambda_0 > \lambda_+^{U,w}$ if,

$$
h > \frac{1}{\lim_{\lambda \searrow \lambda_+^{U,w}} g_\lambda^{U,w}(x, x)}.
$$

Now, by the first inequality of (29) of part (i) of Corollary 2 and by Lemma 2 the right-hand side of the above inequality is larger than $\bar{w}_U + 2d\kappa$. □

3.3 Proof of Theorem 4

Let us now prove Theorem 4. To prove part (i) note that the unique λ_0 which satisfies (30) of Lemma 3 has to be the principal Dirichlet eigenvalue by Theorem 6 of Appendix A. Applying the expansion (26) of Lemma 1 of the Green function we obtain part (i). To prove part (ii) let us first note that by Lemma 3, λ_0 is an isolated point of the spectrum of $H_{U,\omega}$. It follows that there is simple closed curve Γ in the complex plane which contains λ_0 in its interior and the rest of the spectrum of $H_{U,\omega}$ in its exterior. Therefore

$$P := \int_\Gamma R_\lambda d\lambda,$$

where R_λ is the resolvent of $H_{U,\omega}$, is the orthogonal projection onto the eigenspace of λ_0. Now, from Theorem 6 of Appendix A, we can see that

$$P = C_3 (q_{\lambda_0}, \cdot) q_{\lambda_0},$$

for some constant C_3. From here we can deduce part (ii). Part (iii) follows immediately from part (ii). Part (iv) follows from Corollary 3 of Appendix A and Lemma 2.

4 High Peak Statistics of a Weibull Potential

In this section we will derive several results describing the asymptotic behavior as $l \to \infty$ of quantities defined in terms of the order statistics $v_{(1)} \geq v_{(2)} \geq \cdots \geq v_{(N)}$ of an i.i.d. potential v on the box Λ_l with Weibull distribution. We will occasionally also use the notation $v(x_{(j)})$ instead of $v_{(j)}$ to indicate explicitly the site $x_{(j)}$ where the value $v_{(j)}$ is attained.

Lemma 4 *Consider the order statistics $\{v_{(1)}, \ldots, v_{(N)}\}$ of the potential v. Let*

$$a_l := (\rho \log |\Lambda_l|)^{1/\rho} \quad \text{for} \quad l \geq 1.$$

(i) For every $x \in \mathbb{R}$,

$$\lim_{l \to \infty} \mu[v_{(1)} < a_l + b_l x] = \exp\{-e^{-x}\},$$

where $b_l := \frac{1}{(\rho \log |\Lambda_l|)^{1-1/\rho}}$. In particular, μ-a.s.,

$$\lim_{l \to \infty} \frac{v_{(1)}}{a_l} = 1.$$

(ii) For every sequence $(c_l)_{l \geq 1}$ such that $c_l \geq a_l$, we have that

$$\mu[v_{(1)} \geq c_l] \leq |\Lambda_l| e^{-\frac{1}{\rho} c_l^\rho} + o\left(|\Lambda_l| e^{-\frac{1}{\rho} c_l^\rho}\right). \tag{31}$$

and that

$$\mu[v_{(2)} \geq c_l] \leq |\Lambda_l|^4 e^{-\frac{2}{\rho} c_l^\rho} \tag{32}$$

Proof Consider a sequence $\{c_l : l \geq 0\}$ and note that

$$\mu(v_{(1)} < c_l) = (1 - \mu(v_1 \geq c_l))^{|\Lambda_l|} = \left(1 - \frac{1}{|\Lambda_l|^{\left(\frac{c_l}{a_l}\right)^\rho}}\right)^{|\Lambda_l|}.$$

Therefore, using the fact the for all natural n, $e^{-1}(1 - 1/n) \leq (1 - 1/n)^n \leq e^{-1}$, we have that

$$\exp\left\{-|\Lambda_l|^{1-\left(\frac{c_l}{a_l}\right)^\rho}\right\}\left(1 - \frac{1}{|\Lambda_l|^{\left(\frac{c_l}{a_l}\right)^\rho}}\right)^{|\Lambda_l|^{1-\left(\frac{c_l}{a_l}\right)^\rho}} \leq \mu(v_{(1)} < c_l) \leq \exp\left\{-|\Lambda_l|^{1-\left(\frac{c_l}{a_l}\right)^\rho}\right\}.$$

(33)

Choosing $c_l = a_l + b_l x$ in the above inequalities and taking the limit when $l \to \infty$, we prove part (i).

Part (ii). Note that for all $x \geq 0$, $e^{-x} \leq 1 - x + \frac{x^2}{2}$. Therefore, by (33), we have that

$$\mu(v_{(1)} \geq c_l) \leq \left(\frac{1}{|\Lambda_l|^{\left(\frac{c_l}{a_l}\right)^\rho - 1}} - \frac{1}{|\Lambda_l|^{2\left(\frac{c_l}{a_l}\right)^\rho - 2}}\right)\left(1 - \frac{1}{|\Lambda_l|^{\left(\frac{c_l}{a_l}\right)^\rho}}\right)^{|\Lambda_l|^{1-\left(\frac{c_l}{a_l}\right)^\rho}}$$

$$= |\Lambda_l|e^{-\frac{1}{\rho}c_l^\rho} + o\left(|\Lambda_l|e^{-\frac{1}{\rho}c_l^\rho}\right),$$

which gives (31). The proof of (32) is completely similar. □

5 Stable Limit Laws and Structure of the Scaling

Here we will prove Theorem 3. The existence of the scaling function h satisfying Eq. (17) will be proved in Sect. 5.1 and its expansion given in Theorem 3 in Sect. 5.2.

5.1 Existence of the Scaling Function

Here we will prove the existence of the scaling function h satisfying (17) of Theorem 3, under the assumption that $\lim_{t\to\infty} L(t) = \infty$. We first need the following lemma.

Lemma 5 *For every potential $v \in W$ and integer N the function $\zeta_N(s) : [0, \infty) \to (0, e^{-\frac{1}{\rho}B_M^\rho}]$ [c.f. (10)] defined by (13) is a homeomorphism.*

Proof Note that for every fixed potential $v \in W$ and integer N the function $B_N(s, v)$ defined in (8) is strictly increasing in $[0, \infty)$. By the bounded convergence theorem, this implies that ϕ is strictly decreasing and continuous with range $(0, e^{-\frac{1}{\rho}B_M^\rho}]$, which proves the lemma. □

Now note that by the definition of t_0 in (17), since the scale $L(t)$ is an increasing function of t, we have that whenever $t \geq t_0$

$$0 < e^{-\gamma \frac{\tau(t)\rho'}{\rho'}} \leq e^{-\frac{1}{\rho} B_M^\rho}.$$

By Lemma 5 it is clear that for each $t \geq t_0$ there exists a $h(t) \in [0, \infty)$ such that (17) is satisfied. We define now as in the statement of Theorem 2 for $t \geq 0$,

$$g(t) := th(t). \tag{34}$$

5.2 Properties of the Scaling Function

Here we will prove the expansion (19) of Theorem 3, which states that the scaling function $h(t)$ defined in (17) has the expansion specified by (19), (21) and (22).

We will now prove that the functions defined recursively by (20) and (21) of Theorem 3 are such that (19) and (22) are satisfied. Define for $1 \leq j \leq M$, $\mathcal{H}_j := h_0 + \cdots + h_j$. We will need the following lemma.

Lemma 6 Consider the functions $\{B_j : j \geq 1\}$ defined in (8) and assume condition (16). Then following are satisfied.

(i) For all $\epsilon > 0$ and $t \geq 0$ such that and $\max_{e \in E} v(e) \leq (1 - \epsilon) \mathcal{A}_0 \tau^{\rho'-1}(t)$ (recall the that E is the set of points of $\mathbb{Z}^d$ of unit norm [cf. (3)], we have that

$$B_1(h_0(t), v)^\rho = \mathcal{A}_0^\rho \tau^{\rho'}(t) + O\left(\frac{1}{t^{\rho'-2}}\right), \tag{35}$$

where

$$\left| O\left(\frac{1}{t^{\rho'-2}}\right) \right| \leq \frac{\rho \kappa \mathcal{A}_0^{\rho-2}}{\epsilon} \tau^{-(\rho'-2)}(t).$$

and for $j \geq 1$ that

$$B_{j+1}(\mathcal{H}_j, v)^\rho = B_j(\mathcal{H}_{j-1}, v)^\rho + \rho \mathcal{A}_0 \tau h_j + O\left(\frac{1}{t^{(2j+1)\rho'-(2j+2)}}\right). \tag{36}$$

where in (36) the error term satisfies

$$\left| O\left(\frac{1}{t^{(2j+1)\rho'-(2j+2)}}\right) \right| \leq C_8 \frac{1}{t^{(2j+1)\rho'-(2j+2)}},$$

where C_8 is a constant depending only ϵ, κ, γ and ρ.

(ii) *There is a constant $C_{6,1}$ such that*

$$e^{-\left(\gamma\frac{\tau^{\rho'}}{\rho'}+\frac{C_{6,1}}{\tau^{\rho'-2}}\right)} \leq E_\mu\left[e^{-\frac{1}{\rho}B_1(h_0,v)^\rho}\right] \leq e^{-\left(\gamma\frac{\tau^{\rho'}}{\rho'}-\frac{C_{6,1}}{\tau^{\rho'-2}}\right)}. \tag{37}$$

Similarly, for each $j \geq 1$ there is a constant $C_{6,j}$ such that

$$E_\mu\left[e^{-\left(\frac{1}{\rho}B_j(\mathcal{H}_{j-1},v)^\rho+\rho\tau h_j+\frac{C_{6,j}}{\tau^{(2j+1)\rho'-(2j+2)}}\right)}\right] \leq E_\mu\left[e^{-\frac{1}{\rho}B_{j+1}(\mathcal{H}_j,v)^\rho}\right]$$

$$\leq E_\mu\left[e^{-\left(\frac{1}{\rho}B_j(\mathcal{H}_{j-1},v)^\rho+\rho\tau h_j-\frac{C_{6,j}}{\tau^{(2j+1)\rho'-(2j+2)}}\right)}\right]. \tag{38}$$

Proof Part (i) follows from a standard Taylor expansion. We will now prove (37) of part (ii). Define for $\epsilon > 0$ the event

$$A := \left\{\max_{e\in E} v(e) \leq (1-\epsilon)\mathcal{A}_0\tau^{\rho'}\right\}.$$

Then

$$E_\mu\left[e^{-\frac{1}{\rho}B_1(h_0,v)^\rho}\right] = E_\mu\left[e^{-\frac{1}{\rho}B_1(h_0,v)^\rho}, A\right] + E_\mu\left[e^{-\frac{1}{\rho}B_1(h_0,v)^\rho}, A^c\right].$$

But by part (i) we have that

$$E_\mu\left[e^{-\frac{1}{\rho}B_1(h_0,v)^\rho}, A\right] \leq e^{-\left(\gamma\frac{\tau^{\rho'}}{\rho'}-\frac{C_8}{\tau^{\rho'-2}}\right)}. \tag{39}$$

On the other hand, since

$$\frac{1}{\rho}B_1(h_0, v)^\rho \geq \frac{1}{\rho}\tau^{\rho'}\mathcal{A}_0^\rho\frac{B_1}{2d\kappa} = \frac{\gamma}{\rho'}\tau^{\rho'}\frac{B_1}{2d\kappa},$$

we have that for ϵ small enough using (31) of part (ii) of Lemma 4 that

$$E_\mu\left[e^{-\frac{1}{\rho}B_1(h_0,v)^\rho}, A^c\right] \leq e^{-\tau^{\rho'}\frac{\gamma}{\rho'}\frac{B_1}{2d\kappa}}\mu(A^c) \leq 2de^{-\tau^{\rho'}\left(\frac{\gamma}{\rho'}\frac{B_1}{2d\kappa}+\frac{1}{\rho}(1-\epsilon)^\rho\mathcal{A}_0^\rho\right)}$$

$$= 2de^{-\frac{\gamma}{\rho'}\tau^{\rho'}\left(\frac{B_1}{2d\kappa}+(1-\epsilon)^\rho\right)} = O\left(e^{-(1+\epsilon)\frac{\gamma}{\rho'}\tau^{\rho'}}\right), \tag{40}$$

where we used that fact that $\frac{B_1}{2d\kappa} + (1-\epsilon)^\rho \geq 1+\epsilon$ for ϵ small enough. Combining (39) with (40) we see that there is a constant C_6 such that

$$E_\mu\left[e^{-\frac{1}{\rho}B_1(h_0,v)^\rho}\right] \leq e^{-\left(\gamma\frac{\tau^{\rho'}}{\rho'}-\frac{C_6}{\tau^{\rho'-2}}\right)}.$$

On the other hand, from $\frac{1}{\rho}B_1(h_0, v)^\rho \le \gamma \frac{\tau^{\rho'}}{\rho'}$, we immediately get that

$$E_\mu\left[e^{-\frac{1}{\rho}B_1(h_0,v)^\rho}\right] \ge e^{-\gamma\frac{\tau^{\rho'}}{\rho'}},$$

which finishes the proof of (37). The proof of (38) is analogous.   □

Let us now prove (22) of Theorem 3. Recall that $\tau(t) \sim t$ [cf. (16)]. Then, by the definition of h_1 in (21) and by (37) of part (ii) of Lemma 6, we see that

$$
-\frac{C_{6,1}}{A_0^{\rho-1}t^{\rho'-1}} \le \frac{1}{A_0^{\rho-1}}\left(\gamma\frac{\tau^{\rho'-1}}{\rho'} + \frac{1}{\tau}\log e^{-\left(\gamma\frac{\tau^{\rho'}}{\rho'}+\frac{C_{6,1}}{\tau^{\rho'-2}}\right)}\right)
$$

$$
\le \frac{1}{A_0^{\rho-1}}\left(\gamma\frac{\tau^{\rho'-1}}{\rho'} + \frac{1}{\tau}\log E_\mu\left[e^{-\frac{1}{\rho}B_1(h_0,v)^\rho}\right]\right) = h_1(t)
$$

$$
\le \frac{1}{A_0^{\rho-1}}\left(\gamma\frac{\tau^{\rho'-1}}{\rho'} + \frac{1}{\tau}\log e^{-\left(\gamma\frac{\tau^{\rho'}}{\rho'}-\frac{C_{6,1}}{\tau^{\rho'-2}}\right)}\right) \le \frac{C_{6,1}}{A_0^{\rho-1}t^{\rho'-1}}.
$$

Hence

$$h_1(t) = O\left(\frac{1}{t^{\rho'-1}}\right).$$

We will now prove that for $j \ge 1$

$$h_{j+1}(t) = O\left(\frac{1}{t^{(2j+1)(\rho'-1)}}\right). \tag{41}$$

Using the definition (21) for h_j and h_{j+1}, by (38) of part (ii) of Lemma 6 we conclude that

$$
-\frac{C_{6,j-1}}{t^{(2j+1)(\rho'-1)}} \le \frac{1}{A_0^{\rho-1}}\left(\gamma\frac{\tau^{\rho'-1}}{\rho'} + \frac{1}{\tau}\log E_\mu\left[e^{-\left(\frac{1}{\rho}B_j(\mathcal{H}_{j-1},v)^\rho+\rho\mathcal{A}_o\tau h_j+\frac{C_{6,j-1}}{\tau^{(2j+1)\rho'-(2j+2)}}\right)}\right]\right)
$$

$$
\le \frac{1}{A_0^{\rho-1}}\left(\gamma\frac{\tau^{\rho'-1}}{\rho'} + \frac{1}{\tau}\log E_\mu\left[e^{-\frac{1}{\rho}B_{j+1}(\mathcal{H}_j,v)^\rho}\right]\right) = h_{j+1}(t)
$$

$$
\le \frac{1}{A_0^{\rho-1}}\left(\gamma\frac{t^{\rho'-1}}{\rho'} + \frac{1}{t}\log E_\mu\left[e^{-\left(\frac{1}{\rho}B_j(\mathcal{H}_{j-1},v)^\rho+\rho\mathcal{A}_o\tau h_j-\frac{C_{6,j-1}}{\tau^{(2j+1)\rho'-(2j+2)}}\right)}\right]\right) \le \frac{C_{6,j-1}}{t^{(2j+1)(\rho'-1)}},
$$

which proves (41) and hence (22) of Theorem 3. It now follows that

$$
e^{-\gamma\frac{\tau^{\rho'}}{\rho'}} = E_\mu\left[e^{-\frac{1}{\rho}B_M(\mathcal{H}_{M-1},v)^\rho-A_0^{\rho-1}\tau h_M}\right] = E_\mu\left[e^{-\frac{1}{\rho}B_M(\mathcal{H}_M,v)^\rho+O\left(\frac{1}{\tau^{(2M+1)\rho'-(2M+2)}}\right)}\right]
$$

which implies that $h(t) - \mathcal{H}_M(t) = O\left(\frac{1}{t^{(2M+1)(\rho'-1)}}\right)$, which proves (19) of part (ii) of Theorem 3.

6 Convergence to Stable Laws

We will now prove the stable limit of Theorem 2. Some of the computations will be similar to those done by Ben Arous, Bogachev and Molchanov in [6], within the context of sums of i.i.d. random exponential variables. As a first step, we will first recall the coarse graining methods introduced in [7], which will enable us to reduce the problem to a sum of approximately independent random variables in Sect. 6.1. In Sect. 6.2, we will show how to reduce the problem to a sum of independent random variables. In Sect. 6.3 we recall a classical criterion in Theorem 5 for convergence to infinite divisible distributions. These criteria will be verified in Sects. 6.4, 6.5 and 6.6.

6.1 Mesoscopic Scales

Here we will recall the coarse graining methods introduced in [7]. The idea is to make a decomposition of the spatial average sum $m_L(t)$ defined in (4), where for the moment L is a positive real number, in an approximate sum of i.i.d. random variables. Of course, eventually we will choose L as an increasing function of time. To make this decomposition, we will subdivide the box Λ_L into smaller boxes of an appropriate side l, separated by very narrow strips of width r. The contribution on the spatial average of the terms in the strips will be negligible, but nevertheless, the presence of the strips will give enough separation for the boxes of side l to produce independence between them. We call this decomposition the *strip-box partition* as in [7]. In the subsequent sections, we will apply standard theorems which imply the convergence of the corresponding sum of i.i.d. random variables to stable laws.

So choose a parameter l smaller than or equal to L, called the *mesoscopic scale*. Note that, there exist natural numbers q and $\bar{q}$ such that $2L + 1 = ql + \bar{q}$, with $0 \le \bar{q} \le q$. Hence

$$2L + 1 = \sum_{i=1}^{q} l_i,$$

where $l_i = l + \theta_{\bar{q}}(i)$ and $\theta_{\bar{q}}(i) = 1$ for $i \le \bar{q}$ and $\theta_{\bar{q}}(i) = 0$ for $i > \bar{q}$. Given a pair of real numbers a, b, we will use the notation $[[a, b]]$ for $[a, b] \cap \mathbb{Z}$. Now define $I_1 := [[-L, -L + l_1 - 1]]$ and for $1 < i \le q$ let $I_i := \left[\left[-L + \sum_{j=1}^{i-1} l_j, -L + \sum_{j=1}^{i} l_j - 1\right]\right]$. Now, we introduce a second parameter r which is a natural number smaller than or equal to l, called the *fine scale*. Let $r_i := r + \theta_{\bar{q}}(i)$. Define $J_1 := [[-L + r_1, -L + l_1 - 1 - r_1]]$ and for $1 < i \le q$ let $J_i := \left[\left[-L + \sum_{j=1}^{i-1} l_j + r_i, -L + \sum_{j=1}^{i} l_j - 1 - r_i\right]\right]$.

Now, let $\mathcal{I} := \{1, 2, \ldots, q\}^d$. For a given element $\mathbf{i} \in \mathcal{I}$, of the form $\mathbf{i} = (i_1, \ldots, i_d)$ with $1 \le i_k \le q$, $1 \le k \le d$, we define

$$\Lambda_{\mathbf{i}}'' := J_{i_1} \times J_{i_2} \times \cdots \times J_{i_d},$$

called a *main box*. Its cardinality is $|\Lambda_i''| = (l - 2r)^d$. Now let

$$S_L := \Lambda_L - \bigcup_{i \in \mathcal{I}} \Lambda_i'',$$

called the *strip set*. The sets S_L and $\{\Lambda_i'' : i \in \mathcal{I}\}$ define a partition of Λ_L called the *strip-box partition at scale l* of Λ_L.

Let us now write

$$\sum_{x \in \Lambda_L} m(x, t) = \sum_{i \in \mathcal{I}} m_i + \sum_{x \in S_L} m(x, t), \qquad (42)$$

where

$$m_i := \sum_{x \in \Lambda_i''} m(x, t).$$

We will use the notation $\mathbf{0} := (0, \dots, 0)$. Recall that the first assumption of Theorem 3, is the existence of an increasing function $L(t)$ such that $\lim_{t \to \infty} L(t) = \infty$. In what follows we will choose the mesoscopic scale as $L(t)$. Recalling the definition of $\tau(t)$ in (15), we can write

$$L(t) := e^{\gamma \frac{t^{\rho'}}{\tau}}. \qquad (43)$$

We will also make the choose the fine scale l and the separation scale r as functions of time so that

$$l(t) := e^t \qquad (44)$$

and

$$r(t) := t^2. \qquad (45)$$

6.2 Dirichlet Boundary Conditions

Throughout, we will denote by $m(x, t, v)$ the solution $m(x, t)$ of the parabolic Anderson problem (1), emphasizing the dependence on the potential v of it. For each finite set $U \subset \mathbb{Z}^d$ and environment $v \in W$, we define $\tilde{m}_U(x, t) = m_U(x, t, v_U)$ as the solution of the parabolic Anderson equation with potential v with Dirichlet boundary conditions on U and initial condition $\mathbf{1}_U$, so that

$$\frac{\partial \tilde{m}_U(x, t)}{\partial t} = \kappa \Delta \tilde{m}_U(x, t) + v(x) \tilde{m}_U(x, t), \quad \text{for all } t > 0, x \in \mathbb{Z}^d,$$

$$\tilde{m}_U(x, 0) = \mathbf{1}_U(x),$$

where the Laplacian Δ has Dirichlet boundary conditions on U. In other words, $v_U(x) = v(x)$ for $x \in U$ while $v_U(x) = -\infty$ for $x \notin U$, with $m(x, t, v_U)$ the solution of (1). Throughout, for $r > 0$, we will use the notation $\tilde{m}_r(x, t)$ instead of $\tilde{m}_{\Lambda(0,r)}(x, t)$. The following lemma can be proved in the same way as part (iii) of Proposition 4.4 of [8].

Lemma 7 *For each $\beta > 0$, there is a constant $C > 0$ such that for all $R \geq 2\kappa t$*

$$\left\langle |m(x, t) - \tilde{m}_R(x, t)|^\beta \right\rangle \leq C(R + 1)^d e^{-2\beta\kappa t I \left(\frac{R}{2\kappa t} \right)} e^{H(\beta t)},$$

where $I := y \sinh^{-1} y - \sqrt{1 + y^2} + 1$.

We will also define for $\mathbf{i} \in \mathcal{I}$

$$\tilde{m}_\mathbf{i}(t) := \sum_{x \in \Lambda_\mathbf{i}''} \tilde{m}_{\Lambda_\mathbf{i}''}(x, t). \tag{46}$$

Let

$$s(t) := e^{g(t)}, \tag{47}$$

where $g(t)$ is defined in (34). Theorem 2 states that

$$\lim_{t \to \infty} \frac{|\Lambda_L|}{s(t)} (m_L(t) - A(t)) = S_\alpha,$$

in distribution. Recall the assumption (16) of Theorem 2:

$$\tau(t) \sim t.$$

By this, Lemma 7, the decomposition (42) and the choice of scales (43), (44) and (45), it is enough to prove that

$$\lim_{t \to \infty} \frac{1}{s(t)} \left(\sum_{\mathbf{i} \in \mathcal{I}} \tilde{m}_\mathbf{i} - |\Lambda_L| A(t) \right) = S_\alpha. \tag{48}$$

6.3   Criteria for Convergence to Stable Laws

In order to prove part (iii) of Theorem 2, we recall the following result which gives conditions for a triangular array of independent random variables to converge to a given infinite divisible distribution (see for example Theorems 7 and 8 of Chap. 4 of Petrov [14] or [6]).

A random variable is infinite divisible if and only if its characteristic function $\phi(t)$ admits the expression (see for example Theorem 5 of Chap. 2 of Petrov [14])

$$\phi(t) = \exp\left\{i\nu t - \frac{\sigma^2 t^2}{2} + \int_{-\infty}^{\infty}\left(e^{itx} - 1 - \frac{itx}{1+x^2}\right)d\mathcal{L}(x)\right\}, \quad (49)$$

where ν is a real constant, σ^2 a non-negative constant, the function $\mathcal{L}(x)$ is non-decreasing in $(-\infty, 0)$ and $(0, \infty)$ and satisfies $\lim_{x\to\infty}\mathcal{L}(x) = \lim_{x\to-\infty}\mathcal{L}(x) = 0$ and for every $\delta > 0$, $\int_{x:0<|x|\leq\delta} x^2 d\mathcal{L}(x) < \infty$. $\mathcal{L}$ is called the *Lévy-Khintchine spectral function*. We will denote by $X_{\nu,\sigma,\mathcal{L}}$ the infinite divisible random variable with characteristic function (49).

Theorem 5 *For each $t \geq 0$ let $\mathcal{S}(t)$ be a growing set of indexes and consider a set $(Y_i(t))_{i\in\mathcal{S}(t)}$ of independent i.i.d. random variables. Call P_t the law of one of these random variables, say of $Y_0(t)$, where we just call $\mathbf{0}$ an arbitrary element of $\mathcal{S}(t)$. Assume that for every $\epsilon > 0$ it is true that*

$$\lim_{t\to\infty} P_t(Y_0(t) > \epsilon) = 0.$$

Now let $\mathcal{L} : \mathbb{R}\backslash\{0\} \to \mathbb{R}$ be a Lévy-Khintchine spectral function, $\nu \in \mathbb{R}$ and $\sigma > 0$, and let $A : [0, \infty) \to \mathbb{R}$ be some function. Then, if $n(t) := |\mathcal{S}(t)|$ the following statements are equivalent.

(i) We have that

$$\lim_{t\to\infty}\left(\sum_{i\in\mathcal{S}(t)} Y_i(t) - A(t)\right) = X_{\nu,\sigma,\mathcal{L}},$$

where the convergence is in distribution.

(ii) Define for $y > 0$ the truncated random variable at level y as $Z_y(t)$ $:= Y_0(t)\mathbf{1}_{|Y_0(t)|\leq y}$. Also, let $E_t(\cdot)$ and $Var_t(\cdot)$ denote the expectation and variance corresponding to the law P_t. Then if x is a continuity point of $\mathcal{L}$,

$$\mathcal{L}(x) = \begin{cases} \lim_{t\to\infty} n(t)P_t(Y_0(t) \leq x) & \text{for } x < 0, \\ -\lim_{t\to\infty} n(t)P_t(Y_0(t) > x) & \text{for } x > 0, \end{cases} \quad (50)$$

$$\sigma^2 = \lim_{y\to 0}\lim_{t\to\infty} n(t)Var_t(Z_y(t)), \quad (51)$$

and for any $y > 0$ which is a continuity point of $\mathcal{L}(x)$,

$$\nu = \lim_{n\to\infty}\left(n(t)E_t(Z_y(t)) - A(t)\right) + \int_{|x|>y}\frac{x}{1+x^2}d\mathcal{L}(x) - \int_{y\geq|x|>0}\frac{x^3}{1+x^2}d\mathcal{L}(x). \quad (52)$$

We will apply Theorem 5, to the set of i.i.d. random variables $(Y_i)_{i \in \mathcal{I}}$ with

$$Y_i := \frac{1}{s(t)} \tilde{m}_i,\tag{53}$$

$$n(t) := \left(\frac{L(t)}{l(t)}\right)^d = \frac{e^{\gamma \frac{\tau(t)^{\rho'}}{\rho'}}}{e^{dt}}.\tag{54}$$

and

$$\tilde{A}(t) := \begin{cases} 0 & \text{if } 0 < \gamma < \gamma_1 \\ E_\mu[Y_0, Y_0 \le 1] & \text{if } \gamma = \gamma_1 \\ E_\mu[Y_0] & \text{if } \gamma_1 < \gamma < \gamma_2. \end{cases}$$

The first, second and third cases correspond to the definition of $A(t)$ in (18) of Theorem 5.

6.4 Lévy-Khintchine Spectral Function

Here we will verify that the i.i.d. random variables $(Y_i)_{i \in \mathcal{I}}$ defined in (53) satisfy condition (50) of part (ii) of Theorem 5, with the Lévy-Khintchine spectral function

$$\mathcal{L}(x) := \begin{cases} 0 & \text{for } x \le 0 \\ -\frac{1}{x^{\alpha(\gamma,\rho)}} & \text{for } x > 0, \end{cases}\tag{55}$$

where $\alpha(\gamma, \rho)$ is defined in (11). To prove that condition (50) is satisfied with the Lévy-Khintchine spectral function (55) it will be enough to show that the following proposition is satisfied.

Proposition 1 *Consider $\tilde{m}_0$ defined in (46), with v an i.i.d. potential with Weibull law μ. Then, for $s(t)$ defined in (47) and $n(t)$ in (54) we have that for all $u > 0$ it is true that*

$$\lim_{t \to \infty} n(t) \mu \left(\frac{\tilde{m}_0}{s(t)} > u\right) = \frac{1}{u^{\alpha(\gamma,\rho)}}.$$

Proof We will use the expansion

$$\tilde{m}_{\Lambda_0''}(x, t) = \sum_{n=0}^{N} e^{t\lambda_n} \psi_n(x)(\psi_n, 1(\Lambda_0'')),\tag{56}$$

where $\{\lambda_n : 0 \le n \le N\}$ and $\{\psi_n : 0 \le n \le N\}$ are the eigenvalues in decreasing order and the corresponding set of orthonormal eigenfunctions, respectively, of the Schrödinger operator $H_{\Lambda_0'',v}$ on Λ_0'' with Dirichlet boundary conditions. Furthermore, we will need to choose δ_1 and δ_2 so that $0 < \delta_1 < \delta_2$ and

$$2(1 - \delta_2)^\rho > 1 \tag{57}$$

and to consider the events

$$A_1 := \left\{ v_{(1)} \ge (1 - \delta_1) \mathcal{A}_0 t^{\rho'-1} \right\},$$

and

$$A_2 := \left\{ v_{(2)} \le (1 - \delta_2) \mathcal{A}_0 t^{\rho'-1} \right\}.$$

Step 1. Note that

$$\mu\left(\frac{\tilde{m}_0}{s(t)} > u\right) = \mu\left(A_1, A_2, \tilde{m}_0 s(t)^{-1} > u\right)$$
$$+ \mu\left(A_1^c, A_2, \tilde{m}_0 s(t)^{-1} > u\right) + \mu\left(A_1, A_2^c, \tilde{m}_0 s(t)^{-1} > u\right). \tag{58}$$

By inequality (32) of part (ii) of Lemma 4, we have that

$$\mu(A_2^c) = \mu\left(v_{(2)} > (1 - \delta_2)\left(\frac{\gamma\rho}{\rho'}\right)^{1/\rho} t^{\rho'-1}\right) \le e^{-2(1-\delta_2)^\rho \frac{\gamma}{\rho'} t^{\rho'}}. \tag{59}$$

But from (57) we have that

$$\lim_{t\to\infty} n(t)\mu(A_2^c) = 0. \tag{60}$$

It follows from (58) and from (60) that

$$\lim_{t\to\infty} n(t)\mu\left(\frac{\tilde{m}_0}{s(t)} > u\right) = \lim_{t\to\infty} n(t)\mu\left(A_1, A_2, \frac{\tilde{m}_0}{s(t)} > u\right)$$
$$+ \lim_{t\to\infty} n(t)\mu\left(A_1^c, A_2, \frac{\tilde{m}_0}{s(t)} > u\right).$$

Step 2. From the expansion (56) note that on the event A_1^c we have that

$$\tilde{m}_0 = e^{t\lambda_0} \sum_{n=0}^{N} e^{-t(\lambda_0 - \lambda_n)} (\psi_n, 1(\Lambda_0''))^2$$
$$\le e^{t\lambda_0} l^{2d} \le e^{t(\lambda_0 + 2d)} \le e^{t(v_{(1)} + 2d)} \le e^{(1-\delta_1)\mathcal{A}_0 t^{\rho'} + 2dt},$$

where in the second to last inequality we have used (28). Using the fact that $\lim_{t\to\infty} e^{-g(t)} e^{(1-\delta_1)A_0 t^{\rho'} + 2d\beta t} = 0$, we therefore see that $\lim_{t\to\infty} \tilde{m}_0 e^{-g(t)} = 0$ and hence

$$\lim_{t\to\infty} n(t)\mu(A_1^c, \tilde{m}_0 e^{-g(t)} > u) = 0.$$

It follows from (58) that

$$\lim_{t\to\infty} n(t)\mu\left(m_0 s(t)^{-1} > u\right) = \lim_{t\to\infty} n(t)\mu\left(A_1, A_2, m_0 s(t)^{-1} > u\right). \qquad (61)$$

Step 3. By part (iii) of Theorem 4, on the event $A_1 \cap A_2$, the normalized principal Dirichlet eigenfunction ψ_0 is such that for some $x_0 \in \Lambda_0''$ one has that

$$\psi_0(x) = 1 + \varepsilon(x), \qquad (62)$$

where

$$|\varepsilon(x)| \leq \frac{C_4}{t^{(|x-x_0|_1+1)(\rho'-1)}}.$$

for some constant C_4. Therefore, from (62) and the expansion (56) we can see that on the event $A_1 \cap A_2$,

$$\tilde{m}_0 = e^{t\lambda_0}\left(1 + O\left(\frac{1}{t^{\rho'-1}}\right)\right) + \sum_{n=1}^{N} e^{t\lambda_n}(\psi_n, 1(\Lambda_0''))^2.$$

Therefore, from the identity (23) of part (i) of Theorem 4 and part (iv) of the same theorem, we see that on $A_1 \cap A_2$ it is true that

$$e^{t\lambda_0}\left(1 + O\left(\frac{1}{t^{\rho'-1}}\right)\right) \leq \tilde{m}_0$$

$$\leq e^{t\lambda_0}\left(1 + O\left(\frac{1}{t^{\rho'-1}}\right) + (2l+1)^2 e^{-(\delta_2-\delta_1)\left(\frac{2\rho}{\rho'}\right)^{1/\rho} t^{\rho'}}\right)$$

$$= e^{t\lambda_0}\left(1 + O\left(\frac{1}{t^{\rho'-1}}\right)\right).$$

Hence, on $A_1 \cap A_2$,

$$\tilde{m}_0 = e^{t\lambda_0}\left(1 + O\left(\frac{1}{t^{\rho'-1}}\right)\right). \qquad (63)$$

Step 4. Using (63) in display (61) we see that for any sequence $(t_k)_{k\geq 1}$ such that $\lim_{k\to\infty} n(t_k)\mu\left(\frac{\tilde{m}_0(t_k)}{s(t_k)}\right)$ exists (possibly being equal to ∞), one has that

$$\lim_{k\to\infty} n(t_k)\mu\left(\frac{\tilde{m}_0(t_k)}{s(t_k)} > u\right)$$

$$= \lim_{k\to\infty} n(t_k)\mu\left(A_1, A_2, e^{t_k\lambda_0}\left(1 + O\left(t_k^{-(\rho'-1)}\right)\right) \geq u e^{g(t_k)}\right)$$

$$= \lim_{k\to\infty} n(t_k)\mu\left(e^{t_k\lambda_0}\left(1 + O\left(\frac{1}{t_k^{\rho'-1}}\right)\right) \geq u e^{g(t_k)}\right), \tag{64}$$

where in the last step we used (60) of Step 2 and the fact that since on A_1^c it is true that $\lambda_0 \leq (1 - \delta_1)g(t)$, eventually in t

$$e^{t\lambda_0} \leq x e^{(1-\delta_1)g(t)}. \tag{65}$$

From (64) we now get that there is a constant C_1 such that

$$\liminf_{t\to\infty} n(t)\mu\left(\lambda_0 \geq h(t) + \frac{1}{t}\log u + \frac{C_1}{t^{\rho'}}\right) \leq \liminf_{t\to\infty} n(t)\mu\left(\frac{\tilde{m}_0}{s(t)} > u\right)$$

$$\leq \limsup_{t\to\infty} n(t)\mu\left(\frac{\tilde{m}_0}{s(t)} > u\right) \leq \limsup_{t\to\infty} n(t)\mu\left(\lambda_0 \geq h(t) + \frac{1}{t}\log u - \frac{C_1}{t^{\rho'}}\right). \tag{66}$$

Step 5. For each $t > 0$, define W_t as the set of potentials v such that $v(y) = 0$ for $y \notin \Lambda_0''$. For each $v \in W_t$ and $x \in \Lambda_0''$, consider the function

$$B(s, v, x) := \frac{s + 2d\kappa}{1 + \sum_{j=1}^{\infty} \sum_{\gamma \in \mathbb{P}^{2j+1}(x,x)} \prod_{z \in \gamma, z \neq z_1} \frac{\kappa}{s+2d\kappa - v_0(z)}},$$

which is well defined whenever $s > \bar{v} := \max\{v(x) : x \in \Lambda_0''\}$. Using the fact that the function $B(s, v)$ is increasing on $[\bar{v}, \infty)$ and part (i) of Theorem 4, note that on the event $A_1 \cap A_2$, the inequality $\lambda_0 \geq s$ is equivalent to $v_{(1)} \geq B(s, v, x_0)$, where $x_0 \in \Lambda_0''$ is such that $v_{(1)} = v(x_0)$. It follows from (60) of Step 2 and (65) of Step 4 that

$$\liminf_{t\to\infty} n(t)\mu\left(\lambda_0 \geq h(t) + \frac{1}{t}\log u + \frac{C_1}{t^{\rho'}}\right)$$

$$= \liminf_{t\to\infty} n(t)\mu\left(A_1, A_2, \lambda_0 \geq h(t) + \frac{1}{t}\log u + \frac{C_1}{t^{\rho'}}\right)$$

$$= \liminf_{t\to\infty} n(t)\mu\left(A_1, A_2, v_{(1)} \geq B\left(h(t) + \frac{1}{t}\log u + \frac{C_1}{t^{\rho'}}, v, x_0\right)\right). \tag{67}$$

Now, note that

$$A_1 \cap A_2 = \bigcup_{x \in \Lambda_0''} \left\{v(x) \geq (1 - \delta_1)\mathcal{A}_0 t^{\rho'-1}\right\} \cap A_2,$$

where the symbol $\cup$ denotes a disjoint union. Therefore, the probability appearing in the right-most hand side of display (67) can be written as

$$\mu\left(A_1, A_2, v_{(1)} \geq B\left(h(t) + \frac{1}{t}\log u + \frac{C_1}{t^{\rho'}}, v, x_0\right)\right)$$

$$= \sum_{x \in \Lambda_0''} \mu\left(A_2, v(x) \geq (1 - \delta_1)\mathcal{A}_0 t^{\rho'-1}, v(x) \geq B\left(h(t) + \frac{1}{t}\log u + \frac{C_1}{t^{\rho'}}, v, x\right)\right).$$

$$(68)$$

Furthermore, for each $x \in \Lambda_0''$, on the event $A_2 \cap \{v(x) \geq (1 - \delta_1)\mathcal{A}_0 t^{\rho'-1}\}$ whenever $s \geq (1 - \delta_1)t^{\rho'-1}$, we have that

$$\sum_{j=1}^{\infty} \sum_{\gamma \in \mathbb{P}^{2j+1}(x,x)} \prod_{z \in \gamma, z \neq z_1} \frac{\kappa}{s + 2d\kappa - v_0(z)} = O\left(\frac{1}{t^{\rho'-1}}\right),$$

Now, by part (ii) of Theorem 3, we have that

$$h(t) = h_0(t) + O\left(\frac{1}{t^{\rho'-1}}\right). \tag{69}$$

Hence, $h(t) = \mathcal{A}_0 \tau^{\rho'-1} + O(1)$ and on the event $A_2 \cap \{v(x) \geq (1 - \delta_1)\mathcal{A}_0 t^{\rho'-1}\}$ we have

$$B\left(h(t) + \frac{1}{t}\log u + \frac{C_1}{t^{\rho'}}, v, x\right) = h(t) + \frac{1}{t}\log u + \frac{C_1}{t^{\rho'}} + 2d\kappa + O\left(\frac{h(t)}{t^{\rho'-1}}\right)$$

$$\geq \mathcal{A}_0 \tau^{\rho'-1} + O(1) \tag{70}$$

and also that

$$B\left(h(t) + \frac{1}{t}\log u + \frac{C_1}{t^{\rho'}}, v, x\right) = B_M\left(h(t) + \frac{1}{t}\log u, \theta_x v\right) + O\left(\frac{1}{t^{\rho'}}\right) + O\left(\frac{1}{t^{2M(\rho'-1)}}\right)$$

$$= B_M(h(t), \theta_x v) + \frac{1}{t}\log u + O\left(\frac{1}{t^{\rho'}}\right) + O\left(\frac{1}{t^{2M(\rho'-1)}}\right). \tag{71}$$

where $\{\theta_x : x \in \mathbb{Z}^d\}$ is the canonical set of translations acting on the potentials $v \in W_t$ as $\theta_x v(y) := v(x + y)$. It follows from (70) that eventually in t for all $x \in \Lambda_0''$ one has the inclusion

$$\{v(x) \geq (1 - \delta_1)\mathcal{A}_0 t^{\rho'-1}\} \subset \left\{v(x) \geq B\left(h(t) + \frac{1}{t}\log u + \frac{C_1}{t^{\rho'}}, \theta_x v\right)\right\}. \tag{72}$$

Hence, going back to (68) we see after considering (71) and (72) that eventually in t it is true that

$$\mu\left(A_1, A_2, v_{(1)} \geq B\left(h(t) + \frac{1}{t}\log u + \frac{C_1}{t^{\rho'}}, v, x_0\right)\right)$$

$$= \sum_{x \in \Lambda_0''} \mu\left(A_2, v(x) \geq B_M\left(h(t), \theta_x v\right) + \frac{1}{t}\log u + O\left(\frac{1}{t^{\rho'}}\right) + O\left(\frac{1}{t^{2M(\rho'-1)}}\right)\right).$$

$$(73)$$

From the bound (59) of Step 3, using the equality (73) we conclude that in fact there is a constant C_2 such that

$$\liminf_{t\to\infty} n(t)\mu\left(A_1, A_2, v_{(1)} \geq B\left(h(t) + \frac{1}{t}\log u + \frac{C_1}{t^{\rho'}}, v, x_0\right)\right)$$

$$\geq \liminf_{t\to\infty} n(t) \sum_{x \in \Lambda_0''} \mu\left(v(x) \geq B_M\left(h(t), \theta_x v\right) + \frac{1}{t}\log u + \frac{C_2}{t^{\rho'}} + \frac{C_2}{t^{2M(\rho'-1)}}\right)$$

$$= \liminf_{t\to\infty} n(t)|\Lambda_0''|\mu\left(v(0) \geq B_M\left(h(t), v\right) + \frac{1}{t}\log u + \frac{C_2}{t^{\rho'}} + \frac{C_2}{t^{2M(\rho'-1)}}\right),$$

$$(74)$$

where in the last equality we have used the fact that the terms under the summation in (74) are translation invariant when $x \in \Lambda_0$ is far enough of the boundary, and that the points which do not have this property have a negligible cardinality with respect to $|\Lambda_0''|$. Combining this with (67) we get that

$$\liminf_{t\to\infty} n(t)\mu\left(\lambda_0 \geq h(t) + \frac{1}{t}\log u + \frac{C_1}{t^{\rho'}}\right)$$

$$\geq \liminf_{t\to\infty} e^{-\gamma \frac{t^{\rho'}}{\rho'}} E_\mu\left[e^{-\frac{1}{\rho}\left(B_M(h(t),v) + \frac{1}{t}\log u + \frac{C_2}{t^{\rho'}} + \frac{C_2}{t^{2M(\rho'-1)}}\right)^\rho}\right].$$

$$(75)$$

Now, for $x > 1$ and $\epsilon < 1$ one has that

$$(x + \epsilon)^\rho = x^\rho + \rho x^{\rho-1}\epsilon + O\left(\frac{\epsilon^2}{x^{2-\rho}}\right).$$

Calling for the moment $B_M = B_M(h(t), v)$, we see from the lower bound (69) of Step 1 that this implies that

$$\frac{1}{\rho}\left(B_M + \frac{1}{t}\log u + O\left(\frac{1}{t^{\rho'}}\right) + O\left(\frac{1}{t^{2M(\rho'-1)}}\right)\right)^\rho$$

$$= \frac{1}{\rho}B_M^\rho + \frac{B_M^{\rho-1}}{t}\log u + + O\left(\frac{B_M^{\rho-1}}{t^{2M(\rho'-1)}}\right) + \frac{1}{B_M^{2-\rho}}\left(O\left(\frac{1}{t^2}\right) + O\left(\frac{1}{t^{4M(\rho'-1)}}\right)\right).$$

$$(76)$$

Now, by the bounds (69) and (76), and the fact that $B_M \geq h(t) + 2d\kappa$, on the event A_2 we have that

$$\mathcal{A}_0 T^{\rho'-1} \leq B_M \leq \mathcal{A}_0 T^{\rho'-1} + \frac{C_5}{t^{\rho'-1}}. \tag{77}$$

Combining (77) with (76) we conclude that on the event A_2 one has that

$$\frac{1}{\rho}\left(B_M + \frac{1}{t}\log u + O\left(\frac{1}{t^{\rho'}}\right) + O\left(\frac{1}{t^{2M(\rho'-1)}}\right)\right)^{\rho}$$
$$= \frac{1}{\rho}B_M^{\rho} + \alpha\log u + O\left(\frac{1}{t^{\rho'}}\right) + O\left(\frac{1}{t^{2M\rho'-(2M+1)}}\right).$$

Inserting this in (75) we conclude that

$$\liminf_{t\to\infty} n(t)\mu\left(\lambda_0 \geq h(t) + \frac{1}{t}\log u + \frac{C_1}{t^{\rho'}}\right)$$
$$\geq \liminf_{t\to\infty} e^{-\gamma\frac{T^{\rho'}}{\rho'}}\frac{1}{u^{\alpha}}E_{\mu}\left[e^{-\frac{1}{\rho}B_M(h(t),v)+\frac{C_5}{t^{\rho'}}+\frac{C_5}{t^{2M\rho'-(2M+1)}}}\right] = \frac{1}{u^{\alpha}}, \tag{78}$$

where in the last equality we have used the definition of M given in (9), which implies that

$$2M\rho' - (2M+1) > 1,$$

and the definition of B_N given in (8). By a similar argument, where in order to control the terms close to the boundary we have to use the fact that for all $x \in \Lambda_0''$ it is true that for some constant C_6

$$\mu\left(v(x) \geq B_M\left(h(t) + \frac{1}{t}\log u, \theta_x v\right) - \frac{C_6}{t^{\rho'}} - \frac{C_6}{t^{2M\rho'-(2M+1)}}\right)$$
$$\leq \mu\left(v(0) \geq B_M\left(h(t) + \frac{1}{t}\log u, v\right) - \frac{C_6}{t^{\rho'}} - \frac{C_6}{t^{2M\rho'-(2M+1)}}\right),$$

we can get that

$$\limsup_{t\to\infty} n(t)\mu\left(\lambda_0 \geq h(t) + \frac{1}{t}\log u - \frac{C_1}{t^{\rho'}}\right) \leq \frac{1}{u^{\alpha}}. \tag{79}$$

Inserting (78) and (79) in (66) we finish the proof of the proposition. $\square$

6.5 The Truncated Moments

Here we will compute some quantities related to the moments of the random variable Y_0 defined in (53) which will be later used to prove that conditions (51) and (52) are satisfied.

Lemma 8 *Consider the random variable Y_0 given in (53). Assume that $0 < \gamma < \gamma_2$. Then the following statements are satisfied.*

(i) For $y > 0$ we have that

$$\lim_{t \to \infty} n(t) \left(E_\mu \left[Y_0(t), Y_0(t) \leq y \right] - \tilde{A}(t) \right) = \begin{cases} \frac{\alpha}{1-\alpha} y^{1-\alpha} & \text{if } \gamma \in (0, \gamma_1) \cup (\gamma_1, \gamma_2) \\ \log y & \text{if } \gamma = \gamma_1. \end{cases}$$

(80)

(ii) For $y > 0$ we have that

$$\lim_{t \to \infty} n(t) E_\mu \left[Y_0^2(t), Y_0(t) \leq y \right] = \frac{\alpha}{2-\alpha} y^{2-\alpha}.$$

Here $\alpha = \alpha(\gamma, \rho)$ is defined in (11).

Proof Part (i). We will first prove (80) for the case $0 < \gamma < \gamma_1$. Note that for each N we have that

$$E_\mu [Y_0(t) 1(Y_0(t) \leq y)] = \sum_{i=0}^{N-1} E_\mu \left[Y_0(t) 1 \left(\frac{i}{N} y \leq Y_0(t) \leq \frac{i+1}{N} y \right) \right]. \quad (81)$$

Now, for each $1 \leq i \leq N-1$ we have by Proposition 1 that

$$\overline{\lim}_{t \to \infty} n(t) E_\mu \left[Y_0(t) 1 \left(\frac{i}{N} y \leq Y_0(t) \leq \frac{i+1}{N} y \right) \right]$$
$$\leq \frac{i+1}{N} y \, \overline{\lim}_{t \to \infty} n(t) \mu \left(\frac{i}{N} y \leq Y_0(t) \leq \frac{i+1}{N} y \right) = \frac{i+1}{N} y \left(\frac{1}{(iy/N)^\alpha} - \frac{1}{((i+1)y/N)^\alpha} \right).$$

(82)

Similarly, we can conclude that

$$\lim_{t \to \infty} n(t) E_\mu \left[Y_0(t) 1 \left(\frac{i}{N} y \leq Y_0(t) \leq \frac{i+1}{N} y \right) \right] \geq \frac{i}{N} y \left(\frac{1}{(iy/N)^\alpha} - \frac{1}{((i+1)y/N)^\alpha} \right).$$

(83)

Combining (82) and (83) with (81) we get

$$\sum_{i=0}^{N-1} \frac{i}{N} y \left(\frac{1}{(iy/N)^\alpha} - \frac{1}{((i+1)y/N)^\alpha} \right) \leq \underline{\lim}_{t \to \infty} n(t) E_\mu [Y_0(t) 1(Y_0(t) \leq y)]$$

$$\leq \overline{\lim}_{t \to \infty} n(t) E_\mu [Y_0(t) 1(Y_0(t) \leq y)] \leq \sum_{i=0}^{N-1} \frac{i+1}{N} y \left(\frac{1}{(iy/N)^\alpha} - \frac{1}{((i+1)y/N)^\alpha} \right).$$

(84)

Now note that for $0 < \gamma < \gamma_1$, one has that $0 < \alpha < 1$. We can hence take the limit when $N \to \infty$ in (84) to deduce that

$$\lim_{t\to\infty} n(t)E_\mu[Y_0(t)1(Y_0(t) \le y)] = \int_0^y \frac{1}{x^\alpha} dx - y^{1-\alpha} = \frac{\alpha}{1-\alpha} y^{1-\alpha},$$

which proves (80) for the case $0 < \gamma < \gamma_1$. The proof of (80) for the case $\gamma_1 < \gamma < \gamma_2$ has to take into account that

$$E_\mu[Y_0(t), Y_0(t) \le y] - \tilde{A}(t) = -E_\mu[Y_0(t), Y_0(t) > y],$$

and then follows an analysis similar to the previous case. The proof of (80) in the case $\gamma = \gamma_1$ uses the fact that

$$E_\mu[Y_0(t), Y_0(t) \le y] - \tilde{A}(t) = \begin{cases} E_\mu[Y_0(t), 1 < Y_0(t) \le y] & \text{if} \qquad y \ge 1 \\ -E_\mu[Y_0(t), y < Y_0(t) < 1] & \text{if} \qquad y < 1, \end{cases}$$

which then enables one to prove that

$$\lim_{N\to\infty} \left(E_\mu[Y_0(t), Y_0(t) \le y] - \tilde{A}(t) \right) = \int_1^y \frac{1}{x} dx = \log y.$$

Part (ii). In analogy to the inequalities (84), we can conclude that

$$\sum_{i=0}^{N-1} \left(\frac{i}{N}y\right)^2 \left(\frac{1}{(iy/N)^\alpha} - \frac{1}{((i+1)y/N)^\alpha}\right) \le \lim_{t\to\infty} n(t)E_\mu\left[Y_0^2(t)1(Y_0(t) \le y)\right]$$

$$\le \lim_{t\to\infty} n(t)E_\mu\left[Y_0^2(t)1(Y_0(t) \le y)\right] \le \sum_{i=0}^{N-1} \left(\frac{i+1}{N}y\right)^2 \left(\frac{1}{(iy/N)^\alpha} - \frac{1}{((i+1)y/N)^\alpha}\right).$$

As in (84), since $0 < \gamma < \gamma_2$ implies that $0 < \alpha < 2$, we can take the limit as $N \to \infty$ above to deduce that

$$\lim_{t\to\infty} n(t)E_\mu\left[Y_0^2(t)1(Y_0(t) \le y)\right] = \int_0^y \frac{1}{x^{\alpha-1}} dx - y^{2-\alpha} = \frac{\alpha}{2-\alpha} y^{2-\alpha},$$

□

6.6   The Parameters of the Infinite Divisible Law

Here we will show that the i.i.d. random variables $(Y_i)_{i\in\mathcal{I}}$ satisfy conditions (51) and (52) of part (ii) of Theorem 5 with

$$\sigma^2 = \lim_{y\to 0} \lim_{t\to\infty} n(t)Var_t(Z_y(t)) = 0, \tag{85}$$

and

$$\nu := \begin{cases} \frac{\alpha\pi}{2\cos\frac{\alpha\pi}{2}} & \text{if } \gamma \in (0, \gamma_1) \cup (\gamma_1, \gamma_2) \\ 0 & \text{if } \gamma = \gamma_1. \end{cases} \tag{86}$$

The proof of (85) is a direct consequence of part (ii) of Lemma 8. Since the proof of (86) is completely analogous to the proofs of Propositions 6.4 and 6.5 of [6], we will just give an outline here. Note that by (52), part (i) of Lemma 8 and (55), we should have

$$\nu = \frac{\alpha}{1-\alpha} y^{1-\alpha} + \alpha \int_y^\infty \frac{x^{-\alpha}}{1+x^2} dx - \alpha \int_0^y \frac{x^{2-\alpha}}{1+x^2} dx.$$

Using the identity $\int_0^\infty \frac{x^{-\alpha}}{1+x^2} dx = \frac{\pi}{2\cos\frac{\alpha\pi}{2}}$ in the case $\gamma \neq \gamma_1$, we can obtain (86) for that case. A similar analysis gives the case $\gamma = \gamma_1$.

6.7 Conclusion

Gathering (55), (85) and (86), into Theorem 5, and the observation that it is enough to find the limiting law in (48), we conclude that for $0 < \gamma < \gamma_2$,

$$\lim_{t \to \infty} \frac{1}{s(t)} m_L(t) = Z,$$

where the convergence is in distribution and Z is an infinite divisible distribution with characteristic function

$$\phi(t) = \exp\left\{ i\nu u + \alpha \int_0^\infty \left(e^{itx} - 1 - \frac{itx}{1+x^2} \right) \frac{dx}{x^{\alpha+1}} \right\}, \tag{87}$$

with ν defined in (86). Now, by Theorem 6.6 of [6], we know that any infinite divisible distribution corresponding to (87) has the canonical representation given in (14). This finishes the proof of Theorem 2.

7 Annealed Asymptotics

In this section we will prove (7) of Theorem 1.

Proposition 2 *Consider the solution* $(m(x, t))_{x \in \mathbb{Z}^d, t \geq 0}$ *of the parabolic Anderson equation (1) with Weibull potential of parameter* $\rho > 1$. *Then*

$$\langle m(0, t) \rangle \sim \left(\frac{\pi}{\rho - 1} \right)^{1/2} t^{1 - \frac{\rho'}{2}} e^{\frac{t^{\rho'}}{\rho'} - 2d(\kappa t - t^{2 - \rho'})}.$$

Define the *cumulant generating function* by

$$H(t) = \log \langle e^{v(0)t} \rangle, \qquad t \geq 0.$$

Using the independence of the coordinates of the potential v, note that

$$\langle m(0, t) \rangle = E_0 \left[e^{\sum_{x \in \mathbb{Z}^d} H(\mathcal{L}(t,x))} \right],$$

where E_0 is the expectation defined by the law P_0 of a simple symmetric random walk on $\mathbb{Z}^d$, starting from 0, of total jump rate 1 and for each $t \geq 0$ and site $x \in \mathbb{Z}^d$, $\mathcal{L}(t, x)$ is the total time spent by the random walk in the time interval $[0, t]$ at x.

The basis of our proof of Theorem 1 will be the following result, where we recall that $\mathbb{P}^n(0)$ is the set of paths on $\mathbb{Z}^d$ of length n starting from 0 (see Sect. 2 for the precise definition).

Proposition 3 *Consider the solution* $(m(x, t))_{x \in \mathbb{Z}^d, t \geq 0}$ *of the parabolic Anderson equation (1) with Weibull potential of parameter* $\rho > 1$. *Then*

$$\langle m(0, t) \rangle \sim \left(\frac{\pi}{\rho - 1} \right)^{1/2} e^{\frac{t^{\rho'}}{\rho'} - 2d\kappa t} \sum_{n=1}^{\infty} \frac{\kappa^{n-1}}{t^{(\rho'-1)n}} \sum_{\gamma \in \mathbb{P}^n(0)} \sum_{i=1}^{k} \frac{t^{\rho'(n_i - \frac{1}{2})}}{(n_i - 1)!}. \qquad (88)$$

7.1 Preliminary Estimates

To prove Proposition 3, we will need the precise asymptotics of the cumulant generating function, given by the following lemma.

Lemma 9 *For* $t \geq 0$,

$$H(t) = \frac{t^{\rho'}}{\rho'} + \frac{\rho'}{2} \log t + \frac{1}{2} \log \frac{\pi}{\rho - 1} + \varepsilon(t),$$

where $\lim_{t \to \infty} \varepsilon(t) = 0$.

Proof Note that

$$\langle e^{tv(0)} \rangle = \int_0^{\infty} u^{\rho-1} e^{-u^\rho/\rho} e^{ut} \, du.$$

Making the variable change $u = w t^{1/(\rho-1)}$, this becomes

$$t^{\rho'} \int_0^{\infty} w^{\rho-1} e^{-t^{\rho'} \left(\frac{w^\rho}{\rho} - w \right)} \, dw. \qquad (89)$$

The function $f(w) := \frac{w^\rho}{\rho} - w$, attains its minimum value $-1/\alpha'$ at $w = 1$, having the expansion

$$f(w) = -\frac{1}{\rho'} + (w-1)^2(\rho - 1) + (w-1)^3 f^{(3)}(\bar{w}), \qquad (90)$$

where $\bar{w}$ is between 1 and w. Now, let $0 < \epsilon < 1$, and make the decomposition

$$\int_0^\infty w^{\rho-1} e^{-t^{\rho'}\left(\frac{w^\rho}{\rho} - w\right)} dw = A_1 + A_2 + A_3, \qquad (91)$$

where $A_1 = \int_0^{1-\epsilon} w^{\rho-1} e^{-t^{\rho'}\left(\frac{w^\rho}{\rho} - w\right)} dw$, $A_2 = \int_{1-\epsilon}^{1+\epsilon} w^{\rho-1} e^{-t^{\rho'}\left(\frac{w^\rho}{\rho} - w\right)} dw$ and $A_3 = \int_{1+\epsilon}^\infty w^{\rho-1} e^{-t^{\rho'}\left(\frac{w^\rho}{\rho} - w\right)} dw$. It is easy to check that there exists a constant $C > 0$ such that, $A_1 \le C e^{-t^{\rho'} f(1+\epsilon)}$ and $A_2 \le C e^{-t^{\rho'} f(1+\epsilon)}$. Furthermore, from the expansion (90), we see that

$$A_3 \le (1+\epsilon)^{\rho-1} e^{c\epsilon^3} e^{\frac{t^{\rho'}}{\rho'}} \int_{-\infty}^\infty e^{-(\rho-1)t^{\rho'}x^2} dx = (1+\epsilon)^{\rho-1} e^{c\epsilon^3} e^{\frac{t^{\rho'}}{\rho'}} \sqrt{\frac{\pi}{(\rho-1)t^{\rho'}}},$$

for $c = |f^{(3)}(2)|$. Similarly we have that,

$$A_3 \ge (1-\epsilon)^{\rho-1} e^{-c\epsilon^3} e^{\frac{t^{\rho'}}{\rho'}} \left(\sqrt{\frac{\pi}{(\rho-1)t^{\rho'}}} - O(e^{-t^{\rho'}\epsilon^2}) \right).$$

Substituting these estimates for A_1, A_2 and A_3 in (91) and this in (89), and choosing $\epsilon = t^{-\gamma}$ for $\gamma > 0$ small enough, we finish the proof of the lemma. $\qquad \square$

Let us finish this section with the following elementary computation.

Lemma 10 *Let $k \ge 1$ and $n_1, n_2, \ldots, n_k$ be natural numbers larger than 0. For $u \ge 0$, let $J_{n_1,\ldots,n_k}(u) = \int_{\sum_{i=1}^k x_k < u} x_1^{n_1-1} \cdots x_k^{n_k-1} dx_1 \cdots dx_k$. Then*

$$\int_0^\infty J_{n_1,\ldots,n_k}(u) e^{-u} du = \prod_{i=1}^k (n_i - 1)!.$$

7.2 Path Decomposition of Annealed First Moment

For $j \ge 1$, call τ_j the time of the j-th jump of the random walk. Note that these random times are independent exponential random variables of rate $2d\kappa$. Also

$$\langle m(0,t) \rangle = \sum_{\gamma \in \mathbb{P}(0)} E_0 \left[e^{\sum_{x \in \mathbb{Z}^d} H(\mathcal{L}(t,x))} C_\gamma \right], \qquad (92)$$

where C_γ is the event that in the time interval $[0, t]$, the random walk follows the path γ. Let us examine a single term A_γ, of the summation in (92) corresponding to a path $\gamma = (x_1 = 0, x_2, \dots, x_n)$, visiting sites $\{y_1 = 0, y_2, \dots, y_k\}$. Without loss of generality, we assume that $y_k = x_n$, so that y_k is the last visited site. For each $1 \le i \le k$, let us call n_i, the number of times the path visits site y_i. Furthermore, let $i_1, \dots, i_{n_i}$ be the set of indices of the set $\{1, 2, \dots, n - 1\}$ such that $x_{1_1} = \cdots = x_{i_{n_i}} = y_i$. Furthermore, note that $C_\gamma = \bar{C}_\gamma \cap \{\tau_1 + \cdots + \tau_{n-1} < t\} \cap \{\tau_n \ge t - \tau_1 - \cdots \tau_{n-1}\}$, where $\bar{C}_\gamma = \cap_{j=1}^{n-1}\{X_{\tau_j} = x_{j+1}\}$. Therefore, if $S_{n-1} := \{s_1 + \cdots + s_{n-1} < t\} \subset \mathbb{R}^{n-1}$ then

$$A_\gamma = \frac{(2d\kappa)^{n-1}}{(2d)^{n-1}} \int_{S_{n-1}} e^{-2d\kappa(s_1 + \cdots + s_{n-1})} e^{-2d\kappa(t - s_1 - \cdots - s_{n-1})} e^{H_1 + \cdots + H_k} ds_1 \cdots ds_{n-1},$$

where $H_i := H(s_{i_1} + \cdots + s_{i_{n_i}})$, for $1 \le i \le k - 1$, while $H_k := H(t - v_1 - \cdots - v_{k-1})$, with $v_i = s_{i_1} + \cdots + s_{i_{n_i}}$. Thus,

$$A_\gamma = \kappa^{n-1} e^{-2d\kappa t} \int_{T_k} \frac{v_1^{n_1-1}}{(n_1 - 1)!} \cdots \frac{v_k^{n_k-1}}{(n_k - 1)!} e^{H(v_1) + \cdots + H(v_{k-1}) + H_k} dv_1 \cdots dv_k$$

$$= \kappa^{n-1} t^n e^{-2d\kappa t} \frac{1}{\prod_{i=1}^k (n_i - 1)!} \int_{T'_k} u_1^{n_1-1} \cdots u_k^{n_k-1} e^{H(tu_1) + \cdots + H(tu_{k-1}) + H(t\bar{u}_k)} du_1 \cdots du_k,$$

$$(93)$$

where $T_k := \{v_1 + \cdots + v_k < t\}$, $T'_k := \{u_1 + \cdots + u_k < 1\}$, $\bar{u}_k := 1 - u_1 - \cdots - u_{k-1}$, in the second equality we made the variable change $u_i = tv_i$ and we have used the fact that $n_1 + \cdots + n_k = n$. Let now $0 < \delta < 1/2$, and define $W_k = \{u_1 + \cdots + u_k < 1, \max_{1 \le i \le k} u_i < 1 - \delta\}$ and $V_k = T'_k - W_k$.

7.3 Asymptotic Lower Bound

Here we compute an asymptotic lower bound for A_γ. Note that $H(t\bar{u}_k) \ge H(tu_k)$. Hence

$$A_\gamma \ge \kappa^{n-1} t^n e^{-2d\kappa t} \frac{1}{\prod_{i=1}^k (n_i - 1)!} \int_{V_k} u_1^{n_1-1} \cdots u_k^{n_k-1} e^{H(tu_1) + \cdots + H(tu_k)} du_1 \cdots du_k.$$

Now, note that $V_k = \cup_{j=1}^k V_{k,j}$, where the union is disjoint and $V_{k,j} := \{u_1 + \cdots + u_k < 1, u_j \ge 1 - \delta\}$. Therefore, $A_\gamma \ge t^n e^{-2dt} \frac{1}{\prod_{i=1}^k (n_i-1)!} \sum_{j=1}^k I_{k,j}$ where,

$$I_{k,j} := \int_{V_{k,j}} u_1^{n_1-1} \cdots u_k^{n_k-1} e^{H(tu_1) + \cdots + H(tu_k)} du_1 \cdots du_k.$$

By symmetry it is enough to examine $I_{k,1}$. From Lemma 9 and (93) we obtain

$$I_{k,1} \geq t^{\frac{\rho'}{2}} \left(\frac{\pi}{\rho-1}\right)^{1/2} \int_{V_{k,1}} u_1^{n_1-1+\frac{\rho'}{2}} u_2^{n_1-1} \cdots u_k^{n_k-1} e^{\frac{t\rho'}{\rho'} u_1^{\rho'} + \varepsilon(tu_1)} du_1 \cdots du_k.$$

Now, the variable change $u_1' := 1 - u_1$ transforms the integral in the above expression to

$$I_{k,1}' := \int_0^\delta \int_{T_{k,1}} (1-u_1')^{n_1-1+\frac{\rho'}{2}} u_2^{n_2-1} \cdots u_k^{n_k-1} e^{\frac{t\rho'}{\rho'}(1-u_1')^{\rho'} + \varepsilon(t(1-u_1'))} du_1' du_2 \cdots du_k.$$

where $T_{k,1} := \{u_2 + \cdots u_k < u_1'\}$. Note that, $(1-u_1')^{\rho'} = 1 - \rho' u_1' + \rho'(\rho'-1)\bar{u}^2/2$, for some $0 \leq \bar{u} \leq u_1'$. Therefore,

$$1 - \rho' u_1' \leq (1-u_1')^{\rho'} \leq 1 - \rho' u_1' + \rho'(\rho'-1) u_1'^2/2. \tag{94}$$

Then, we get a lower bound

$$I_{k,1}' \geq e^{\frac{t\rho'}{\rho'}} \int_0^\delta J_{\gamma,1}(x)(1-x)^{n_1-1+\frac{\rho'}{2}} e^{-t\rho' x + \varepsilon(t(1-x))} dx$$
$$= \frac{1}{t^{\rho'}} e^{\frac{t\rho'}{\rho'}} \int_0^{\delta t\rho'} J_{\gamma,1}\left(\frac{y}{t^{\rho'}}\right) \left(1 - \frac{y}{t^{\rho'}}\right)^{n_1-1+\frac{\rho'}{2}} e^{-y+\varepsilon(t(1-y/t^{\rho'}))} dy. \tag{95}$$

where for $1 \leq j \leq k$, we define $J_{\gamma,j}(x) := \int_{\sum_{i\neq j} u_i < x} \prod_{i\neq j} u_i^{n_i-1} du_i$. Now,

$$J_{\gamma,1}\left(\frac{y}{t^{\rho'}}\right) = t^{-\rho'(n-n_1)} J_{\gamma}(y). \tag{96}$$

Therefore, by the dominated convergence theorem, we see that the right-hand side of (95) is asymptotically equal to, $\frac{1}{t^{\rho'(n+1-n_1)}} e^{\frac{t\rho'}{\rho'}} \int_0^\infty J_{\gamma,1}(x) e^{-x} dx$. We therefore, conclude that $I_{k,1}$ is asymptotically lower bounded by

$$\left(\frac{\pi}{\rho-1}\right)^{1/2} \frac{1}{t^{\rho'(n+\frac{1}{2}-n_1)}} e^{\frac{t\rho'}{\rho'}} \int_0^\infty J_{\gamma,1}(x) e^{-x} dx.$$

and that A_γ is asymptotically lower bounded by

$$\frac{\left(\frac{\pi}{\rho-1}\right)^{1/2}}{\prod_{i=1}^k (n_i-1)!} t^n e^{-2dt} \sum_{j=1}^k \frac{1}{t^{\rho'(n+\frac{1}{2}-n_j)}} e^{\frac{t\rho'}{\rho'}} K_{\gamma,j},$$

where $K_{\gamma,j} := \int_0^\infty J_{\gamma,j}(u) e^{-u} du$. Using Lemma 10, we obtain the desired asymptotic lower bound.

7.4 Asymptotic Upper Bound

Here we will obtain an asymptotic upper bound for A_γ. First, let us examine the integral

$$\bar{I}_k := \int_{\bar{W}_k} u_1^{n_1-1} \cdots u_k^{n_k-1} e^{H(tu_1)+\cdots+H(tu_{k-1})+H(t\bar{u}_k)} du_1 \cdots du_k,$$

where $\bar{W}_k := \{u_1 + \cdots + u_k < 1, \max_{1 \le i \le k-1} u_i < 1 - \delta, \bar{u}_k < 1 - \delta\}$. Since on $\bar{W}_k$ we have $\max_{1 \le j \le k-1} u_j < 1 - \delta$ and $\bar{u}_k < 1 - \delta$, the following inequality is satisfied: $\frac{1}{\rho'} t^{\rho'} (u_1^{\rho'} + \cdots + u_{k-1}^{\rho'} + \bar{u}_k^{\rho'}) \le \frac{1}{\rho'} (1 - \delta)^{\rho'-1} t^{\rho'}$. Hence, the integral $\bar{I}_k$ is upper bounded by

$$(t(1-\delta))^k \frac{\rho'}{2} \left(\frac{\pi}{\rho-1} \right)^{k/2} e^{(1-\delta)^{\rho'-1} \frac{t\rho'}{\rho'} + k\varepsilon(t(1-\delta))} \int_{W_k} u_1^{n_1-1} \cdots u_k^{n_k-1} du_1 \cdots du_k.$$

Therefore, $\bar{I}_k \le c(k) e^{(1-\delta)^{\rho'-1} \frac{t\rho'}{\rho'}}$, for some constant $c(k)$. Let us now define $\bar{V}_k := T'_k - \bar{W}_k$ and note that $\bar{V}_k = \cup_{j=1}^k \bar{V}_{k,j}$ where the union is disjoint and $\bar{V}_{k,j} := \{u_1 + \cdots + u_k < 1, u_j \ge 1 - \delta\}$ for $1 \le j \le k - 1$ while $\bar{V}_{k,k} := \{u_1 + \cdots + u_k < 1, \bar{u}_k \ge 1 - \delta\}$. Then if

$$\bar{I}_{k,j} := \int_{\bar{V}_{k,j}} u_1^{n_1-1} \cdots u_k^{n_k-1} e^{H(tu_1)+\cdots+H(tu_{k-1})+H(t\bar{u}_k)} du_1 \cdots du_k,$$

it follows that

$$A_\gamma \le c(k) e^{(1-\delta)^{\rho'-1} \frac{t\rho'}{\rho'}} + t^n e^{-2dt} \frac{1}{\prod_{i=1}^k (n_i - 1)!} \sum_{j=1}^k \bar{I}_{k,j}. \tag{97}$$

Then, we need to upper bound the integrals $\bar{I}_{k,j}$. Define $u'_1 = 1 - u_1$. By the subadditivity of the cumulant generating function, note that $H(tu_2) + \cdots + H(tu_{k-1}) + H(t\bar{u}_k) \le H(t(u_2 + \cdots + u_{k-1} + \bar{u}_k)) = H(t(1 - u_1))$. Hence, $\bar{I}_{k,1}$ is upper bounded by

$$t^{\frac{\rho'}{2}} \left(\frac{\pi}{\rho-1} \right)^{1/2} \int_{\bar{V}_{k,1}} u_1^{n_1-1+\frac{\rho'}{2}} u_2^{n_1-1} \cdots u_k^{n_k-1} e^{\frac{t\rho'}{\rho'} u_1^{\rho'} + H(t(1-u_1)) + \varepsilon(tu_1)} du_1 \cdots du_k.$$

But by the second inequality of display (94), the integral in the above expression is upper bounded by

$$e^{\frac{t^{\rho'}}{\rho'}} \int_0^\delta J_{\gamma,1}(x)(1-x)^{n_1-1+\frac{\rho'}{2}} e^{-t^{\rho'}x+t^{\rho'}\rho'(\rho'-1)\frac{x^2}{2}+H(tx)+\varepsilon(t(1-x))} dx$$

$$= \frac{1}{t^{\rho'}} e^{\frac{t^{\rho'}}{\rho'}} \int_0^{\delta t^{\rho'}} J_{\gamma,1}\left(\frac{y}{t^{\rho'}}\right) e^{-y+\rho'(\rho'-1)\frac{y^2}{2t^{\rho'}}+H(t^{1-\rho'})+\varepsilon(t(1-y/t^{\rho'}))} dy.$$

By (96) and the dominated convergence theorem, this is asymptotically equivalent as $t \to \infty$ to $\frac{1}{t^{\rho'(n+1-n_1)}} e^{\frac{t^{\rho'}}{\rho'}} K_{\gamma,1}$. This provides an asymptotic upper bound for $\bar{I}_{k,1}$. A similar argument gives us the asymptotic upper bounds $\frac{1}{t^{\rho'(n+1-n_1)}} e^{\frac{t^{\rho'}}{\rho'}} K_{\gamma,j}$, for $\bar{I}_{k,j}$, $2 \le j \le k$. Combining these estimates with (97), and using Lemma 10, finishes the proof of the asymptotic upper bound.

7.5 Proof of Proposition 2

Define $\mathbb{Q}^n(0,0)$ as the set of paths $\gamma \in \mathbb{P}^n(0,0)$ such that 0 is visited n_1 times with $n_1 > n/2$. Note that if n is even this set is empty, whereas for $n > 1$ odd all of these paths start and end at 0, and they are of the form $\gamma = (x_1 = 0, x_2, x_3, \ldots, x_{n-1}, x_n = 0)$, with $x_i = 0$ for i odd, $1 \le i \le n$. Let us express the series

$$\sum_{n=1}^{\infty} \frac{1}{t^{(\rho'-1)n}} \sum_{\gamma \in \mathbb{P}^n(0)} \sum_{i=1}^{k} \frac{t^{\rho'(n_i-\frac{1}{2})}}{(n_i-1)!},$$

of the right-hand side of display (88) as, $S_1 + S_2 + S_3$, where

$$S_1 := \sum_{n=1}^{\infty} \frac{1}{t^{(\rho'-1)n}} \sum_{\gamma \in \mathbb{Q}^n(0)} \frac{t^{\rho'(n_1-\frac{1}{2})}}{(n_1-1)!}$$

$$S_2 := \sum_{n=2}^{\infty} \frac{1}{t^{(\rho'-1)n}} \sum_{\gamma \in \mathbb{Q}^n(0)} \sum_{i=2}^{k} \frac{t^{\rho'(n_i-\frac{1}{2})}}{(n_i-1)!}$$

$$S_3 := \sum_{n=2}^{\infty} \frac{1}{t^{(\rho'-1)n}} \sum_{\gamma \in \mathbb{R}^n(0)} \sum_{i=1}^{k} \frac{t^{\rho'(n_i-\frac{1}{2})}}{(n_i-1)!},$$

where $\mathbb{R}^n(0) := \mathbb{P}^n(0) - \mathbb{Q}^n(0)$. We will show that only S_1 contributes to the final result.

By our previous remarks, note that the summation in S_1 runs only over odd values of $n = 2m + 1$, $m \ge 0$. Furthermore, $n_1 = m + 1$ and $|\mathbb{Q}^n(0)| = (2d)^m$. Therefore

$$S_1 = \sum_{m=0}^{\infty} \frac{1}{t^{(\rho'-1)(2m+1)}} (2d)^m \frac{t^{\rho'(m+\frac{1}{2})}}{m!} = t^{1-\frac{\rho'}{2}} \sum_{m=0}^{\infty} \frac{(2dt^{2-\rho'})^m}{m!} = t^{1-\frac{\rho'}{2}} e^{2dt^{2-\rho'}}.$$

Let us next examine S_2. Note that for $\gamma \in \mathbb{Q}^n(0)$, we have $n_i = 1$ for $2 \leq i \leq k$. Again, only the odd terms in the series count, and we have

$$S_2 = t^{1-\frac{\rho'}{2}} \sum_{m=1}^{\infty} \left(\frac{2d}{t^{2(\rho'-1)}} \right)^m = t^{1-\frac{\rho'}{2}} \frac{2d}{t^{2(\rho'-1)} - 2d} \ll S_1.$$

Now, let $c < 1/\rho'$. Let us write $S_3 = S_3' + S_3''$, where

$$S_3' := \sum_{n=2}^{\infty} \frac{1}{t^{(\rho'-1)n}} \sum_{\gamma \in \mathbb{R}^n(0)} \sum_{i=1}^{k} 1(n_i \geq cn) \frac{t^{\rho'\left(n_i - \frac{1}{2}\right)}}{(n_i - 1)!}$$

$$S_3'' := \sum_{n=2}^{\infty} \frac{1}{t^{(\rho'-1)n}} \sum_{\gamma \in \mathbb{R}^n(0)} \sum_{i=1}^{k} 1(n_i < cn) \frac{t^{\rho'\left(n_i - \frac{1}{2}\right)}}{(n_i - 1)!}.$$

Using the bounds $k \leq n$ and $|\mathbb{Q}^n(0)| \leq (2d)^n$, note that

$$S_3'' \leq \sum_{n=2}^{\infty} \frac{1}{t^{(\rho'-1)n}} n(2d)^n t^{\rho'\left(cn-\frac{1}{2}\right)} = t^{-\frac{\rho'}{2}} \sum_{n=2}^{\infty} n \left(2dt^{2(1+\rho(c-1))} \right)^n \ll S_1,$$

since $1 + \rho(c - 1) < 0$. To estimate S_3', note that all the paths in $\mathbb{R}^n(0)$, are such that $n_i \leq n/2$, for $1 \leq i \leq k$. Then

$$S_3' \leq t^{-\frac{\rho'}{2}} \sum_{n=2}^{\infty} \frac{(2d)^n}{t^{(\rho'-1)n}} \sum_{m=\lfloor cn \rfloor}^{\lfloor n/2 \rfloor} \frac{t^{\rho'm}}{(m-1)!} \leq t^{-\frac{\rho'}{2}} \sum_{m=0}^{\infty} \sum_{n=f(m)}^{\infty} \frac{(2d)^n}{t^{(\rho'-1)n}} \frac{t^{\rho'm}}{(m-1)!}.$$

Now, performing summation by parts, we see that

$$\sum_{m=0}^{\infty} \sum_{n=f(m)}^{\infty} \frac{(2d)^n}{t^{(\rho'-1)n}} \frac{t^{\rho'm}}{(m-1)!} = \left(\frac{2d}{t^{\rho'-1}} \right)^2 + \sum_{m=1}^{\infty} \left(\frac{2d}{t^{\rho'-1}} \right)^{2m} \frac{t^{\rho'm}}{(m-1)!}$$
$$= (2d)^2 t^{2-2\rho'} + (2d)^2 t^{2-\rho'} e^{2dt^{2-\rho'}}.$$

Hence

$$S_3' \leq (2d)^2 t^{2-5\rho'/2} + (2d)^2 t^{2-3\rho'/2} e^{2dt^{2-\rho'}} \ll t^{1-\rho'/2} e^{2dt^{2-\rho'}}.$$

8 Asymptotic Expansion of the Scaling Function for $1 < \rho < (3 + \sqrt{17})/2$

Here we will prove Corollary 1, giving an expansion of the scaling function h defined in (17) for $1 < \rho < (3 + \sqrt{17})/2$. We make the calculations only up to the value $(3 + \sqrt{17})/2$ because for ρ larger than this number extra terms in the expansion of h have to be computed, and in order to keep the length of this section limited, we have decided to stop there. There seems to be no straightforward interpretation on the appearance of this number. On the other hand, as it will be shown, transitions in the behavior of h occur for $\rho = 2, 3$ and the number $(3 + \sqrt{17})/2$. Above this last value, transitions should appear at integer values of ρ, and additionally we expect that for some other non-integer values of ρ. To prove part (i) of Corollary 1, note that when $1 < \rho < 2$, we have that $\rho' > 2$. Therefore, $M = 1$ [c.f. (9)]. It follows from (22) of Theorem 3, that $h_1(t) = O\left(\frac{1}{t}\right)$. This together with (19) of Theorem 3, shows that

$$h(t) = A_0 \tau^{\rho'-1} - 2d\kappa + O\left(\frac{1}{t^{\rho'-1}}\right),$$

from where using that $\rho' > 2$, part (i) of Corollary 1 follows. Let us now prove parts (ii) and (iii) of Corollary 1. Note that for ϵ small enough

$$E_\mu\left[e^{-\frac{1}{\rho}B_1(h_0, v)^\rho} 1(\max_{e \in E} v(e) > (1-\epsilon)A_0\tau^{\rho'-1})\right]$$

$$\le e^{-\frac{A_0^\rho \tau^{\rho'}}{1+\frac{1}{2d}}} \mu(\max_{e \in E} v(e) > (1-\epsilon)A_0\tau^{\rho'-1}) = e^{-\frac{A_0^\rho \tau^{\rho'}}{1+\frac{1}{2d}}} e^{-\frac{\gamma}{\rho'}(1-\epsilon)^\rho \tau^{\rho'}}$$

$$= o\left(e^{-\frac{\gamma}{\rho'}\tau^{\rho'}}\right), \qquad (98)$$

where in the inequality we have used the fact that $B_1(h_0, v) \ge \frac{1}{1+\frac{1}{2d}} A_0 \tau^{\rho'-1}$ and in the last equality that ϵ is small enough.

It follows from (19) that

$$h(t) = A_0\tau^{\rho'-1} - 2d\kappa + h_1(t) + O\left(\frac{1}{t^{3(\rho'-1)}}\right).$$

Now

$$h_1(t) := \frac{1}{A_0^{\rho-1}}\left(\frac{\gamma}{\rho'}\tau^{\rho'-1}(t) + \frac{1}{\tau(t)}\log E_\mu\left[e^{-\frac{1}{\rho}B_1(h_0, v)^\rho}\right]\right) \qquad (99)$$

and for $s \ge 0$,

$$B_1(s, v) = \frac{s + 2d\kappa}{1 + \frac{\kappa}{2d\kappa+s}\sum_{e:|e|_1=1}\frac{\kappa}{2d\kappa+(s-v(e))_+}}.$$

In analogy with the proof of part (i) of Lemma 6, but going to a higher order Taylor expansion, we see that when $\sup_{e \in E} v_0(e) \le A_0\tau^{\rho'-1}(1-\epsilon)$ for some $\epsilon > 0$, we have

$$\frac{1}{\rho}B_1^\rho(h_0(t), v) = \frac{\gamma}{\rho'}\tau^{\rho'} - \frac{\gamma\rho}{\rho'A_0}\kappa^2\sum_{e \in E}\frac{\tau}{A_0\tau^{\rho'-1} - v_0(e)} + O\left(\frac{1}{\tau^{3\rho'-4}}\right).$$

Now, note that for all $x > 0$ one has that

$$\frac{\tau}{A_0 \tau^{\rho'-1} - x} = A_0^{-1} \tau^{2-\rho'} + A_0^{-2} x \tau^{3-2\rho'} + A_0^{-3} x^2 \frac{\tau^{4-3\rho'}}{1 - \frac{x}{\tau^{\rho'-1}}}$$

$$= A_0^{-1} \tau^{2-\rho'} + A_0^{-2} x \tau^{3-2\rho'} + x^2 O\left(\tau^{4-3\rho'}\right)$$

It follows that

$$E_\mu\left[e^{-\frac{1}{\rho}B_1(h_0,v)^\rho} 1(\max_{e \in E} v(e) \leq (1-\epsilon)A_0 \tau^{\rho'-1})\right] = e^{-\frac{\gamma}{\rho'}\tau^{\rho'} + 2d\frac{\gamma\rho}{\rho'}A_0^{-2}\kappa^2\tau^{2-\rho'}}$$

$$\times \left(\int_0^{(1-\epsilon)A_0 \tau^{\rho'-1}} e^{K_1 x \tau^{3-2\rho'} + x^2 O\left(\tau^{4-3\rho'}\right)} e^{-\frac{1}{\rho}x^\rho} x^{\rho-1} dx\right)^{2d}, \qquad (100)$$

where

$$K_1 := \frac{\gamma\rho}{\rho'}\kappa^2 A_0^{-3}$$

Now, for $2 \leq \rho < 3$ we claim that

$$A := \int_0^{(1-\epsilon)A_0 \tau^{\rho'-1}} e^{K_1 x \tau^{3-2\rho'} + x^2 O\left(\tau^{4-3\rho'}\right)} e^{-\frac{1}{\rho}x^\rho} x^{\rho-1} dx = 1 + o(1). \qquad (101)$$

To prove (101), for $\delta < 2\rho' - 3$ and t large enough, write

$$A = \int_0^{\tau^\delta} g(x) dx + \int_{(1-\epsilon)A_0 \tau^{\rho'-1}}^{\tau^\delta} g(x) dx \qquad (102)$$

where

$$g(x) := e^{K_1 x \tau^{3-2\rho'} + x^2 O\left(\tau^{4-3\rho'}\right)} e^{-\frac{1}{\rho}x^\rho} x^{\rho-1} dx.$$

Note that for $0 \leq x \leq \tau^\delta$, since $x\tau^{3-2\rho'} = o(1)$ and also $x^2 \tau^{4-3\rho'} = o(1)$, with $\lim_{t\to\infty} \sup_{0 \leq x \leq \tau^\delta} o(1) = 0$, we have that

$$g(x) = e^{o(1)} e^{-\frac{1}{\rho}x^\rho} x^{\rho-1} = e^{-\frac{1}{\rho}x^\rho} x^{\rho-1} + o(1).$$

Therefore the first integral in (102) satisfies

$$\int_0^{\tau^\delta} g(x) dx = 1 + o(1).$$

For the second integral in (102), remark that

$$\sup_{\tau^\delta \le x \le (1-\epsilon)\mathcal{A}_0 \tau^{\rho'-1}} x^2 O\left(\tau^{4-3\rho'}\right) = o(1).$$

On the other hand, the function

$$u(x) := K_1 x \tau^{3-2\rho'} - \frac{1}{\rho} x^\rho,$$

is decreasing in the interval $[\tau^\delta, (1-\epsilon)\mathcal{A}_0 \tau^{\rho'-1}]$, so that

$$\int_{(1-\epsilon)\mathcal{A}_0 \tau^{\rho'-1}}^{\tau^\delta} g(x)dx = O\left(e^{-C\tau^{\delta\rho}}\right) = o(1),$$

for some constant $C > 0$. This finishes the proof of (101). Substituting now (101) into (100) we conclude that

$$E_\mu \left[e^{-\frac{1}{\rho}B_1(h_0,v)^\rho} 1(\max_{e \in E} v(e) \le (1-\epsilon)\mathcal{A}_0 \tau^{\rho'-1}) \right] = e^{-\frac{\gamma}{\rho'}\tau^{\rho'} + 2d\frac{\gamma\rho}{\rho'}A_0^{-2}\kappa^2\tau^{2-\rho'}} (1 + o(1)). \tag{103}$$

Combining (103) with (98) we conclude that

$$E_\mu \left[e^{-\frac{1}{\rho}B_1(h_0,v)^\rho} \right] = e^{-\frac{\gamma}{\rho'}\tau^{\rho'} + 2d\frac{\gamma\rho}{\rho'}A_0^{-2}\kappa^2\tau^{2-\rho'}} (1 + o(1)). \tag{104}$$

Substituting (104) back into (99) we see that for $2 \le \rho < 3$,

$$h_1(t) = 2d\kappa^2 A_0^3 \tau^{1-\rho'}(t) + o\left(\frac{1}{t}\right),$$

which proves part (ii).

To prove part (iii), consider the function

$$r(x) := A_0^2 K_1 \frac{\tau}{\mathcal{A}_0 \tau^{\rho'-1} - x} - \frac{1}{\rho} x^\rho + (\rho - 1) \log x, \tag{105}$$

defined for $x > 0$. We will establish the following lemma.

Lemma 11 Let $3 < \rho \le 4$. Then, the following are satisfied.

(i) The function r defined in (105) has a global maximum x_t on the interval $[0, (1 - \epsilon)\mathcal{A}_0 \tau^{\rho'-1}]$ where its derivative vanishes and such that

$$x_t = (A_0 K_1)^{\frac{1}{\rho-1}} \tau^{\frac{1}{\rho-1}(3-2\rho')} - \frac{1}{A_0 K_1} \tau^{-(3-2\rho')} + o\left(\tau^{-(3-2\rho')}\right). \tag{106}$$

(ii) The function r satisfies

$$r(x_t) = A_0 K_1 \tau^{2-\rho'} + K_1 (A_0 K_1)^{\frac{1}{\rho-1}} \left(1 - \frac{1}{\rho} A_0\right) \tau^{\frac{\rho}{\rho-1}(3-2\rho')}$$
$$+ (3 - 2\rho') \log A_o \tau - A_0^2 K_1^2 + o(1).$$

(iii) For every $t > 0$ we have that

$$E_\mu \left[e^{-\frac{1}{\rho} B_1 (h_0, v)^\rho} \right] = e^{r(x_t)} \sqrt{\frac{\pi(\rho - 1)}{\tau^{\frac{\rho-2}{\rho-1}(3-2\rho')}}} \, (1 + o(1)). \tag{107}$$

Proof Proof of parts (i) and (ii). Note that

$$r'(x) = A_0^2 K_1 \frac{\tau}{(A_0 \tau^{\rho'-1} - x)^2} - x^{\rho-1} + (\rho - 1)\frac{1}{x} \quad \text{and}$$
$$r''(x) = 2 A_0^2 K_1 \frac{\tau}{(A_0 \tau^{\rho'-1} - x)^3} - (\rho - 1)x^{\rho-2} - (\rho - 1)\frac{1}{x^2}.$$

Now, for every $\epsilon' > 0$ we have that $r'(x) > 0$ whenever $x \leq \bar{x} := \tau^{\frac{1-\epsilon'}{\rho-1}(3-2\rho')}$, while $r'(x) < 0$ whenever $x \geq (1 - \epsilon)A_0 \tau^{\rho'-1}$. On the other hand, it is easy to check that $r''(x) < 0$ for $\bar{x} \leq x \leq (1 - \epsilon)A_0 \tau^{\rho'-1}$. It follows that there exists only one root x_t of the equation $r'(x) = 0$ on the interval $[0, (1 - \epsilon)A_0 \tau^{\rho'-1})$. To prove (106), as a first step, we note that

$$x_t = (A_0 K_1)^{\frac{1}{\rho-1}} \tau^{\frac{1}{\rho-1}(3-2\rho')} + y_t,$$

where $y_t = o\left(\tau^{\frac{1}{\rho-1}(3-2\rho')}\right)$. Furthermore

$$2 K_1 \tau^{4-3\rho'} x_t - (\rho - 1)x_t^{\rho-2} y_t - (\rho - 1)\frac{1}{x_t} = u(t), \tag{108}$$

where $u(t)$ is of smaller order in t than the three terms of the left-hand side of (108). Now, for $\rho < \frac{3+\sqrt{17}}{2}$, the last term of the left-hand side of (108) has a higher order than the first term. This implies that

$$y_t = -\frac{1}{x_t^{\rho-1}} + o\left(\frac{1}{x_t^{\rho-1}}\right) = -\frac{1}{A_0 K_1 \tau^{(3-2\rho')}} + o\left(\frac{1}{\tau^{(3-2\rho')}}\right),$$

which proves (106) of part (i). The proof of part (ii) now follows using the expansion (106) of x_t of part (i).

Proof of part (iii). By a standard Taylor expansion, we see that for every real y such that $x_t - |y| > 0$, there is a $\vartheta \in [x_t - |y|, x_t + |y|]$ such that

$$r(x_t + y) = r(x_t) + \frac{y^2}{2} r''(x_t) + \frac{y^3}{6} r'''(\vartheta).$$

Note that

$$r'''(x) = 6A_0^2 K_1 \frac{\tau}{(A_0 \tau^{\rho'-1} - x)^4} - (\rho - 2)(\rho - 1)x^{\rho-3} + 2(\rho - 1)\frac{1}{x^3}.$$

Therefore

$$r''(x_t) = -(\rho - 1)x_t^{\rho-2} + O\left(t^{4-3\rho'}\right)$$

and for $|\vartheta| \leq 2x_t$,

$$r'''(\vartheta) = -(\rho - 1)(\rho - 2)\vartheta^{\rho-3} + O\left(t^{5-4\rho'}\right). \tag{109}$$

It follows that

$$E_\mu\left[e^{-\frac{1}{\rho}B_1(h_0,v)^\rho} 1(\max_{e \in E} v(e) \leq (1-\epsilon)A_0\tau^{\rho'-1})\right] = \int_0^{(1-\epsilon)A_0\tau^{\rho'-1}} e^{r(x)} dx$$

$$= e^{r(x_t)} \int_{-x_t}^{(1-\epsilon)A_0\tau^{\rho'-1}-x_t} e^{\frac{1}{2}y^2 r''(x_t) + \frac{1}{6}y^3 r'''(\vartheta)} dy$$

$$= e^{r(x_t)} \int_{-x_t}^{(1-\epsilon)A_0\tau^{\rho'-1}-x_t} e^{-\frac{\rho-1}{2}y^2 x_t^{\rho-2} + +y^2 O\left(t^{4-3\rho'}\right) + \frac{1}{6}y^3 r'''(\vartheta)} dy. \tag{110}$$

For δ such that $\tau^\delta \leq x_t$ write

$$\int_{-x_t}^{(1-\epsilon)A_0\tau^{\rho'-1}-x_t} e^{-\frac{\rho-1}{2}y^2 x_t^{\rho-2} + +y^2 O\left(t^{4-3\rho'}\right) + \frac{1}{6}y^3 r'''(\vartheta)} dy$$

$$= \int_{-\tau^\delta}^{\tau^\delta} e^{-\frac{\rho-1}{2}y^2 x_t^{\rho-2} + +y^2 O\left(t^{4-3\rho'}\right) + \frac{1}{6}y^3 r'''(\vartheta)} dy + \int_{B_\delta} e^{-\frac{\rho-1}{2}y^2 x_t^{\rho-2} + +y^2 O\left(t^{4-3\rho'}\right) + \frac{1}{6}y^3 r'''(\vartheta)} dy, \tag{111}$$

where $B_\delta := \{y : |y| \geq \tau^\delta, -x_t \leq y \leq (1-\epsilon)A_0\tau^{\rho'-1} - x_t\}$. For the first integral in the right-hand side of (111), we have that

$$\int_{-\tau^\delta}^{\tau^\delta} e^{-\frac{\rho-1}{2}y^2 x_t^{\rho-2} + +y^2 O\left(t^{4-3\rho'}\right) + \frac{1}{6}y^3 r'''(\vartheta)} dy$$

$$= \frac{1}{\sqrt{x_t^{\rho-2}}} \int_{-\tau^\delta \sqrt{x_t^{\rho-2}}}^{\tau^\delta \sqrt{x_t^{\rho-2}}} e^{-\frac{\rho-1}{2}y^2 + y^2 O\left(\frac{t^{4-3\rho'}}{x_t^{\rho-2}}\right) + \frac{1}{6x_t^{3(\rho-2)/2}}y^3 r'''(\vartheta)} dy \tag{112}$$

$$= (1 + o(1)) \frac{1}{\sqrt{x_t^{\rho-2}}} \int_{-\tau^\delta \sqrt{x_t^{\rho-2}}}^{\tau^\delta \sqrt{x_t^{\rho-2}}} e^{-\frac{\rho-1}{2}y^2 + \frac{1}{6x_t^{3(\rho-2)/2}}y^3 r'''(\vartheta)} dy. \tag{113}$$

where we have used the fact that when $\rho > 3$, one has that $y^2 O \left(t^{4-3\rho'} / x_t^{\rho-2} \right) \leq C t^{2\delta+4-3\rho'} = o(1)$. Now

$$
\frac{1}{\sqrt{x_t^{\rho-2}}} \int_{-\tau^\delta \sqrt{x_t^{\rho-2}}}^{\tau^\delta \sqrt{x_t^{\rho-2}}} e^{-\frac{\rho-1}{2} y^2 + \frac{1}{6x_t^{3(\rho-2)/2}} y^3 r'''(\vartheta)} \, dy
$$

$$
= \frac{1}{\sqrt{x_t^{\rho-2}}} \int_{-\tau^\delta}^{\tau^\delta} e^{-\frac{\rho-1}{2} y^2 + \frac{1}{6x_t^{3(\rho-2)/2}} y^3 r''(\vartheta)} \, dy
$$

$$
+ \frac{1}{\sqrt{x_t^{\rho-2}}} \int_{D_\delta} e^{-\frac{\rho-1}{2} y^2 + \frac{1}{6x_t^{3(\rho-2)/2}} y^3 r'''(\vartheta)} \, dy, \tag{114}
$$

where $D_\delta := \{y : |y| \geq \tau^\delta, -\tau^\delta \sqrt{x_t^{\rho-2}} \leq y \leq \tau^\delta \sqrt{x_t^{\rho-2}}\}$. For the first integral of the right-hand side of (114), we have that

$$
\frac{1}{\sqrt{x_t^{\rho-2}}} \int_{-\tau^\delta}^{\tau^\delta} e^{-\frac{\rho-1}{2} y^2 + \frac{1}{6x_t^{3(\rho-2)/2}} y^3 r'''(\vartheta)} \, dy = \sqrt{\frac{\pi(\rho-1)}{x_t^{\rho-2}}} + o\left(x_t^{-(\rho-2)/2}\right), \tag{115}
$$

where we have used the fact that by (109) we have that $|y^3| x_t^{-3(\rho-2)/2} r'''(\vartheta) \leq C\tau^{3\delta} x_t^{\rho-3-3(\rho-2)/2} \leq C t^{3\delta-1/2} = o(1)$ for $\delta < 1/6$. For the second integral on the right-hand side of (114), note that since $\frac{1}{x_t^{3(\rho-2)/2}} y^3 r'''(\vartheta) \leq y^2 o(t)$ uniformly for $y \in D_\delta$, we have

$$
\frac{1}{\sqrt{x_t^{\rho-2}}} \int_{D_\delta} e^{-\frac{\rho-1}{2} y^2 + \frac{1}{6x_t^{3(\rho-2)/2}} y^3 r'''(\vartheta)} \, dy = \frac{1}{\sqrt{x_t^{\rho-2}}} O\left(e^{-\tau^{2\delta}}\right) = o\left(x_t^{-(\rho-2)/2}\right). \tag{116}
$$

Substituting (115) and (116) into (114), (113), (111) and (110), and using (98) together with (106) we conclude that (107) of part (ii) of Lemma 11 is satisfied. □

Appendix A
Abstract Rank-One Perturbation Theory

For the sake of completeness, we review here the standard rank-one perturbation theory (see [12, 15] for an overview). Let H_0 be a bounded self-adjoint operator in a Hilbert space $\mathcal{H}$. Here we want to establish some cases under which a rank-one self-adjoint perturbation H of H_0, has a principal eigenvalue and eigenfunction possibly with a series expansion on some small parameter. We will need the resolvents $R_\lambda := (\lambda I - H)^{-1}$ and $R_\lambda^0 := (\lambda I - H_0)^{-1}$, of H and H_0 respectively, defined for λ not in the corresponding spectrums $\sigma(H)$ and $\sigma(H_0)$. Let us denote by $res(H_0)$ and $res(H)$ the respective resolvent sets. The top of the spectrum of H_0, will be denoted by

$$\lambda_+^0 := \sup\{\lambda : \lambda \in \sigma(H_0)\}.$$

A.1 Definitions

Let us consider a rank one perturbation of H_0 depending on a *large* parameter $h > 0$:

$$H := H_0 + hB, \qquad B := (\phi, \cdot)\phi$$

for some normalized $\phi \in \mathcal{H}$. Note that H is also bounded and self-adjoint. We will show that if h is large enough, H has a principal eigenvalue and eigenfunction with a Laurent series expansion on h. Define then the following two families of elements of $\mathcal{H}$,

$$r_\lambda := (\lambda I - H)^{-1}\phi, \qquad \lambda \notin \sigma(H),$$

$$q_\lambda := (\lambda I - H_0)^{-1}\phi, \qquad \lambda \notin \sigma(H_0).$$

A.2 The Aronszajn-Krein Formula

Here we will state and prove the famous Aronzajn-Krein formula (see for example [15]), in our particular context. Let us first define the following set

$$S := \{\lambda \in res(H_0) : h(\phi, q_\lambda) = 1\}.$$

and the quantity

$$h_0 := \frac{1}{\lim_{\lambda \searrow \lambda_+^0}(\phi, q_\lambda)}. \tag{117}$$

Note that (ϕ, q_λ) is decreasing in λ for $\lambda > \lambda_+^0$. Indeed, $\frac{d(\phi, q_\lambda)}{d\lambda} = -||q_\lambda||^2 < 0$, since $\phi \neq 0$. Hence, the limit in display (117) exists, possibly having the value ∞. In the sequel, we will interpret the quantity h_0 as 0 when the limit in the denominator of the right hand side of (117) is ∞.

Lemma 12 $S \subset \mathbb{R}$, and has only isolated points. Furthermore, there is a $\lambda \in S$ such that $\lambda > \lambda_+^0$ if and only if $h > h_0$. In this case it is unique.

Proof Note that $h(\phi, q_\lambda) - 1$ is an analytic function on the open set $res(H_0)$. Therefore, its zeros are isolated. On the other hand, since H is self-adjoint, they have to be real. The last statement follows from the fact that (ϕ, q_λ) is decreasing if $\lambda > \lambda_+^0$ and $\lim_{\lambda \to \infty}(\phi, q_\lambda) = 0$. $\qquad \square$

Theorem 6 *Consider the bounded selfadjoint operators H_0 and H.*

(i) **Aronszajn-Krein formula.** *If* $\lambda \notin \sigma(H_0) \cup \sigma(H)$ *then*

$$R_\lambda = R_\lambda^0 + \frac{h}{1 - (\phi, q_\lambda)h}(q_\lambda, \cdot)q_\lambda. \tag{118}$$

(ii) **Spectrum of** *H*.

$$\mathcal{S} \subset \sigma(H) \subset \mathcal{S} \cup \sigma(H_0). \tag{119}$$

Proof Let us first prove part *(i)* and the first inclusion of part *(ii)*. Assume that $\lambda \in res(H_0) \cap res(H)$. By definition we have, $(\lambda I - H)r_\lambda = \phi$. Hence

$$(\lambda I - H_0)r_\lambda = (1 + (\phi, r_\lambda)h)\phi.$$

Making the resolvent $(\lambda I - H_0)^{-1}$ act on both sides of this equality, we get

$$r_\lambda = (1 + (\phi, r_\lambda)h)q_\lambda. \tag{120}$$

Taking the scalar product with ϕ, we see that $(1 - (\phi, q_\lambda)h)(\phi, r_\lambda) = (\phi, q_\lambda)$. This shows that $(\phi, q_\lambda)h \neq 1$ and hence

$$(\phi, r_\lambda) = \frac{(\phi, q_\lambda)}{1 - (\phi, q_\lambda)h}. \tag{121}$$

Therefore, $\mathcal{S} \subset \sigma(H)$. Substituting (121) back in the identity (120) and using $R_\lambda = R_\lambda^0 + hr_\lambda(q_\lambda, \cdot)$ proves (118). Now, assume that $\lambda \notin \mathcal{S} \cap \sigma(H_0)$. Then the right hand side of (118) is well defined as a bounded selfadjoint operator in $\mathcal{H}$. A simple computation shows that it is the inverse of the operator $(\lambda I - H)$. □

From Theorem 6 we can now deduce the following corollary.

Corollary 3 *Either of the following is true:*

(i) *If* $h > h_0$, *H has a unique simple eigenvalue* $\lambda_{max} > \lambda_+^0$ *and* $\sigma(H)/\{\lambda_{max}\} \subset (-\infty, \lambda_+^0]$.
(ii) *If* $h \leq h_0$, *then* $\sigma(H) \subset (-\infty, \lambda_+^0]$.

Furthermore, if *(i)* *is satisfied the eigenfunction of* λ_{max} *is proportional to* $q_{\lambda_{max}}$ *and there exist an* $r_0 > h_0$ *such that* λ_{max} *admits a Laurent series expansion for* $h > r_0$,

$$\lambda_{max} = h + \sum_{k=0}^{\infty} \frac{b_k}{h^k}.$$

Proof If $h \leq h_0$, by Lemma 12 the equation $h(\phi, q_\lambda) = 1$ does not have any solution $\lambda > \lambda_+^0$. By Theorem 6, there is no $\lambda \in \sigma(H)$ such that $\lambda > \lambda_+^0$. On the other hand, by Lemma 12, if $h > h_0$, there is a unique $\lambda_{max} > \lambda_+^0$ such that $h(\phi, q_{\lambda_{max}}) = 1$. By Theorem 6, $\lambda_{max} \in \sigma(H_0)$ and the spectral projector of H on λ_{max} is given by

$$P = \frac{1}{||q_{\lambda_{max}}||^2}(q_{\lambda_{max}}, \cdot)q_{\lambda_{max}}.$$

This shows that the eigenfunction of λ_{max} is proportional to q_λ. Finally, defining $u := 1/h$, we see that if $\lambda(u)$ satisfies $(\phi, q_{\lambda(u)}) = u$, then

$$\frac{d(\phi, q_{\lambda(u)})}{du} = 1.$$

By the implicit function theorem, this implies that there is a neighborhood of the point $u = 0$, where the function $1/\lambda(u)$ is analytic. $\qquad\square$

Appendix B
Comments on the Article "Stable Limit Laws for the Parabolic Anderson Model Between Quenched and Annealed Behaviour" of Gärtner and Schnitzler [11]

As a way of clarifying possible confusions that may arise in regard with the article [11] of Gärtner-Schnitzler, here we point out the main passages of the proof of Theorem 1 where we believe the arguments are incomplete:

(1) In the statement of Theorem 1 of the quoted article it is claimed that the spatial average of the mean number of particles starting at points in a box with its size increasing with time, properly centered and normalized by a function $B_\alpha(t)$, converges to stable distribution. Nevertheless, the function $B_\alpha(t)$ is not even properly defined. It is in fact said before the statement of Theorem 1 that

$$B_\alpha(t) = \exp\{t(h_{\alpha t} - \chi + o(1))\}$$

saying that "the error term ... is chosen in a suitable way".
(2) The proof of Lemma 4 of the quoted article is incorrect. In fact, in the expression of page 200 of the quoted article,

$$P\left(\xi(0) > \frac{\log B_\alpha(t)}{t} + \frac{\log x}{t} + \chi + \varepsilon(t)\right),$$

since we only know by Lemma 3 of the quoted article that

$$\varepsilon(t) = o(1),$$

this error term might be larger than the term $\frac{\log x}{t}$. Since the equality in the statement of Lemma 4 gives the value of a limit which is a function of x, without more information about $\varepsilon(t)$ it is therefore impossible to prove it. Furthermore, in the first display of page 197 the authors choose

$$B_\alpha(t) := \exp\{t(h_{\alpha t} - \chi + o(1))\},$$

for a "suitable" error term $o(1)$ (which necessarily has to be deterministic). Now, the function $h_{\alpha t}$ is related to the scales giving the stable laws through

$$L_\alpha(t) := \exp\{\phi(h_{\alpha t})\}.$$

Then, at the end of the proof of Lemma 4 it is clamed that "by our choice of $B_\alpha(t)$ it follows that"

$$\log|Q_{L_\alpha(t)}| = \phi\left(\frac{\log B_\alpha(t)}{t} + \chi + o(1)\right).$$

So the authors claims that the choice of the error term of $B_\alpha(t)$ cancels the error term appearing inside ϕ. But the first one is deterministic and the second one is random, so it is not possible to make such a choice. In the proof of part (i) of Theorem 1 here, we show that there is indeed a choice of $B_\alpha(t)$ with a deterministic term $o(1)$. Nevertheless, again the use of rank-one perturbation theory for this proof is decisive.

(3) The proof of Lemma 3 in the quoted article is incorrect. In the second display of page 199 of the quoted article the authors claim that

$$P(\xi^{(1)}_{Q_{l(t)}} > h(t) + \chi - \varepsilon(t)) = 1 - (1 - P(\xi(0) > h(t) + \chi - \varepsilon(t))^{|Q_{l(t)}|}.$$

To justify this obviously the authors had in mind the fact that the random variables $\xi = (\xi(x))_{x \in \mathbb{Z}^d}$ are independent. Nevertheless, $\varepsilon(\xi)$ is a function of the whole field ξ, so that this independence property cannot be used in this way. One really needs to have more precise information about the field, its extremes and the term $\varepsilon(t)$ in order to show that

$$P(\mu_t > h(t)) \sim |Q_{l(t)}| P(\xi(0) > h(t) - \chi - \varepsilon(t))$$

(which is part of the claim in the proof of Lemma 3, and which is true). For the proof of Theorem 3 here, we solve this difficulty through the use of rank-one perturbation theory.

References

1. Astrauskas, A.: Poisson-type limit theorems for eigenvalues of finite-volume Anderson Hamiltonians. Acta Appl. Math. **96**(1–3), 3–15 (2007)
2. Astrauskas, A.: Extremal theory for spectrum of random discrete Schrödinger operator. I. Asymptotic expansion formulas. J. Stat. Phys. **131**(5), 867–916 (2008)
3. Astrauskas, A.: Extremal theory for spectrum of random discrete Schrödinger operator. II. Distributions with heavy tails. J. Stat. Phys. **146**(1), 98–117 (2012)
4. Astrauskas, A.: Extremal theory for spectrum of random discrete Schrödinger operator. III. Localization properties. J. Stat. Phys. **150**(5), 889–907 (2013)
5. Astrauskas, A.: From extreme values of i.i.d. random fields to extreme eigenvalues of finite-volume Anderson Hamiltonian. Probab. Surv. **13**, 156–244 (2016)
6. Ben Arous, G., Bogachev, L., Molchanov, S.: Limit theorems for sums of random exponentials. Probab. Theory Relat. Fields **132**(4), 579–612 (2005)
7. Ben Arous, G., Molchanov, S., Ramírez, A.F.: Transition from the annealed to the quenched asymptotics for a random walk on random obstacles. Ann. Probab. **33** 6, 2149–2187 (2005)
8. Ben Arous, G., Molchanov, S., Ramírez, A.F.: Transition asymptotics for reaction-diffusion in random media. Probability and mathematical physics. CRM Proceedings of Lecture Notes, vol. 42, pp. 1–40. American Mathematical Society, Providence, RI (2007)
9. Biskup, M., König, W.: Eigenvalue order statistics for random Schrödinger operators with doubly-exponential tails. Commun. Math. Phys. **341**, 179–218 (2016)
10. Gärtner, J., Molchanov, S.: Parabolic problems for the Anderson model I, Intermittency and related topics. Commun. Math. Phys. **132**, 613–655 (1990)
11. Gärtner, J., Schnitzler, A.: Stable limit laws for the parabolic Anderson model between quenched and annealed behaviour. Ann. Inst. Henri Poincaré Probab. Stat. **51**(1), 194–206 (2015)
12. Kato, T.: Perturbation Theory for Linear Operators. Springer (1995)
13. König, W.: The Parabolic Anderson Model (Random Walk in Random Potential). Pathways in Mathematics. Birkhäuser, Springer (2016)
14. Petrov, V.V.: Sums of Independent Random Variables. Springer, Berlin (1975)
15. Simon, B.: Spectral Analysis of Rank One Perturbations and Applications. CRM Proceeding and Lecture Notes, vol. 8, pp. 109–149 (1995)

Quenched Central Limit Theorem for the Stochastic Heat Equation in Weak Disorder

Yannic Bröker and Chiranjib Mukherjee

Abstract We continue with the study of the mollified stochastic heat equation in $d \geq 3$ given by $\mathrm{d}u_{\varepsilon,t} = \frac{1}{2}\Delta u_{\varepsilon,t}\mathrm{d}t + \beta\varepsilon^{(d-2)/2} u_{\varepsilon,t} \, \mathrm{d}B_{\varepsilon,t}$ with spatially smoothened cylindrical Wiener process B, whose (renormalized) Feynman-Kac solution describes the partition function of the continuous directed polymer. This partition function defines a (quenched) polymer path measure for every realization of the noise and we prove that as long as $\beta > 0$ stays small enough, the distribution of the diffusively rescaled Brownian path converges under the aforementioned polymer path measure to the standard Gaussian distribution.

Keywords Stochastic heat equation · Continuous directed polymer · Quenched central limit theorem · Weak disorder

AMS Subject Classification. 60J65 · 60J55 · 60F10

1 Introduction and the Result

1.1 Motivation

We continue the study of the *stochastic heat equation* (SHE) with multiplicative space-time white noise, formally written as

$$\partial_t u(t, x) = \frac{1}{2}\Delta u(t, x) + u(t, x)\eta(t, x) \tag{1.1}$$

Y. Bröker · C. Mukherjee (✉)
University of Muenster, Einsteinstrasse 62, 48149 Muenster, Germany
e-mail: chiranjib.mukherjee@uni-muenster.de

Y. Bröker
e-mail: yannic.broeker@uni-muenster.de

© Springer Nature Switzerland AG 2019
P. Friz et al. (eds.), *Probability and Analysis in Interacting Physical Systems*, Springer Proceedings in Mathematics & Statistics 283,
https://doi.org/10.1007/978-3-030-15338-0_6

with η being a centered Gaussian process with covariance $\mathbf{E}[\eta(t, x)\eta(s, y)] = \delta_0(s - t)\delta_0(x - y)$ for $s, t > 0$ and $x, y \in \mathbb{R}^d$. Note that the Cole-Hopf transformation $h := -\log u$ translates the SHE to the *Kardar-Parisi-Zhang* (KPZ) equation, which is a non-linear stochastic partial differential equation also written formally as

$$\partial_t h(t, x) = \frac{1}{2}\Delta h(t, x) - \frac{1}{2}(\partial_x h(t, x))^2 + \eta(t, x).$$

Note that both SHE and KPZ are a-priori ill-posed, as only distribution valued solutions are expected for both equations which carry fundamental obstacles arising from multiplying or squaring distributions. When the spatial dimension is one, both equations can be analyzed on a rigorous level, as they turn out to be the scaling limit of front propagation of some exclusion processes [2, 4, 22]. An intrinsic precise construction of their solutions also yields to the powerful theories of *regularity structures* [16] as well as *paracontrolled distributions* [14].

In the discrete lattice $\mathbb{Z}^d$, the solution of the SHE is directly related to the *partition function* of the *discrete directed polymer*, which is a well-studied model in statistical mechanics (see [1, 10]). The directed polymer measure is defined as

$$\mu_n(d\omega) = \frac{1}{Z_n}\exp\left\{\beta \sum_{i=1}^{n} \eta(i, \omega(i))\right\}d\mathbb{P}_0, \qquad (1.2)$$

and in this scenario, the space-time white noise potential is replaced by i.i.d. random variables $\eta = \{\eta(n, x): n \in \mathbb{N}, x \in \mathbb{Z}^d\}$ and the strength of the noise is captured by the disorder strength $\beta > 0$. If $\mathbf{P}$ denotes the law of the potential η with $\mathbb{P}_0$ denoting the distribution of a simple random walk $\omega_n = (\omega(i))_{i \leq n}$ starting at the origin and independent of the noise η, and $Z_n = \mathbb{E}_0[\exp\{\beta \sum_{i=1}^{n} \eta(i, \omega(i))\}]$ denotes the normalizing constant, or the *partition function* of the discrete directed polymer, it is well-known that, when $d \geq 3$, the renormalized partition function $Z_n/\mathbf{E}[Z_n]$ converges almost surely to a random variable Z_∞, which, when β is small enough, is positive almost surely (i.e., *weak disorder* persists [3, 17]), and in this case, the distribution $\mu_n\left(\frac{\omega(n)}{\sqrt{n}}\right)^{-1}$ of the rescaled paths converges, for any realization of the noise η, to a centered non-degenerate Gaussian distribution [12]. On the other hand, for β large enough, the limiting random variable satisfies $Z_\infty = 0$ (i.e., *strong disorder holds* [9, 10]).

1.2 The Result

We turn to the scenario in the continuum in $d \geq 3$. Note that the Eq. (1.1) can also be written (formally) as an SDE

$$du_t = \frac{1}{2}\Delta u_t dt + \beta u_t dB_t, \qquad (1.3)$$

where B_t is now a *cylindrical Wiener process*. That is, the family $\{B_t(f)\}_{f \in \mathcal{S}(\mathbb{R}^d)}$ is a centered Gaussian process with covariance given by $\mathbf{E}[B_t(f) B_s(g)] = (t \wedge s)\langle f, g \rangle_{L^2(\mathbb{R}^d)}$ for Schwartz functions $f, g \in \mathcal{S}(\mathbb{R}^d)$. Defining (1.3) precisely requires studying a spatially smoothened version

$$B_\varepsilon(t, x) = B_t(\phi_\varepsilon(x - \cdot))$$

of B_t for any smooth mollifier $\phi_\varepsilon(x) = \varepsilon^{-d} \phi(x/\varepsilon)$. Here ϕ is chosen to be positive, even, smooth and compactly supported and normalized to have total mass $\int \phi = 1$. If we write

$$V_\varepsilon(x) = (\phi_\varepsilon \star \phi_\varepsilon)(x) \qquad V(x) = (\phi \star \phi)(x), \tag{1.4}$$

note that, for any fixed $\varepsilon > 0$, B_ε is also a centered Gaussian process with covariance kernel $\mathbf{E}[B_\varepsilon(t, x) B_\varepsilon(s, y)] = (t \wedge s) V_\varepsilon(x - y)$. If we denote by $\mathbb{P}_x$ the law of a Brownian motion $W = (W_t)_t$ started at $x \in \mathbb{R}^d$ and independent of the process B, then

$$u_\varepsilon(t, x) = \mathbb{E}_x \left[\exp\left\{ \beta \varepsilon^{(d-2)/2} \int_0^t \int_{\mathbb{R}^d} \phi_\varepsilon(W_s - y) \dot{B}(t - s, dy) \, ds - \frac{\beta^2 \varepsilon^{d-2}}{2} t V_\varepsilon(0) \right\} \right] \tag{1.5}$$

represents the renormalized Feynman-Kac solution of the mollified stochastic heat equation

$$du_{\varepsilon,t} = \frac{1}{2} \Delta u_{\varepsilon,t} dt + \beta \varepsilon^{\frac{d-2}{2}} u_{\varepsilon,t} \, dB_{\varepsilon,t} \qquad d \geq 3, \ \beta > 0$$

$$u_{\varepsilon,0} = 1. \tag{1.6}$$

Clearly, $\mathbf{E}[u_\varepsilon(t, x)] = 1$. By time reversal, for any fixed $t > 0$ and $\varepsilon > 0$,

$$u_\varepsilon(t, \cdot) \overset{(d)}{=} M_{\varepsilon^{-2}t}(\varepsilon^{-1} \cdot), \tag{1.7}$$

where

$$M_T(x) = M_{\beta,T}(x) = \mathbb{E}_x \left[\exp\left\{ \beta \int_0^T \int_{\mathbb{R}^d} \phi(W_s - y) \dot{B}(s, dy) ds - \frac{\beta^2 T}{2} V(0) \right\} \right]. \tag{1.8}$$

See (Eq. (2.6) in [18]) for details. Then it was shown [18, Theorem 2.1 and Remark 2.2] that for $\beta > 0$ sufficiently small and any test function $f \in C_c^\infty(\mathbb{R}^d)$,

$$\int_{\mathbb{R}^d} u_\varepsilon(t, x) f(x) \, dx \to \int_{\mathbb{R}^d} \bar{u}(t, x) f(x) \, dx \tag{1.9}$$

as $\varepsilon \to 0$ in probability, with $\bar{u}$ solving the heat equation $\partial_t \bar{u} = \frac{1}{2} \Delta \bar{u}$ with unperturbed diffusion coefficient. Furthermore, it was also shown in [18] that, with β small enough, and for any $t > 0$ and $x \in \mathbb{R}^d$, $u_{\varepsilon,t}(x)$ converges in law to a non-degenerate random variable M_∞ which is almost surely strictly positive, while $u_{\varepsilon,t}(x)$ converges in probability to zero if β is chosen large. The pointwise fluctuations of $M_T(x)$ and $u_{\varepsilon,t}(x)$ were studied also in a recent article [8] when β is sufficiently small (in particular when M_∞ is strictly positive). In particular, it was shown that (see [8, Theorem 1.1 and Corollary 1.2]), in this regime,

$$T^{\frac{d-2}{4}}\left(\frac{M_T(x) - M_\infty(x)}{M_T(x)}\right) \Rightarrow N\left(0, \sigma^2(\beta)\right) \quad \text{and} \quad \sigma^2(\beta) \neq 1.$$

Note that in view of the Feynman-Kac representation (1.5), $u_\varepsilon(t,x)$ and $M_T(x)$ are directly related to the (renormalized) *partition function of the continuous directed polymer*, and following the terminology for discrete directed polymer, the strictly positive limit M_∞ for small disorder strength β is referred to as the *weak-disorder regime*, while for β large, a vanishing partition function $\lim_{T\to\infty} M_T$ underlines the *strong disorder phase*. In fact, the polymer model corresponding to (1.8) is known as the *Brownian directed polymer in a Gaussian environment* and the reader is refered to [7] for a review of a similar model driven by a Poissonian noise (see also [6, 11]).

Despite the aforementioned recent results pertaining to the partition function for the continuous directed polymer, the investigation of the actual *polymer path measure* had remained open. Note that the (quenched) polymer path measure is defined as

$$d\widehat{\mathbb{Q}}_{\beta,T}^{(x)} = \frac{1}{Z_{\beta,T}} \exp\left\{\beta \int_0^T \int_{\mathbb{R}^d} \phi(W_s - y)\, \dot{B}(s, dy)\, ds\right\} d\mathbb{P}_x, \qquad (1.10)$$

for every realization of the noise B. Here $Z_{\beta,T}$ is the un-normalized partition function, i.e.,

$$Z_{\beta,T} = \mathbb{E}_x\left[\exp\left\{\beta \int_0^T \int_{\mathbb{R}^d} \phi(W_s - y)\, \dot{B}(s, dy)\, ds\right\}\right] = M_T(x)\, \mathbb{E}[Z_{\beta,T}]$$

$$= M_T(x)\exp\left\{\frac{\beta^2 T}{2} V(0)\right\}.$$

Throughout this article we will assume that $d \geq 3$ and

$$\beta < \beta_{L^2} := \sup\left\{b > 0: \sup_T \|M_T\|_{L^2(\mathbf{P})} < \infty\right\} \in (0, \infty)$$

Then the goal of the present article is to show[1] that, for almost every realization of the noise B, the law of the diffusively rescaled Brownian path under $\widehat{\mathbb{Q}}_{\beta,T}^{(x)}$ converges to the standard Gaussian law. We turn to a precise statement of our main result.

---

[1] Note that a standard Gaussian computation implies that $\mathbf{E}[M_T^2] = \mathbb{E}_0\left[\exp\{\beta^2 \int_0^T V(\sqrt{2}W_s)ds\}\right] \leq \mathbb{E}_0\left[\exp\{\beta^2 \int_0^\infty V(\sqrt{2}W_s)ds\}\right]$. Since $d \geq 3$ and $V \geq 0$ is a continuous function with compact support, $\beta^2 \sup_{x\in\mathbb{R}^d} \mathbb{E}_x[\int_0^\infty V(\sqrt{2}W_s)ds] = \eta < 1$ as soon as $\beta > 0$ is chosen small enough. Then by Khas'minskii's lemma, $\sup_{x\in\mathbb{R}^d} \mathbb{E}_x[\exp\{\beta^2 \int_0^\infty V(\sqrt{2}W_s)ds\}] = \frac{\eta}{1-\eta} < \infty$, and hence $\beta_{L^2} \in (0, \infty)$.

Theorem 1.1 *Let us assume that $d \geq 3$ and $\beta < \beta_{L^2}$. Then for any $x \in \mathbb{R}^d$ and* **P**-*almost surely, the distribution* $\widehat{\mathbb{Q}}_{\beta,T}^{(x)} \left(\frac{W_T}{\sqrt{T}}\right)^{-1}$ *converges weakly to a d-dimensional centered Gaussian measure with identity covariance matrix.*

Remark 1 For the discrete directed polymer (recall (1.2)), a stronger version of Theorem 1.1 was obtained ([12, Theorem 1.2]) for $d \geq 3$ and disorder strength $\beta < \beta_c(d)$ such that the renormalized partition function $Z_n/\mathbf{E}[Z_n]$ converges to a strictly positive random variable (i.e., the *whole weak disorder region* is considered where $Z_n/\mathbf{E}[Z_n]$ is not necessarily $L^2(\mathbf{P})$-bounded). Also, the result in [12] covers not only convergence of the one-time marginal but also convergence of the process, i.e., it is shown that for any suitable test function F on the path space, $\mathbb{E}^{\mu_n}[F(\omega(n\cdot)/\sqrt{n})] \to \mathbb{E}_0[F(W)]$ in probability with respect to $\mathbf{P}$ and W is a Brownian motion in $\mathbb{R}^d$. However, the convergence assertion in [12] holds in probability w.r.t $\mathbf{P}$ unlike the almost sure statement in Theorem 1.1. We believe that using the approach in [12], Theorem 1.1 can be extended to the whole weak disorder region (i.e., when M_∞ is a non-degenerate strictly positive random variable). □

Theorem 1.1 also implies a central limit theorem for the path measures for the mollified stochastic heat equation (1.6). Recall the relation (1.7), and note that, for any fixed $t > 0$,

$$d\widetilde{\mathbb{Q}}_{\beta,\varepsilon,t} = \frac{1}{Z_{\beta,\varepsilon,t}} \exp\left\{ \beta\varepsilon^{\frac{d-2}{2}} \int_0^t \int_{\mathbb{R}^d} \phi_\varepsilon(W_s - y)\, \dot{B}(t - s, dy)\, ds \right\} d\mathbb{P}_0, \quad (1.11)$$

defines the (quenched) path measure for (1.6), and here

$$Z_{\beta,\varepsilon,t} = u_{\varepsilon,t}(0) \exp\left\{ \frac{t\beta^2}{2} \varepsilon^{(d-2)/2}\, V_\varepsilon(0) \right\}.$$

The following result then will be a direct consequence of Theorem 1.1.

Corollary 1.2 *Let $d \geq 3$ and $\beta < \beta_{L^2}$ as in Theorem 1.1. Then for any fixed $t > 0$,*

$$\widetilde{\mathbb{Q}}_{\beta,\varepsilon,t} \left(\varepsilon W_{\varepsilon^{-2}}\right)^{-1} \Rightarrow N(\mathbf{0}, \mathbf{I_d})$$

and the above convergence holds in probability with respect to the law **P** *of the noise B.*

Proof of Corollary 1.2 (assuming Theorem 1.1) Since $t > 0$ is fixed, for simplicity we will take $t = 1$ and prove the result for $\widetilde{\mathbb{Q}}_{\beta,\varepsilon} = \widetilde{\mathbb{Q}}_{\beta,\varepsilon,1}$. It suffices to show that

$$\mathbb{E}^{\widetilde{\mathbb{Q}}_{\beta,\varepsilon}}\left[e^{\varepsilon\langle \lambda, W_{\varepsilon^{-2}}\rangle} \right] \to e^{|\lambda|^2/2} \quad \text{in probability w.r.t } \mathbf{P}.$$

Then for $\varepsilon < 1$,

$$
\mathbb{E}_0\left[\exp\left\{\beta\varepsilon^{(d-2)/2}\int_0^1\int_{\mathbb{R}^d}\phi_\varepsilon(W_s - y)\dot{B}(1-s,\mathrm{d}y)\mathrm{d}s\right\}e^{\varepsilon\langle\lambda, W_{\varepsilon^{-2}}\rangle}\right]
$$

$$
= \mathbb{E}_0\left[\exp\left\{\beta\varepsilon^{(d-2)/2}\int_0^1\int_{\mathbb{R}^d}\phi_\varepsilon(W_s - y)\dot{B}(1-s,\mathrm{d}y)\mathrm{d}s\right\}e^{\varepsilon\langle\lambda, W_1\rangle}e^{\varepsilon\langle\lambda, W_{\varepsilon^{-2}-W_1}\rangle}\right]
$$

$$
= \mathbb{E}_0\left[\exp\left\{\beta\varepsilon^{(d-2)/2}\int_0^1\int_{\mathbb{R}^d}\phi_\varepsilon(W_s - y)\dot{B}(1-s,\mathrm{d}y)\mathrm{d}s\right\}e^{\varepsilon\langle\lambda, W_1\rangle}\right]\mathbb{E}_0\left[e^{\varepsilon\langle\lambda, W_{\varepsilon^{-2}}-W_1\rangle}\right]
$$

$$
= \mathbb{E}_0\left[\exp\left\{\beta\varepsilon^{(d-2)/2}\int_0^1\int_{\mathbb{R}^d}\phi_\varepsilon(W_s - y)\dot{B}(1-s,\mathrm{d}y)\mathrm{d}s\right\}e^{\varepsilon\langle\lambda, W_1\rangle}\right]e^{\frac{|\lambda|^2}{2}(1-\varepsilon^2)},
$$

and we have used Markov property for W at time 1. If we now recall (1.7),

$$
\mathbb{E}_0\left[\exp\left\{\beta\varepsilon^{(d-2)/2}\int_0^1\int_{\mathbb{R}^d}\phi_\varepsilon(W_s - y)\dot{B}(1-s,\mathrm{d}y)\mathrm{d}s\right\}e^{\varepsilon\langle\lambda, W_1\rangle}\right]
$$

$$
\overset{(d)}{=} \mathbb{E}_0\left[\exp\left\{\beta\int_0^{\varepsilon^{-2}}\int_{\mathbb{R}^d}\phi(y - W_s)\dot{B}(s,\mathrm{d}y)\mathrm{d}s\right\}e^{\varepsilon^2\langle\lambda, W_{\varepsilon^{-2}}\rangle}\right].
$$

From now on we will abbreviate

$$
H_{\beta,\varepsilon}(W, B) = \beta\int_0^{\varepsilon^{-2}}\int_{\mathbb{R}^d}\phi(y - W_s)\dot{B}(s,\mathrm{d}y)\mathrm{d}s - \frac{\beta^2}{2\varepsilon^2}V(0).
$$

Then, we have

$$
\mathbb{E}^{\widetilde{\mathbb{Q}}_{\beta,\varepsilon}}\left[e^{\varepsilon\langle\lambda, W_{\varepsilon^{-2}}\rangle}\right] \overset{(d)}{=} e^{\frac{|\lambda|^2}{2}(1-\varepsilon^2)}\frac{\mathbb{E}_0\left[\exp\{H_{\beta,\varepsilon}(W, B) + \varepsilon\langle\lambda, \ \varepsilon\ W_{\varepsilon^{-2}}\rangle\}\right]}{\mathbb{E}_0\left[\exp\{H_{\beta,\varepsilon}(W, B)\}\right]}
$$

$$
= e^{\frac{|\lambda|^2}{2}\left(1-\frac{1}{T}\right)}\mathbb{E}^{\widehat{\mathbb{Q}}_{\beta,T}}\left[e^{\frac{1}{\sqrt{T}}\langle\lambda, \frac{W_T}{\sqrt{T}}\rangle}\right]
$$

$$
\to e^{|\lambda|^2/2},
$$

where, in the second identity above, we wrote $T := \varepsilon^{-2}$ and the last statement holds true for **P**-almost every realization of B and follows from Theorem 1.1. Indeed, we can expand the exponential $\exp\{\frac{1}{\sqrt{T}}\langle\lambda, \frac{W_T}{\sqrt{T}}\rangle\}$ into a power series, and since all moments $\mathbb{E}^{\widehat{\mathbb{Q}}_{\beta,T}}[\langle\lambda, \frac{W_T}{\sqrt{T}}\rangle^n]$ converge, according to Theorem 1.1, to the moments $E[\langle\lambda, X\rangle^n]$ of a Gaussian $X \sim N(0, \mathbf{I_d})$, the expectation $\mathbb{E}^{\widehat{\mathbb{Q}}_{\beta,T}}\left[\exp\{\frac{1}{\sqrt{T}}\langle\lambda, \frac{W_T}{\sqrt{T}}\rangle\}\right]$ converges **P**-a.s. to 1. This concludes the proof of Corollary 1.2. $\qquad\square$

Remark 2 As remarked earlier, in [18] it was shown that the solution $u_{\varepsilon,t}(x)$ of the stochastic heat equation (1.6) with constant initial condition 1 converges in probability w.r.t. **P** in the weak disorder phase (i.e., for $\beta > 0$ small enough) to a strictly positive random variable $M_\infty(x)$. Note that the argument for the proof of Corollary 1.2 also implies the convergence of the ratio

$$\frac{\widehat{u}_\varepsilon^{(\lambda)}}{u_\varepsilon} \to 1 \qquad \text{as } \varepsilon \to 0,$$

where $\widehat{u}_\varepsilon^{(\lambda)}$ denotes the solution of the stochastic heat equation (1.6) with initial condition $u_{\varepsilon,0}^{(\lambda)}(x) = f(x) = e^{\varepsilon\langle\lambda,x\rangle}$. Although a proof of Corollary 1.2 could probably be given by first proving the convergence of the above ratio following analytical tools, for our purposes we choose to rely on more probabilistic arguments based on $L^2(\mathbf{P})$ computations and martingale methods as in [3, 18]. □

Remark 3 The case when the noise B is smoothened both in time and space has also recently been considered. If $F(t, x) = \int_{\mathbb{R}^d} \int_0^\infty \phi_1(t - s)\phi_2(x - y)\mathrm{d}B(s, y)$ is the mollified noise, and $\hat{u}_\varepsilon(t, x) = u(\varepsilon^{-2}t, \varepsilon^{-1}x)$ with u solving $\partial_t u = \frac{1}{2}\Delta u + \beta F(t, x)u$ with initial condition $u(0, x) = u_0(x)$ with $u_0 \in C_b(\mathbb{R}^d)$, then it was shown in [19] that for any $\beta > 0$ and $x \in \mathbb{R}^d$, $\frac{1}{\kappa(\varepsilon,t)} \mathbb{E}[\hat{u}_\varepsilon(t, x)] \to \hat{u}(t, x)$ as $\varepsilon \to 0$, where $\kappa(\varepsilon, t)$ is a divergent constant and $\hat{u}(t, x)$ solves the homogenized heat equation

$$\partial_t \hat{u} = \frac{1}{2}\mathrm{div}\left(\mathrm{a_{fi}}\nabla\hat{u}\right) \qquad (1.12)$$

with diffusion coefficient $\mathrm{a}_\beta \neq \mathbf{I}_d$. It was shown in [15, 20] that, for $\beta > 0$ small enough, the rescaled and spatially averaged fluctuations converge, i.e.,

$$\varepsilon^{1-d/2} \int \mathrm{d}x f(x)[\hat{u}_\varepsilon(t, x) - \mathbf{E}(\hat{u}_\varepsilon(t, x))] \Rightarrow \int f(x)\mathscr{U}(t, x)\,\mathrm{d}x$$

and the limit $\mathscr{U}$ satisfies the additive noise stochastic heat equation

$$\partial_t \mathscr{U} = \frac{1}{2}\mathrm{div}\left(\mathrm{a_{fi}}\nabla\mathscr{U}\right) + \beta\nu^2(\beta)\,\hat{u}\,\mathrm{d}B$$

with diffusivity a_β and variance $\nu^2(\beta)$, and $\hat{u}$ solves (1.12).

Remark 4 Finally we remark on the $T \to \infty$-asymptotic behavior of the polymer measures $\widehat{\mathbb{Q}}_{\beta,T}$ when β is large. Recall that, for large β, it was shown in [18] that the renormalized partition function M_T converges in distribution to 0 (i.e., we are in the strong disorder regime). In a recent article [5], based on the compactness theory developed in [21], we show that for large enough β, the distribution of the endpoint of the path W_T under $\widehat{\mathbb{Q}}_{\beta,T}$ is concentrated in random spatial regions, leading to a strong localization effect.

The rest of the article is devoted to the proof of Theorem 1.1.

2 Proof of Theorem 1.1

We remind the reader that $V = \phi \star \phi$ where ϕ is a smooth, positive, even function and has support in a ball of radius K around the origin. Moreover, $\int_{\mathbb{R}^d} \phi = 1$. Furthermore, $\mathbb{P}_0$ denotes the law of a Brownian motion W in $\mathbb{R}^d$, starting from 0, with $\mathbb{E}_0$ denoting the corresponding expectation, while $\mathbf{P}$ denotes the law of the white noise B which is independent of W and $\mathbf{E}$ denotes the corresponding expectation. We will also denote by $\mathcal{F}_T$ the σ-field generated by the noise B up to time T.

For simplicity, we will fix $x = 0$ and we will show that, for $\beta < \beta_{L^2}$ and any $\lambda \in \mathbb{R}^d$, $\mathbf{P}$-almost surely,

$$\lim_{T \to \infty} \mathbb{E}^{\widehat{\mathbb{Q}}_{\beta,T}} \left[e^{\langle \lambda, W_T / \sqrt{T} \rangle} \right] = e^{|\lambda|^2/2} \tag{2.1}$$

where $\widehat{\mathbb{Q}}_{\beta,T} = \widehat{\mathbb{Q}}_{\beta,T}^{(0)}$.

Lemma 2.1 *In $d \geq 3$ and for $\beta > 0$ small enough, M_T converges almost surely to a random variable M_∞ that satisfies*

$$\mathbf{E}[M_\infty] = 1 \quad and \quad \mathbf{P}(M_\infty = 0) = 0.$$

Proof Let us set

$$H_{\beta,T}(W, B) = \beta \int_0^T \int_{\mathbb{R}^d} \phi(y - W_s) \dot{B}(s, \mathrm{d}y) \mathrm{d}s - \frac{\beta^2 T}{2} V(0) \tag{2.2}$$

Then the proof follows from the fact that $M_T = \exp\{H_{\beta,T}\}$ is a nonnegative $(\mathcal{F}_T)$-martingale, which remains bounded in $L^2(\mathbf{P})$ for $\beta < \beta_{L^2}$. □

Lemma 2.2 *Let $\mathcal{G}_T$ be the σ-algebra generated by the Brownian path $(W_s)_{0 \leq s \leq T}$ until time T. Then, for $n = (n_1, \ldots, n_d) \in \mathbb{N}_0^d$ with $|n| := \sum_{i=1}^d n_i$, and $\lambda = (\lambda_1, \ldots, \lambda_d) \in \mathbb{R}^d$,*

$$I_n(T, W_T) = \frac{\partial^{|n|}}{\prod_{j=1}^d \partial \lambda_j^{n_j}} \left[\exp\left\{ \sum_{i=1}^d \lambda_i W_T^{(i)} - \frac{1}{2} |\lambda|^2 T \right\} \right] \Bigg|_{\lambda=0} \tag{2.3}$$

is a $(\mathcal{G}_T)$-martingale.

Proof We first note that

$$U_\lambda(T) = \exp\left\{ \sum_{i=1}^d \lambda_i W_T^{(i)} - \frac{1}{2} |\lambda|^2 T \right\}$$

is a $(\mathcal{G}_T)$-martingale for every $\lambda \in \mathbb{R}^d$. Now if we write $U_\lambda(T)$ in its Taylor series at $\lambda = 0$ as

$$U_\lambda(T) = \sum_{n_1=0}^{\infty} \cdots \sum_{n_d=0}^{\infty} I_n(T, W_T) \frac{\prod_{i=1}^{d} \lambda_i^{n_i}}{\prod_{i=1}^{d} n_i!},$$

it follows that $I_n(T, W_T)$ is a $(\mathcal{G}_T)$-martingale for every $n \in \mathbb{N}_0^d$. □

Lemma 2.3 *The sequence*

$$Y_n(T) = \mathbb{E}_0\left[\exp\left\{H_{\beta,T}(W, B)\right\} I_n(T, W_T)\right]$$

is a martingale with respect to the filtration $\mathcal{F}_T$ generated by $(B_s)_{0 \le s \le T}$.

Proof Since $I_n(T, W_T)$ is a $\mathcal{G}_T = \sigma(W_s : 0 \le s \le T)$-martingale, we get for $S \le T$

$$\mathbf{E}[Y_n(T)|\mathcal{F}_S] = \mathbb{E}_0\left[\mathbf{E}\left[\exp\{H_{\beta,T}(W, B)\}|\mathcal{F}_S\right] I_n(T, W_T)\right]$$
$$= \mathbb{E}_0\left[\exp\{H_{\beta,S}(W, B)\} \mathbb{E}_0\left[I_n(T, W_T)|\mathcal{G}_S\right]\right]$$
$$= Y_n(S).$$

Furthermore, since $I_n(T, W_T)$ is integrable w.r.t $\mathbb{P}_0$, $\mathbf{E}[\exp\{H_{\beta,T}(W, B)\}] = 1$ proves the lemma as $\mathbf{E}[Y_n(T)] = \mathbb{E}_0[I_n(T, W_T)] < \infty$. □

We will now control the martingale difference sequence $Y_n(T) - Y_n(T - 1)$ in $L^2(\mathbf{P})$ which will provide some estimate on the decay on correlations.

Lemma 2.4 *If $d \ge 3$ and $\beta < \beta_{L^2}$ as in Theorem 1.1, then there exists $C = C(\beta, d, \phi)$ such that for any $T \ge 1$ and $|n| \ge 1$,*

$$\mathbf{E}\left[(Y_n(T) - Y_n(T - 1))^2\right] \le CT^{|n|-p}.$$

for $p = 9/8$.

Proof We compute the martingale difference as follows:

$$Y_n(T) - Y_n(T - 1)$$
$$= \mathbb{E}_0\left[\exp\{H_{\beta,T}(W, B)\} I_n(T, W_T) - \exp\{H_{\beta,T-1}(W, B)\} I_n(T, W_T)\right.$$
$$\left. + \exp\{H_{\beta,T-1}(W, B)\}\left(I_n(T, W_T) - I_n(T - 1, W_{T-1})\right)\right]$$
$$= \mathbb{E}_0\left[\exp\{H_{\beta,T-1}(W, B)\}\left(\exp\{H_{\beta,T-1,T}(W, B)\} - 1\right) I_n(T, W_T)\right],$$

where $H_{\beta,T-1,T}(W,B)$ is nothing but $H_{\beta,T}(W,B)$ with the integral starting in $T-1$ instead of 0 and we have used that $I_n(T,W_T)$ is a martingale. Then

$$\mathbf{E}\Big[(Y_n(T)-Y_n(T-1))^2\Big] \tag{2.4}$$

$$= (\mathbb{E}_0 \otimes \mathbb{E}_0)\Big[\mathbf{E}\Big[\exp\Big\{\beta\int_0^{T-1}\int_{\mathbb{R}^d}\big(\phi(y-W_s)+\phi(y-W_s')\big)\dot{B}(s,dy)ds - \beta^2(T-1)V(0)\Big\}\Big]$$

$$\times \mathbf{E}\Big[\exp\Big\{\beta\int_{T-1}^{T}\int_{\mathbb{R}^d}\big(\phi(y-W_s)+\phi(y-W_s')\big)\dot{B}(s,dy)ds - \beta^2 V(0)\Big\} - 1\Big]I_n(T,W_T)I_n(T,W_T')\Big] \tag{2.5}$$

with W and W' being two independent copies of the Brownian path. If $K < \infty$ denotes the radius of the support of ϕ, and if we assume $|W_s - W_s'| > 2K$, either $\phi(\cdot - W_s)$ or $\phi(\cdot - W_s')$ must vanish, everywhere in $\mathbb{R}^d$. Therefore, on the event $\{|W_s - W_s'| > 2K \ \forall \ s \in [T-1,T]\}$,

$$\mathbf{E}\Big[\exp\Big\{\beta\int_{T-1}^{T}\int_{\mathbb{R}^d}\big(\phi(y-W_s)+\phi(y-W_s')\big)\dot{B}(s,dy)s\Big\}\Big] = \Big[\mathbf{E}\Big[\exp\Big\{\beta\int_{T-1}^{T}\int_{\mathbb{R}^d}\phi(y-W_s)\dot{B}(s,dy)ds\Big\}\Big]\Big]^2$$

$$= e^{\beta^2 V(0)}.$$

Hence, we can estimate

$$(\mathbb{E}_0 \otimes \mathbb{E}_0)\Big[\mathbf{E}\Big[\exp\Big\{\beta\int_{T-1}^{T}\int_{\mathbb{R}^d}\big(\phi(y-W_s)+\phi(y-W_s')\big)\dot{B}(s,dy)ds - \beta^2 V(0)\Big\} - 1\Big]\Big]$$

$$= (\mathbb{E}_0 \otimes \mathbb{E}_0)\Big[\mathbb{1}_{\{|W_s-W_s'|\le 2K \text{ for some } s\in[T-1,T]\}}$$

$$\times \mathbf{E}\Big[\exp\Big\{\beta\int_{T-1}^{T}\int_{\mathbb{R}^d}\big(\phi(y-W_s)+\phi(y-W_s')\big)\dot{B}(s,dy)ds - \beta^2 V(0)\Big\} - 1\Big]\Big]$$

$$= (\mathbb{E}_0 \otimes \mathbb{E}_0)\Big[\mathbb{1}_{\{|W_s-W_s'|\le 2K \text{ for some } s\in[T-1,T]\}}\Big(\exp\Big\{\beta^2\int_{T-1}^{T}V(W_s-W_s')ds\Big\} - 1\Big)\Big]$$

$$\le \Big(e^{\beta^2\|V\|_\infty} - 1\Big)(\mathbb{P}_0 \otimes \mathbb{P}_0)\big(|W_s - W_s'| \le 2K \text{ for some } s \in [T-1,T]\big),$$

where the inequality holds as $V = \phi \star \phi$ is bounded.

Since

$$(\mathbb{E}_0 \otimes \mathbb{E}_0)\Big[\mathbf{E}\Big[\exp\Big\{\beta\int_0^{T-1}\int_{\mathbb{R}^d}\big(\phi(y-W_s)+\phi(y-W_s')\big)\dot{B}(s,dy)ds - \beta^2(T-1)V(0)\Big\}\Big]\Big]$$

$$= (\mathbb{E}_0 \otimes \mathbb{E}_0)\Big[\exp\Big\{\beta^2\int_0^{T-1}V(W_s-W_s')ds\Big\}\Big]$$

and $W_s - W_s' \stackrel{(d)}{=} W_{2s}$, if we now invoke Hölder's inequality with $q = 4/3$ and $p = r = 8$ to (2.5), we get

$$\mathbf{E}\left[(Y_n(T) - Y_n(T-1))^2\right] \le \left(\mathbb{E}_0\left[\exp\left\{8\beta^2 \int_0^\infty V(W_{2s})ds\right\}\right]\right)^{1/8} \tag{2.6}$$

$$\times \left(c(\mathbb{P}_0 \otimes \mathbb{P}_0)(|W_s - W_s'| \le 2K \text{ for some } s \in [T-1, T])\right)^{3/4} \tag{2.7}$$

$$\times \left((\mathbb{E}_0 \otimes \mathbb{E}_0)\left[(I_n(T, W_T)I_n(T, W_T'))^8\right]\right)^{1/8}. \tag{2.8}$$

As $\beta > 0$ is chosen small enough and $d \ge 3$, the factor on the right hand side of (2.6) is finite (see Lemma 3.1 in [18]), while for the factor in (2.7) we use

$$(\mathbb{P}_0 \otimes \mathbb{P}_0)(|W_s - W_s'| \le 2K \text{ for some } s \in [T-1, T])$$

$$\le \int_{T-1}^T \mathbb{P}_0(W_{2s} \in B_0(2K)) \le c'(T - (T-1))(T-1)^{-d/2} \tag{2.9}$$

for T large enough and a proper constant c'. For the factor in (2.8), we need some facts regarding the polynomial

$$I_n(T, x) = \frac{\partial^{|n|}}{\prod_{j=1}^d \partial \lambda_j^{n_j}} \left[\exp\left\{\sum_{i=1}^d \lambda_i x^{(i)} - \frac{1}{2}|\lambda|^2 T\right\}\right]\Bigg|_{\lambda=0}$$

for $x = (x^{(1)}, \ldots, x^{(d)})$, $\lambda = (\lambda_1, \ldots, \lambda_d) \in \mathbb{R}^d$ and it is useful to collect them now:

Lemma 2.5 *Let $T > 0$, the polynomial $I_n(T, X_T)$ can be rewritten as*

$$I_n(T, X_T) = \sum_{i_1,\ldots,i_d,j} A_n^{X_T}(i_1, \ldots, i_d, j) X_T^{(1)^{i_1}} \cdots X_T^{(d)^{i_d}} T^j, \tag{2.10}$$

where the coefficients $A_n^{X_T}(i_1, \ldots, i_d, j)$ satisfy the properties

(a) $A_n^{X_T}(i_1, \ldots, i_d, j) = 0$ for $i_1 + \ldots + i_d + 2j \ne |n|$
(b) $A_n^{X_T}(i_1, \ldots, i_d, j) = A_n^{X_1}(i_1, \ldots, i_d, j)$
(c) $A_n^{X_T}(i_1, \ldots, i_d, 0) = \delta_{i_1 n_1} \cdots \delta_{i_d n_d}$ for $i_1 + \ldots + i_d = |n|$.

We assume Lemma 2.5 for now and continue with the proof of Lemma 2.4. Note that we only have to estimate the factor in (2.8), for which we can now apply Lemma 2.5(a) and use that W_T and W_T' are independent such that

$$\left((\mathbb{E}_0 \otimes \mathbb{E}_0)\left[(I_n(T, W_T)I_n(T, W_T'))^8\right]\right)^{1/8} = \left(\mathbb{E}_0\left[I_n(T, W_T)^8\right]\right)^{1/4}.$$

We claim that $E_0[I_n(T, W_T)^8] = O(T^{4|n|})$. Indeed, note that by Lemma 2.5(a) and (2.10),

$$I_n(T, W_T) = \sum_{i_1+\cdots+i_d+2j=|n|} A_n^{W_T}(i_1, \ldots, i_d, j) W_T^{(1)^{i_1}} \cdots W_T^{(d)^{i_d}} T^j.$$

Now we apply multinomial theorem for $I_n(T, W_T)^8$ in the above display whence the expansion yields

$$I_n(T, W_T)^8 = \sum_{\substack{i_1+\cdots+i_d+2j=|n| \\ \sum_\ell r_\ell=8}} \frac{8!}{\prod_\ell r_\ell!} \prod_\ell \left(A_n^{W_T}(i_1, ..., i_d, j) W_T^{(1)^{i_1}} \cdots W_T^{(d)^{i_d}} T^j \right)^{r_\ell}.$$

Since $W_T^{(1)}, \ldots, W_T^{(d)}$ are independent and $\mathbb{E}_0[(W_T^{(i)})^k] = T^{k/2} C_k$ for any $k \in \mathbb{N}$ and $i = 1, \ldots, d$, and furthermore by Lemma 2.5(b), the coefficients $A_n^{W_T}(i_1, ..., i_d, j)$ do not depend on W_T or T, we have

$$\mathbb{E}_0[I_n(T, W_T)^8] = \sum_{\substack{i_1+\cdots+i_d+2j=|n| \\ \sum_\ell r_\ell=8}} C(r_\ell, i_1, \ldots, i_d, j) T^{\left(\frac{i_1}{2}+\cdots+\frac{i_d}{2}+j\right)\sum_\ell r_\ell} = O(T^{4|n|}),$$

proving that

$$\left((\mathbb{E}_0 \otimes \mathbb{E}_0)\left[(I_n(T, W_T) I_n(T, W_T'))^8\right]\right)^{1/8} = O(T^{|n|}). \tag{2.11}$$

Then (2.9) and (2.11), together with (2.6)–(2.8) imply that for $d \geq 3$

$$\mathbf{E}\left[(Y_n(T) - Y_n(T-1))^2\right] = O(T^{|n|-p}).$$

and $p = 9/8$. We now owe the reader only the proof of Lemma 2.5.

Proof of Lemma 2.5: We make two simple observations:

(1)

$$\frac{\partial}{\partial \lambda_k}\left[\exp\left\{\sum_{i=1}^d \lambda_i X_T^{(i)} - \frac{1}{2}|\lambda|^2 T\right\}\right] = (X_T^{(k)} - \lambda_k T)\exp\left\{\sum_{i=1}^d \lambda_i X_T^{(i)} - \frac{1}{2}|\lambda|^2 T\right\},$$

(2)

$$\frac{\partial}{\partial \lambda_k}\left[(X_T^{(k)} - \lambda_k T)^{i_k}\right] = -i_k T (X_T^{(k)} - \lambda_k T)^{i_k-1}.$$

We will now prove Lemma 2.5 by induction as follows:

For $|n| = 0$ we have $I_n(T, X_T) = 1$ and thus $i_1 + \ldots + i_d + 2j = 0 = |n|$. For the induction step we assume $i_1 + \ldots + i_d + 2j = |n|$ for every summand in $I_n(T, X_T)$ which is non-zero, where $n \in \mathbb{N}_0^d$ is fixed. If we differentiate $I_n(T, X_T)$ w.r.t. λ_k, every summand of the new polynomial changes as written in (1) or (2). Without loss of generality we assume $k = 1$. In case (1), only the exponent of $X_T^{(1)}$ increases by 1 and since the assumption yields $(i_1 + 1) + i_2 + \ldots + i_d + 2j = |n| + 1$, no summand influences the induction step.

For a summand that follows case (2) the exponent of T increases by 1, while the exponent of $X_T^{(1)}$ decreases by 1. The assumption yields $(i_1 - 1) + i_2 + ... + i_d + 2(j + 1) = |n| + 1$. From both cases together one can conclude Lemma 2.5(a).

Furthermore, as the coefficients do not depend on the values of X_T or T, part (b) follows immediately.

For the statement in (c) we again analyze the cases (1) and (2). If we again assume k to be one, in case (2) the exponent of $X_T^{(1)}$ decreases by 1 and in case (1) only i_1 increases by 1. This means that the requirement $i_1 + ... + i_d = |n|$ can only be fulfilled by that summand that always follows case (1). Since $A_0(0, ..., 0) = 1$, the statement in Lemma 2.5(c) now also follows.

This concludes the proof of Lemma 2.5. Hence Lemma 2.4 is also proven. □

Lemma 2.6 *Under the assumptions imposed in Lemma 2.4, the process $(M_{\tau+N,n})_{N \in \mathbb{N}}$ is an $(\mathcal{F}_{\tau+N})_{N \in \mathbb{N}}$-martingale bounded in $L^2(\mathbf{P})$, where*

$$M_{\tau+N,n} = \sum_{S=1}^{N} S^{-|n|/2}(Y_n(\tau + S) - Y_n(\tau + S - 1))$$

and $\tau = T - \lfloor T \rfloor$.

Proof By Lemma 2.3, $(Y_n(T))_{T \geq 0}$ is an $(\mathcal{F}_T)_{T \geq 0}$-martingale, and we know that the process $(M_{\tau+N,n})_{N \in \mathbb{N}}$ is an $(\mathcal{F}_{\tau+N})_{N \in \mathbb{N}}$-martingale. Thus

$$\mathbf{E}[M_{\tau+N,n}^2] = \sum_{S=1}^{N} S^{-|n|}\mathbf{E}[(Y_n(\tau + S) - Y_n(\tau + S - 1))^2]$$

$$+ \sum_{S=1}^{N} \sum_{S \neq R=1}^{N} S^{-|n|/2}R^{-|n|/2}\mathbf{E}[(Y_n(\tau + S) - Y_n(\tau + S - 1))(Y_n(\tau + R) - Y_n(\tau + R - 1))]$$

$$= \sum_{S=1}^{N} S^{-|n|}\mathbf{E}[(Y_n(\tau + S) - Y_n(\tau + S - 1))^2]$$

and by Lemma 2.4

$$\mathbf{E}[M_{\tau+N,n}^2] \leq C \sum_{S=1}^{N} S^{-9/8}\left(\frac{S+\tau}{S}\right)^{|n|} \leq 2^{|n|}C \sum_{S=1}^{N} S^{-9/8} \leq 2^{|n|}C \sum_{S=1}^{\infty} S^{-9/8} < \infty$$

for a proper constant C and $T = \tau + N$ large enough. From this we can conclude that $M_{\tau+N,n}$ is an L^2-bounded martingale for every $n \in \mathbb{N}^d \setminus \{0\}$. □

Lemma 2.7 *Under the assumptions imposed in Lemma 2.4, if $n \in \mathbb{N}^d \setminus \{0\}$, then*

$$\lim_{T \to \infty} T^{-|n|/2}Y_n(T) = 0$$

holds $\mathbf{P}$-almost surely.

Proof From Lemma 2.6 we deduce that $M_{\tau+N,n}$ converges almost surely w.r.t. **P**. Then

$$\lim_{N\to\infty} N^{-|n|/2} \sum_{S=1}^{N} Y_n(\tau+S) - Y_n(\tau+S-1) = 0$$

for all $\tau \in [0, 1)$ and it follows

$$\lim_{T\to\infty} T^{-|n|/2} \sum_{S=1}^{\lfloor T\rfloor} Y_n(\tau+S) - Y_n(\tau+S-1) = 0$$

for $\tau = T - \lfloor T\rfloor$. Since

$$Y_n(T) = \sum_{S=1}^{\lfloor T\rfloor} Y_n(\tau+S) - Y_n(\tau+S-1) + Y_n(\tau),$$

the claim in Lemma 2.7 follows if we show that $T^{-|n|/2} Y_n(\tau)$ goes almost surely to zero. But by the Cauchy-Schwarz inequality

$$Y_n(\tau)^2 \le \mathbb{E}_0\Big[I_n(\tau, W_\tau)^2\Big]\mathbb{E}_0\Big[\exp\Big\{2\beta \int_0^\tau \int_{\mathbb{R}^d} \phi(y-W_s)\dot{B}(s,\mathrm{d}y)\mathrm{d}s - \beta^2\tau V(0)\Big\}\Big].$$

The first factor on the right hand side is finite as τ is bounded by zero and one and all moments of a Gaussian random variable are finite. Again, because τ is bounded and as ϕ is a bounded function with compact support,

$$T^{-|n|/2}\Big(\mathbb{E}_0\Big[\exp\Big\{2\beta \int_0^\tau \int_{\mathbb{R}^d} \phi(y-W_s)\dot{B}(s,\mathrm{d}y)\mathrm{d}s - \beta^2\tau V(0)\Big\}\Big]\Big)^{1/2} \to 0$$

almost surely, as $T \to \infty$. $\square$

We finally turn to the proof of Theorem 1.1.

Proof of Theorem 1.1 By Lemma 2.5(a) we can rewrite the almost sure convergence statement in Lemma 2.7 as

$$\lim_{T\to\infty} \mathbb{E}_0\Big[\sum_{i_1,\dots,i_d} A_n^{W_T}\Big(i_1,\dots,i_d, \frac{|n|-i_1-\dots-i_d}{2}\Big)\Big(\frac{W_T^{(1)}}{\sqrt{T}}\Big)^{i_1}\cdots\Big(\frac{W_T^{(d)}}{\sqrt{T}}\Big)^{i_d}$$

$$\times \exp\Big\{\beta \int_0^T \int_{\mathbb{R}^d} \phi(y-W_s)\dot{B}(s,\mathrm{d}y)\mathrm{d}s - \frac{\beta^2 T}{2}V(0)\Big\}\Big] = 0 \quad (2.12)$$

where we implicitly assume $(|n|-i_1-\dots-i_d)/2 \in \mathbb{N}_0$.

We now consider the sum

$$\sum_{i_1,...,i_d} A_n^X\left(i_1,..,i_d,\frac{|n|-i_1-...-i_d}{2}\right) X^{(1)^{i_1}} \cdots X^{(d)^{i_d}}$$

with the i.i.d. random variables $X, X^{(1)}, ..., X^{(d)}$, where $X \sim \text{Normal}(0, 1)$. The random variable X has mean zero so that the odd moments of X are zero. Since the random variables are independent, this means if at least one i_k is odd, the expectation of that summand is zero. A summand with no odd exponents arises from a summand where $i_1, ..., i_{k-1}, i_{k+1}, ..., i_d$ are even and i_k is odd in the $|n| - 1$-th derivative of $\exp\{\sum_{i=1}^d \lambda_i X^{(i)} - \frac{1}{2}\sum_{i=1}^d \lambda_i^2\}$. Note that by the product rule

$$\frac{\partial}{\partial \lambda_j}\left[(X^{(j)} - \lambda_j)^{2k+1}\exp\left\{\sum_{i=1}^d \lambda_i X^{(i)} - \frac{1}{2}|\lambda|^2\right\}\right]\Bigg|_{\lambda=0} \tag{2.13}$$

$$= \left[(X^{(j)} - \lambda_j)^{2k+2} - (2k+1)(X^{(j)} - \lambda_j)^{2k}\right]\exp\left\{\sum_{i=1}^d \lambda_i X^{(i)} - \frac{1}{2}|\lambda|^2\right\}\Bigg|_{\lambda=0}.$$

We denote by E the expectation w.r.t the random variables $X, X^{(1)}, ..., X^{(d)}$. As X is Normal$(0, 1)$ distributed, $E[X^{2k+2}] = (2k+1)E[X^{2k}]$ for all $k \in \mathbb{N}_0$ and so by (2.13) one can obtain that the expectation of the summands with no odd exponents cancel each other. Both statements together (that for all exponents are even and that for at least one is odd) yield

$$E\left[\sum_{i_1,...,i_d} A_n^X\left(i_1,..,i_d,\frac{|n|-i_1-...-i_d}{2}\right) X^{(1)^{i_1}} \cdots X^{(d)^{i_d}}\right] = 0 \tag{2.14}$$

which shows that the sequence on the left hand side of (2.12) converges almost surely to the left hand side of (2.14) for every $n \in \mathbb{N}_0^d \setminus \{0\}$. By Lemma 2.1 this holds true after dividing the sequence by M_T and thus

$$\mathbb{E}^{\hat{\mathbb{Q}}_{\beta,T}}\left[\sum_{i_1,...,i_d} A_n^{W_T}\left(i_1,...,i_d,\frac{|n|-i_1-...-i_d}{2}\right)\left(\frac{W_T^{(1)}}{\sqrt{T}}\right)^{i_1} \cdots \left(\frac{W_T^{(d)}}{\sqrt{T}}\right)^{i_d}\right] \tag{2.15}$$

$$\longrightarrow E\left[\sum_{i_1,...,i_d} A_n^X\left(i_1,..,i_d,\frac{|n|-i_1-...-i_d}{2}\right) X^{(1)^{i_1}} \cdots X^{(d)^{i_d}}\right]$$

almost surely.

By Lemma 2.5(b), the coefficients $A_n^X\left(i_1,..,i_d,\frac{|n|-i_1-...-i_d}{2}\right)$ coming from

$$\frac{\partial^{|n|}}{\partial \lambda_1^{n_1} \cdots \partial \lambda_d^{n_d}}\left[\exp\left\{\sum_{i=1}^d \lambda_i X^{(i)} - \frac{1}{2}\sum_{i=1}^d \lambda_i^2\right\}\right]\Bigg|_{\lambda=0}$$

are equal to the coefficients $A_n^{W_T}\left(i_1, .., i_d, \frac{|n|-i_1-...-i_d}{2}\right)$. By induction, this statement yields for all $n \in \mathbb{N}_0^d \setminus \{0\}$

$$\lim_{T \to \infty} \mathbb{E}^{\widehat{Q}_{\beta,T}}\left[\left(\frac{W_T^{(1)}}{\sqrt{T}}\right)^{n_1} \cdots \left(\frac{W_T^{(d)}}{\sqrt{T}}\right)^{n_d}\right] = E\left[X^{(1)^{n_1}} \cdots X^{(d)^{n_d}}\right] \quad \text{a.s.} : \quad (2.16)$$

Indeed, for $|n| = 1$ this is obvious. We check the induction step for $n \to \tilde{n}$, where $\tilde{n} = (n_1 + 1, n_2, ..., n_d)$. The induction hypothesis is that (2.16) holds for all $m \in \mathbb{N}_0^d \setminus \{0\}$ that satisfy $m_1 \leq n_1 + 1, ..., m_d \leq n_d$ and at least one of these inequalities is strictly. By Lemma 2.5(c) we know that the coefficient $A_{\tilde{n}}^{W_T}(n_1 + 1, n_2, ..., n_d, 0)$ has value 1. This means that by the induction hypothesis

$$\mathbb{E}^{\widehat{Q}_{\beta,T}}\left[\sum_{i_1,...,i_d} A_{\tilde{n}}^{W_T}\left(i_1, ..., i_d, \frac{|n| + 1 - i_1 - ... - i_d}{2}\right)\left(\frac{W_T^{(1)}}{\sqrt{T}}\right)^{i_1} \cdots \left(\frac{W_T^{(1)}}{\sqrt{T}}\right)^{i_d}\right]$$

$$- \mathbb{E}^{\widehat{Q}_{\beta,T}}\left[\left(\frac{W_T^{(1)}}{\sqrt{T}}\right)^{n_1+1}\left(\frac{W_T^{(2)}}{\sqrt{T}}\right)^{n_2} \cdots \left(\frac{W_T^{(d)}}{\sqrt{T}}\right)^{n_d}\right] \qquad (2.17)$$

$$\longrightarrow E\left[\sum_{i_1,...,i_d} A_{\tilde{n}}^X\left(i_1, ..., i_d, \frac{|n| + 1 - i_1 - ... - i_d}{2}\right)X^{(1)^{i_1}} \cdots X^{(d)^{i_d}}\right]$$

$$- E\left[X^{(1)^{n_1+1}} X^{(2)^{n_2}} \cdots X^{(d)^{n_d}}\right]$$

almost surely as $T \to \infty$. If we set n to be $\tilde{n}$ in (2.15), the induction step follows by combining (2.15) and (2.17). Since (2.16) shows that all moments converge to the moments of the standard normal distribution, we have proved (2.1) and thus Theorem 1.1.

Acknowledgements The authors would like to thank an anonymous referee for a very careful reading of the manuscript and many valuable comments.

References

1. Alberts, T., Khanin, K., Quastel, J.: The continuum directed random polymer. J. Stat. Phys. **154**, 305–326 (2014)
2. Amir, G., Corwin, I., Quastel, J.: Probability distribution of the free energy of the continuum directed random polymer in 1+1 dimensions. Commun. Pure Appl. Math. **64**(4), 466–537 (2011)
3. Bolthausen, E.: A note on the diffusion of directed polymers in a random environment. Commun. Math. Phys. **123**, 529–534 (1989)
4. Bertini, L., Giacomin, G.: Stochastic Burgers and KPZ equations from particle systems. Commun. Math. Phys. **183**(3), 571–607 (1997)
5. Bröker, Y., Mukherjee, C.: On the localization of the stochastic heat equation in strong disorder (2018). arXiv: 1808.05202
6. Cosco, C.: The intermediate disorder regime for Brownian directed polymers in poisson environment (2018). arXiv:1804.09571

7. Comets, F., Cosco, C.: Brownian polymers in poissonian environment: a survey (2018). arXiv:1805.10899
8. Comets, F., Cosco, C., Mukherjee, C.: Fluctuation and rate of convergence of the stochastic heat equation in weak disorder. arXiv: 1807.03902
9. Comets, F., Shiga, T., Yoshida, N.: Directed polymers in a random environment: path localization and strong disorder. Bernoulli **9**, 705–728 (2003)
10. Comets, F., Shiga, T., Yoshida, N.: Probabilistic analysis of directed polymers in a random environment: a review. Stochastic analysis on large scale interacting systems. Adv. Stud. Pure Math. **39**, 115–142 (2004)
11. Comets, F., Yoshida, N.: Brownian directed polymers in random environment. Comm. Math. Phys. **254**, 257–287 (2005)
12. Comets, F., Yoshida, N.: Directed polymers in random environment are diffusive at weak disorder. Ann. Probab. **34**, 1746–1770 (2006)
13. Feng, Z.C.: Rescaled directed random polymer in random environment in dimension 1+ 2. Ph.D. thesis, University of Toronto (2015)
14. Gubinelli, M., Perkowski N.: KPZ reloaded. Comm. Math. Phys. **349**(1), 165–269 (2017)
15. Gu, Y., Ryzhik, L., Zeitouni, O.: The Edwards-Wilkinson limit of the random heat equation in dimensions three and higher (2017). arXiv:1710.00344
16. Hairer, M.: Solving the KPZ equation. Ann. Math. **178**, 559–664 (2013)
17. Imbrie, J.Z., Spencer, T.: Diffusion of directed polymers in a random environment. J. Stat. Phys. **52**(3/4) (1988)
18. Mukherjee, C., Shamov, A., Zeitouni, O.: Weak and strong disorder for the stochastic heat equation and the continuous directed polymer in $d \geq 3$. Electr. Comm. Prob. **21** (2016). arXiv: 1601.01652
19. Mukherjee, C.: Central limit theorem for the annealed path measures for the stochastic heat equation and the continuous directed polymer in $d \geq 3$ (2017). arXiv: 1706.09345
20. Magnen, J., Unterberger, J.: The scaling limit of the KPZ equation in space dimension 3 and higher. J. Stat. Phys. **171**(4), 543–598 (2018)
21. Mukherjee, C., Varadhan, S.R.S.: Brownian occupation measures, compactness and large deviations. Ann. Prob. **44**(6), 3934–3964 (2016). arXiv: 1404.5259
22. Sasamoto, T., Spohn, H.: The one-dimensional KPZ equation: an exact solution and its universality. Phys. Rev. Lett. **104**, 230602 (2010)

GOE and Airy$_{2 \to 1}$ Marginal Distribution via Symplectic Schur Functions

Elia Bisi and Nikos Zygouras

To Raghu Varadhan with great respect on the occasion of his 75th birthday

Abstract We derive Sasamoto's Fredholm determinant formula for the Tracy-Widom GOE distribution, as well as the one-point marginal distribution of the Airy$_{2 \to 1}$ process, originally derived by Borodin-Ferrari-Sasamoto, as scaling limits of point-to-line and point-to-half-line directed last passage percolation with exponentially distributed waiting times. The asymptotic analysis goes through new expressions for the last passage times in terms of integrals of (the continuous analog of) symplectic and classical Schur functions, obtained recently in [6].

Keywords Point-to-line directed last passage percolation ·
Tracy-Widom GOE distribution · Airy$_{2 \to 1}$ · Schur functions ·
Symplectic Schur functions

2010 Mathematics Subject Classification. Primary: 60Cxx · 05E05 · 82B23 ·
Secondary: 11Fxx · 82D60

1 Introduction

The goal of this contribution is to provide a new route to the Tracy-Widom GOE distribution and to the one-point marginal of the Airy$_{2 \to 1}$ process through asymp-

E. Bisi
School of Mathematics and Statistics, University College Dublin, Belfield Dublin 4, Ireland
e-mail: elia.bisi@ucd.ie

N. Zygouras (✉)
Department of Statistics, University of Warwick, Coventry CV4 7AL, UK
e-mail: n.zygouras@warwick.ac.uk

© Springer Nature Switzerland AG 2019
P. Friz et al. (eds.), *Probability and Analysis in Interacting Physical Systems*, Springer Proceedings in Mathematics & Statistics 283,
https://doi.org/10.1007/978-3-030-15338-0_7

totics of directed last passage percolation (dLPP), starting from new exact formulae involving not only the usual Schur functions (that have been appearing so far in treatments of dLPP) but also symplectic Schur functions. The first systematic study on the dLPP model goes back to Johansson [12], who derived the Tracy-Widom GUE limiting distribution for the model with point-to-point path geometry and geometrically distributed waiting times. For an overview on dLPP and other similar integrable probabilistic models related to random matrix theory and determinantal structures, the reader is referred to [11, Chap. 10] and [15].

Let us start by introducing the details of the directed last passage percolation model that we will work with. On the $(1 + 1)$-dimensional lattice we consider paths with fixed starting point and with ending point lying free on a line or half-line. We also consider a random field on the lattice, which is a collection $\{W_{i,j} : (i, j) \in \mathbb{N}^2\}$ of independent random variables, usually called *weights* or *waiting times* and which we will assume to be exponentially distributed. We will consider two different path geometries (see Fig. 1 for a graphical representation):

(i) The *point-to-line* directed last passage percolation, defined by

$$\tau_N^{\text{flat}} := \max_{\pi \in \Pi_N^{\text{flat}}} \sum_{(i,j) \in \pi} W_{i,j}, \tag{1.1}$$

where Π_N^{flat} is the set of directed (down-right) lattice paths of length N (namely, made up of N vertices) starting from $(1, 1)$ and ending at any point of the line $\{(i, j) \in \mathbb{N}^2 : i + j = N + 1\}$.

(ii) The *point-to-half-line* directed last passage percolation, defined by

$$\tau_N^{\text{h-flat}} := \max_{\pi \in \Pi_N^{\text{h-flat}}} \sum_{(i,j) \in \pi} W_{i,j}, \tag{1.2}$$

where $\Pi_N^{\text{h-flat}}$ is the set of directed lattice paths of length N starting from $(1, 1)$ and ending on the half-line $\{(i, j) \in \mathbb{N}^2 : i + j = N + 1, \ i \le j\}$.

As already mentioned, in our treatment we will assume the weights $W_{i,j}$'s to be exponentially distributed. More specifically, for the point-to-line dLPP τ_{2N}^{flat}, we consider a triangular array of independent weights $W = \{W_{i,j} : (i, j) \in \mathbb{N}^2, \ i + j \le 2N + 1\}$ distributed as

$$W_{i,j} \sim \begin{cases} \text{Exp}(\alpha_i + \beta_j) & 1 \le i, j \le N, \\ \text{Exp}(\alpha_i + \alpha_{2N-j+1}) & 1 \le i \le N, \ N < j \le 2N - i + 1, \\ \text{Exp}(\beta_{2N-i+1} + \beta_j) & 1 \le j \le N, \ N < i \le 2N - j + 1, \end{cases} \tag{1.3}$$

for fixed parameters $\alpha_1, \ldots, \alpha_N, \beta_1, \ldots, \beta_N \in \mathbb{R}_+$. Similarly, for the point-to-half-line dLPP $\tau_{2N}^{\text{h-flat}}$, we consider a trapezoidal array of independent weights $W = \{W_{i,j} : (i, j) \in \mathbb{N}^2, \ i + j \le 2N + 1, \ i \le N\}$ distributed according to (1.3) for $i \le N$.

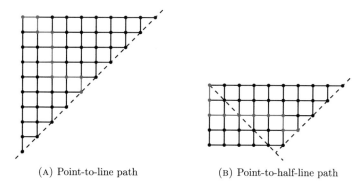

(A) Point-to-line path   (B) Point-to-half-line path

Fig. 1 Directed paths in $\mathbb{N}^2$ of length 10 from the point $(1, 1)$ to the line $i + j - 1 = 10$. The paths, highlighted in red, correspond to the two geometries specified. The picture is rotated by 90° clockwise w.r.t. the Cartesian coordinate system, to adapt it to the usual matrix/array indexing

Let us now present our results starting with the flat case. We will prove that τ_{2N}^{flat} has fluctuations of order $N^{1/3}$, and its scaling limit is given by the Tracy-Widom GOE distribution, whose cumulative distribution function F_1 can be expressed (among other expressions) via a Fredholm determinant [20] as

$$F_1(s) = \det(I - \mathcal{K}_1)_{L^2([s,\infty))} \tag{1.4}$$

for $s \in \mathbb{R}$, where $\mathcal{K}_1$ is the operator on $L^2([s, \infty))$ defined through the kernel

$$\mathcal{K}_1(\lambda, \xi) := \frac{1}{2} \operatorname{Ai}\left(\frac{\lambda + \xi}{2}\right). \tag{1.5}$$

Theorem 1.1 *If the waiting times are independent and exponentially distributed with rate 2γ, the limiting distribution of the point-to-line directed last passage percolation τ_{2N}^{flat} is given, for $r \in \mathbb{R}$, by*

$$\lim_{N\to\infty} \mathbb{P}\left(\tau_{2N}^{\text{flat}} \leq \frac{2N}{\gamma} + rN^{1/3}\right) = F_1\left(2^{1/3}\gamma r\right). \tag{1.6}$$

Regarding the point-to-half-line dLPP, we will prove that, for exponentially distributed i.i.d. environment, $\tau_{2N}^{\text{h-flat}}$ has fluctuations of order $N^{1/3}$, and its scaling limit is given by the one-point distribution $F_{2\to1}$ of the Airy$_{2\to1}$ process. The expression we will arrive at is the following [2]:

$$F_{2\to1}(s) = \det(I - \mathcal{K}_{2\to1})_{L^2([s,\infty))}, \tag{1.7}$$

where $\mathcal{K}_{2\to1}$ is the operator on $L^2([s, \infty))$ defined through the kernel

$$\mathcal{K}_{2\to1}(\lambda,\xi) := \int_0^\infty \mathrm{Ai}(\lambda+x)\,\mathrm{Ai}(\xi+x)\,\mathrm{d}x + \int_0^\infty \mathrm{Ai}(\lambda+x)\,\mathrm{Ai}(\xi-x)\,\mathrm{d}x\,.$$

(1.8)

Theorem 1.2 *If the waiting times are independent and exponentially distributed with rate 2γ, the limiting distribution of the point-to-half-line directed last passage percolation $\tau_{2N}^{\text{h-flat}}$ is given, for $r \in \mathbb{R}$, by*

$$\lim_{N\to\infty} \mathbb{P}\left(\tau_{2N}^{\text{h-flat}} \le \frac{2N}{\gamma} + rN^{1/3}\right) = F_{2\to1}\left(2^{-1/3}\gamma r\right).$$

Expression (1.4)–(1.5) for the GOE distribution is different from the one originally derived by Tracy and Widom, first expressed in terms of Painlevé functions [23] and then in terms of a block-Fredholm Pfaffian [24, 25]. Sasamoto's original derivation of (1.5) came through the analysis of the Totally Asymmetric Simple Exclusion Process (TASEP), with an initial condition of the form $\cdots0101010000\cdots$, where 1 denotes a particle and 0 a hole. The presence of the semi infinite sequence of holes is technical and the actual focus of the asymptotic analysis in [20] is on the alternating particle-hole regime, which simulates *flat* initial conditions. The starting point for this derivation in [20] was Schütz's determinantal formula [22] for the occupation probabilities in TASEP, obtained via Bethe ansatz methods. A proof that Sasamoto's formula actually provides a different expression for the Tracy-Widom GOE distribution was provided in [9]. Subsequently to [20], the $F_{2\to1}$ distribution was derived in [2] by studying, again via Schütz's and Sasamoto's formulae, the asymptotic distribution of TASEP particles with initial configuration $\cdots0101010000\cdots$, but now at the interface between the right half end $000\cdots$ and the left half alternating (flat) configuration $\cdots010101$.

Asymptotics recovering the Tracy-Widom GOE distribution as a limiting law have been performed in [1, 8] for directed last passage percolation and polynuclear growth models, and more recently in [16], at a nonrigorous level, for the KPZ equation with flat initial data. All these works derive Painlevé expressions or various forms of block-Fredholm Pfaffian formulae for the Tracy-Widom GOE. In contrast, our approach leads directly to Sasamoto's Fredholm determinant formula (1.4)–(1.5), as well as to the Airy$_{2\to1}$ formula (1.7)–(1.8).

Methodologically, both [1, 8] use a symmetrization argument that amounts to considering point-to-point dLPP on a square array with waiting times symmetric along the antidiagonal. We do not use such an argument but we rather start from the exact integral formulae obtained in [6] for the cumulative distribution function of (1.1) and (1.2) with the exponential distribution specified above, in terms of integrals of (the continuous analog of) both standard and symplectic Schur functions, cf. (1.11), (1.12). In representation theory, Schur functions appear as characters of irreducible representations of classical groups [10] (in our specific case, general linear groups and symplectic groups), and are defined as certain sums on Gelfand-Tsetlin patterns. Continuous Schur functions, which are of our interest here, are continuous limits of rescaled Schur functions, and by Riemann sum approximation turn out to be integrals on continuous Gelfand-Tsetlin patterns (see for example [6, Proposition 4.2] for the symplectic case). Due to the determinantal structure of Schur functions, which in turn arises from the Weyl character formula [10], continuous Schur functions can also be shown to have a determinantal form. The continuous analogs of standard and symplectic Schur functions, denoted by $s_\alpha^{\text{cont}}(x)$ and $sp_\alpha^{\text{cont}}(x)$ respectively for $\alpha, x \in \mathbb{R}^N$, thus have the following representation:

$$s_\alpha^{\text{cont}}(x) = \frac{\det(e^{\alpha_j x_i})_{1 \leq i, j \leq N}}{\prod_{1 \leq i < j \leq N}(\alpha_i - \alpha_j)}, \tag{1.9}$$

$$sp_\alpha^{\text{cont}}(x) = \frac{\det\left(e^{\alpha_j x_i} - e^{-\alpha_j x_i}\right)_{1 \leq i, j \leq N}}{\prod_{1 \leq i < j \leq N}(\alpha_i - \alpha_j)\prod_{1 \leq i \leq j \leq N}(\alpha_i + \alpha_j)}. \tag{1.10}$$

Our formulae for the two dLPP models then read as follows.

Proposition 1.3 *([6]) Ignoring the normalization constants that depend on the parameters α_i's and β_j's only, the distribution functions of τ_{2N}^{flat} and $\tau_{2N}^{\text{h-flat}}$ are given by*

$$\mathbb{P}(\tau_{2N}^{\text{flat}} \leq u) \propto e^{-u \sum_{k=1}^N (\alpha_k + \beta_k)} \int_{\{0 \leq x_N \leq \cdots \leq x_1 \leq u\}} sp_\alpha^{\text{cont}}(x)\, sp_\beta^{\text{cont}}(x) \prod_{i=1}^N dx_i, \tag{1.11}$$

$$\mathbb{P}(\tau_{2N}^{\text{h-flat}} \leq u) \propto e^{-u \sum_{k=1}^N (\alpha_k + \beta_k)} \int_{\{0 \leq x_N \leq \cdots \leq x_1 \leq u\}} sp_\alpha^{\text{cont}}(x) s_\beta^{\text{cont}}(x) \prod_{i=1}^N dx_i. \tag{1.12}$$

Using the determinantal form of Schur functions (1.9) and (1.10) and the well-known Cauchy-Binet identity, which expresses the multiple integral of two determinantal functions as the determinant of an integral, formulae (1.11) and (1.12) can be turned to ratios of determinants and then to Fredholm determinants, amenable to asymptotic analysis via steepest descent.

It is worth noting that the above formulae have been derived in [6] as the zero temperature limit of certain integrals of Whittaker functions that represent the Laplace transform of the log-gamma polymer partition function [7, 18, 21] in the same path geometries. The derivation of the Laplace transforms for the log-gamma polymer in [6] used combinatorial arguments through the geometric Robinson-Schensted-Knuth correspondence [14, 17].[1] Continuous classical and symplectic Schur functions appear then in our context as scaling limits of Whittaker functions associated to the groups $GL_N(\mathbb{R})$ and $SO_{2N+1}(\mathbb{R})$ respectively.

Organization of the paper: In Sect. 2 we present a general scheme to turn a ratio of determinants to a Fredholm determinant. In Sect. 3 we perform the steepest descent analysis of a central integral. In Sects. 4 and 5 we prove Theorems 1.1 and 1.2 respectively.

---

[1](1.11) and (1.12) have now been obtained directly via the standard Robinson-Schensted-Knuth correspondence, avoiding the route of the zero temperature limit, see [4, 5].

2   From Determinants to Fredholm Determinants

In this section we present a general scheme to turn ratios of determinants into a
Fredholm determinant. Such a scheme is not new (see for example [3, 13, 19]), but
we present it here in a fashion adapted to our framework. Let us start by briefly
recalling the notion of a Fredholm determinant: Given a measure space $(\mathcal{X}, \mu)$, any
linear operator $K : L^2(\mathcal{X}) \to L^2(\mathcal{X})$ can be defined in terms of its integral kernel
$K(x, y)$ by

$$(Kh)(x) := \int_{\mathcal{X}} K(x, y)h(y)\mu(\mathrm{d}y), \qquad h \in L^2(\mathcal{X}).$$

The *Fredholm determinant* of K can then be defined through its series expansion:

$$\det(I + K)_{L^2(\mathcal{X})} := 1 + \sum_{n=1}^{\infty} \frac{1}{n!} \int_{\mathcal{X}^n} \det(K(x_i, x_j))_{i,j=1}^{n} \, \mu(\mathrm{d}x_1) \dots \mu(\mathrm{d}x_n), \quad (2.1)$$

assuming the series converge. Denoting from now on by $\mathbb{C}_+ := \{z \in \mathbb{C} : \Re(z) > 0\}$
the complex right half-plane, we can state:

Theorem 2.1 *Let*

$$H(z, w) := C(z, w) - \widehat{H}(z, w), \tag{2.2}$$

where C is the function

$$C(z, w) = \frac{1}{z + w}, \qquad for \ (z, w) \in \mathbb{C}^2,$$

*and $\widehat{H}$ is a holomorphic function in the region $\mathbb{C}_+ \times \mathbb{C}_+$. For any choice of positive
parameters $\alpha_1, \dots, \alpha_N, \beta_1, \dots, \beta_N$, define the operator K_N on $L^2(\mathbb{R}_+)$ through the
kernel*

$$K_N(\lambda, \xi) := \frac{1}{(2\pi i)^2} \int_{\Gamma_1} \mathrm{d}z \int_{\Gamma_2} \mathrm{d}w \ e^{-\lambda z - \xi w} \ \widehat{H}(z, w) \prod_{m=1}^{N} \left[\frac{(z + \beta_m)(w + \alpha_m)}{(z - \alpha_m)(w - \beta_m)} \right], \tag{2.3}$$

*where $\Gamma_1, \Gamma_2 \subset \mathbb{C}_+$ are any positively oriented simple closed contours such that Γ_1
encloses $\alpha_1, \dots, \alpha_N$ and Γ_2 encloses $\beta_1, \dots, \beta_N$. Then*

$$\frac{\det(H(\alpha_i, \beta_j))_{1 \le i, j \le N}}{\det(C(\alpha_i, \beta_j))_{1 \le i, j \le N}} = \det(I - K_N)_{L^2(\mathbb{R}_+)}. \tag{2.4}$$

Note that the denominator on the left hand side of (2.4) is a Cauchy determinant:

$$\det(C(\alpha_i, \beta_j))_{i,j} = \det \left(\frac{1}{\alpha_i + \beta_j} \right)_{i,j} = \frac{\prod_{1 \le i < j \le N} (\alpha_i - \alpha_j)(\beta_i - \beta_j)}{\prod_{1 \le i, j \le N} (\alpha_i + \beta_j)}. \tag{2.5}$$

Proof For convenience, let us denote by $\mathcal{C}$, $\mathcal{H}$ and $\widehat{\mathcal{H}}$ the $N \times N$ matrices $(C(\alpha_i, \beta_j))_{1 \le i, j \le N}$, $(H(\alpha_i, \beta_j))_{1 \le i, j \le N}$ and $(\widehat{H}(\alpha_i, \beta_j))_{1 \le i, j \le N}$ respectively. We then have:

$$\frac{\det(\mathcal{H})}{\det(\mathcal{C})} = \det\left(\mathcal{C}^{-1}(\mathcal{C} - \widehat{\mathcal{H}})\right) = \det\left(I - \mathcal{C}^{-1}\widehat{\mathcal{H}}\right), \tag{2.6}$$

where I is the identity matrix of order N. To invert $\mathcal{C}$, we use Cramer's formula:

$$\mathcal{C}^{-1}(i, k) = (-1)^{i+k} \frac{\det\left(\mathcal{C}^{(k,i)}\right)}{\det(\mathcal{C})},$$

where $\mathcal{C}^{(k,i)}$ is the matrix of order $N - 1$ obtained from $\mathcal{C}$ by removing its k-th row and i-th column. In our case, both determinants in the above formula are of Cauchy type:

$$\det(\mathcal{C}) = \prod_{l<m}(\alpha_l - \alpha_m) \prod_{l<m}(\beta_l - \beta_m) \prod_{l,m}(\alpha_l + \beta_m)^{-1}$$

$$\det\left(\mathcal{C}^{(k,i)}\right) = \prod_{\substack{l<m \\ l,m \ne k}}(\alpha_l - \alpha_m) \prod_{\substack{l<m \\ l,m \ne i}}(\beta_l - \beta_m) \prod_{\substack{l \ne k \\ m \ne i}}(\alpha_l + \beta_m)^{-1},$$

where indices l and m run in $\{1, \ldots, N\}$. The inverse of $\mathcal{C}$ is thus readily computed:

$$\mathcal{C}^{-1}(i, k) = \frac{\prod_{m=1}^{N}(\alpha_k + \beta_m)(\beta_i + \alpha_m)}{(\alpha_k + \beta_i)\prod_{m \ne k}(\alpha_k - \alpha_m)\prod_{m \ne i}(\beta_i - \beta_m)}.$$

Writing $(\alpha_k + \beta_i)^{-1} = \int_0^\infty \exp(-(\alpha_k + \beta_i)\lambda)\,d\lambda$, we obtain:

$$\left(\mathcal{C}^{-1}\widehat{\mathcal{H}}\right)(i, j) = \sum_{k=1}^{N} \mathcal{C}^{-1}(i, k)\widehat{\mathcal{H}}(k, j) = \int_0^\infty f(i, \lambda)g(\lambda, j)\,d\lambda,$$

where for all $\lambda > 0$

$$f(i, \lambda) := e^{-\beta_i \lambda} \frac{\prod_{m=1}^{N}(\beta_i + \alpha_m)}{\prod_{m \ne i}(\beta_i - \beta_m)}, \qquad g(\lambda, j) := \sum_{k=1}^{N} e^{-\alpha_k \lambda} \frac{\prod_{m=1}^{N}(\alpha_k + \beta_m)}{\prod_{m \ne k}(\alpha_k - \alpha_m)} \widehat{H}(\alpha_k, \beta_j).$$

This proves that matrix $\mathcal{C}^{-1}\widehat{\mathcal{H}}$, viewed as a linear operator on $\mathbb{R}^N$, equals the composition FG, where F and G are the linear operators

$$F: L^2(\mathbb{R}_+) \to \mathbb{R}^N, \qquad \phi \mapsto \left[\int_0^\infty f(i, \lambda)\phi(\lambda) \, d\lambda \right]_{i=1}^N,$$

$$G: \mathbb{R}^N \to L^2(\mathbb{R}_+), \qquad (a(j))_{j=1}^N \mapsto \sum_{j=1}^N g(\lambda, j)a(j).$$

We note that these are well-defined operators, as $f(i, \cdot)$ and $g(\cdot, j)$ are square integrable functions on $\mathbb{R}_+$, for all i and j. We will next use Sylvester's identity, which states that

$$\det(I + K_1 K_2)_{L^2(\mathcal{X}_2)} = \det(I + K_2 K_1)_{L^2(\mathcal{X}_1)} \tag{2.7}$$

for any trace class operators $K_1: L^2(\mathcal{X}_1) \to L^2(\mathcal{X}_2)$ and $K_2: L^2(\mathcal{X}_2) \to L^2(\mathcal{X}_1)$. By applying this identity, we obtain

$$\det\left(I - C^{-1}\widehat{\mathcal{H}}\right)_{\mathbb{R}^N} = \det(I - FG)_{\mathbb{R}^N} = \det(I - K_N)_{L^2(\mathbb{R}_+)}, \tag{2.8}$$

where $K_N := GF$ is the operator on $L^2(\mathbb{R}_+)$ defined through the kernel

$$K_N(\lambda, \xi) := \sum_{i,k=1}^N e^{-\alpha_k \lambda - \beta_i \xi} \, \widehat{H}(\alpha_k, \beta_i) \frac{\prod_{m=1}^N (\alpha_k + \beta_m)(\beta_i + \alpha_m)}{\prod_{m \neq k}(\alpha_k - \alpha_m) \prod_{m \neq i}(\beta_i - \beta_m)}. \tag{2.9}$$

From the latter formula, it is clear that $|K_N(\lambda, \xi)| \le c_1 e^{-c_2 \lambda}$ for all $\lambda \in [0, \infty)$, where the positive constants c_1 and c_2 depend on N and on the parameters. Hadamard's bound then implies that

$$\left| \det(K_N(\lambda_i, \lambda_j))_{i,j=1}^n \right| \le n^{n/2} \prod_{i=1}^n c_1 e^{-c_2 \lambda_i}.$$

It follows that

$$\left| \det(I - K_N)_{L^2(\mathbb{R}_+)} \right| \le 1 + \sum_{n=1}^\infty \frac{n^{n/2}}{n!} \left(\int_0^\infty c_1 e^{-c_2 \lambda} \, d\lambda \right)^n < \infty,$$

hence the right-hand side of (2.8) is indeed a converging Fredholm determinant. By applying the residue theorem (recalling the assumption that $\widehat{H}(z, w)$ is holomorphic in $\mathbb{C}_+ \times \mathbb{C}_+$), the double sum in (2.9) can be turned into a double contour integral, yielding representation (2.3) for the kernel. By combining (2.6) and (2.8) we obtain (2.4). $\qquad\square$

3 Steepest Descent Analysis

Thanks to the results mentioned in Sect. 1 (formulae (1.11) and (1.12)) and to the general scheme introduced in Sect. 2, the distribution function of our two dLPP models can be expressed as a Fredholm determinant on $L^2(\mathbb{R}_+)$ with kernel of type (2.3), as we will state precisely in the next two sections (see Theorems 4.2 and 5.2). As we will see, in the limit $N \to \infty$ all these kernels converge, after rescaling, to expressions involving Airy functions. In order to see this, one needs to perform the asymptotic analysis of a few contour integrals via steepest descent. This procedure is very similar in all cases, as it always involves the same functions. Therefore, we will carry it out in detail only for one of such contour integrals, arguably the most archetypal one as it just approximates the Airy function. Other very similar steepest descent analyses are sketched where needed, specifically in the proof of Theorem 1.2.

Let us first recall that the Airy function Ai has the following contour integral representation:

$$\text{Ai}(x) := \frac{1}{2\pi i} \int_{e^{-i\pi/3}\infty}^{e^{i\pi/3}\infty} \exp\left\{\frac{z^3}{3} - xz\right\} dz \,, \tag{3.1}$$

where the integration path starts at infinity with argument $-\pi/3$ and ends at infinity with argument $\pi/3$ (see for example the red contour in Fig. 2).

Proposition 3.1 *For any fixed $\gamma > 0$ and $f := 2/\gamma$, let us define*

$$J_N(x) := -\frac{1}{2\pi i} \int_\Gamma e^{-z(fN+x)} \left[\frac{\gamma+z}{\gamma-z}\right]^N dz \,, \tag{3.2}$$

where $\Gamma \subset \mathbb{C}$ is a positively oriented contour enclosing γ. Then, for all $x \in \mathbb{R}$,

$$\tilde{J}_N(x) := \frac{\sqrt[3]{2N}}{\gamma} J_N\left(\frac{\sqrt[3]{2N}}{\gamma}x\right) \xrightarrow{N\to\infty} \text{Ai}(x) \,. \tag{3.3}$$

Proof The proof is based on the steepest descent analysis of the integral

$$J_N(x) = -\frac{1}{2\pi i} \int_\Gamma \exp\left\{NF(z) - xz\right\} dz \,,$$

where $F(z) := \log(\gamma + z) - \log(\gamma - z) - fz$. We need to compute the critical points of the function F, whose first three derivatives are given by:

$$F'(z) = \frac{1}{\gamma+z} + \frac{1}{\gamma-z} - f \,,$$

$$F''(z) = -\frac{1}{(\gamma+z)^2} + \frac{1}{(\gamma-z)^2} \,,$$

$$F'''(z) = \frac{2}{(\gamma+z)^3} + \frac{2}{(\gamma-z)^3} \,.$$

The second derivative vanishes if and only if $z = 0$. As in the statement of the theorem, we then set $f := 2/\gamma$, which is the only value of f such that the first derivative also vanishes at $z = 0$. The first non-vanishing derivative of F at the critical point $z = 0$ is then the third one. In particular, we have that

$$F(0) = F'(0) = F''(0) = 0, \qquad F'''(0) = \frac{4}{\gamma^3},$$

hence the Taylor expansion of F near the critical point is

$$F(z) = \frac{2}{3\gamma^3}z^3 + R(z), \qquad (3.4)$$

where $R(z) = o(z^3)$ as $z \to 0$. Since the directions of steepest descent of F from $z = 0$ correspond to the angles $\pm\pi/3$, we deform the *positively* oriented contour Γ into the *negatively* oriented triangular path T_a with vertices 0, $2a\,e^{i\pi/3}$ and $2a\,e^{-i\pi/3}$ for some $a > \gamma$ (so that the pole $z = \gamma$ is still enclosed, see Fig. 2). This only implies a change of sign in the integral, corresponding to the change of orientation of the contour. Indeed, in order to obtain the right estimates in the proof of Corollary 3.2, it is convenient to consider an infinitesimal shift of T_a, by setting the contour to be $T_a + \epsilon\gamma/\sqrt[3]{2N}$, where $\epsilon > 0$ is an arbitrary constant. Moreover, we split the integral into two regions, i.e. a neighborhood of the critical point, where the main contribution of the integral is expected to come from, and its exterior (we choose the neighborhood to be a ball centered at $\epsilon\gamma/\sqrt[3]{2N}$ with radius $\gamma N^{-\alpha}$, where $\alpha > 0$ will be suitably specified later on):

$$J_N(x) = J_N^{\text{in}}(x) + J_N^{\text{ex}}(x),$$

where

$$J_N^{\text{in}}(x) := \frac{1}{2\pi i} \int_{T_a + \frac{\epsilon\gamma}{\sqrt[3]{2N}}} \exp\left\{F(z)N - xz\right\} \mathbb{1}_{\{|z - \epsilon\gamma/\sqrt[3]{2N}| \leq \gamma N^{-\alpha}\}}\, dz,$$

$$J_N^{\text{ex}}(x) := \frac{1}{2\pi i} \int_{T_a + \frac{\epsilon\gamma}{\sqrt[3]{2N}}} \exp\left\{F(z)N - xz\right\} \mathbb{1}_{\{|z - \epsilon\gamma/\sqrt[3]{2N}| \geq \gamma N^{-\alpha}\}}\, dz.$$

Let us first focus on the former integral and denote by C the piecewise linear path going from the point at infinity with argument $-\pi/3$ to the origin to the point at infinity with argument $\pi/3$ (see Fig. 2). We then have that

$$J_N^{\text{in}}(x) = \frac{1}{2\pi i} \int_{C + \frac{\epsilon\gamma}{\sqrt[3]{2N}}} \exp\left\{\frac{2N}{3\gamma^3}z^3 - xz + R(z)N\right\} \mathbb{1}_{\{|z - \epsilon\gamma/\sqrt[3]{2N}| \leq \gamma N^{-\alpha}\}}\, dz,$$

where $R(z)$ is defined by (3.4). If we now rescale both the integration variable and the function J_N^{in} by the factor $\sqrt[3]{2N}/\gamma$, by setting $w := z\sqrt[3]{2N}/\gamma$ and defining $\tilde{J}_N^{\text{in}}$ as in (3.3), we obtain:

$$\tilde{J}_N^{in}(x) = \frac{1}{2\pi i} \int_{C+\epsilon} \exp\left\{\frac{w^3}{3} - xw + R\left(\frac{\gamma}{\sqrt[3]{2N}}w\right)N\right\} \mathbb{1}_{\{|w-\epsilon|\le\sqrt[3]{2}N^{1/3-\alpha}\}}\,dw\,.$$

(3.5)

A standard estimate of the remainder in the Taylor expansion (3.4) yields

$$\left|R\left(\frac{\gamma}{\sqrt[3]{2N}}w\right)N\right| \le \frac{m}{4!}\left|\frac{\gamma}{\sqrt[3]{2N}}w\right|^4 N \le \frac{m}{4!}\left(\gamma N^{-\alpha} + \frac{\epsilon\gamma}{\sqrt[3]{2N}}\right)^4 N$$

for $|w - \epsilon| \le \sqrt[3]{2}N^{1/3-\alpha}$, where the constant m is the maximum modulus of $F^{(4)}$ in some fixed neighborhood of the origin. If we take $\alpha > 1/4$, the above expression vanishes as $N \to \infty$. If we further choose $\alpha < 1/3$, the indicator function in (3.5) converges to 1, yielding

$$\exp\left\{R_N\left(\frac{\gamma}{\sqrt[3]{2N}}w\right)N\right\} \mathbb{1}_{\{|w-\epsilon|\le\sqrt[3]{2}N^{1/3-\alpha}\}} \xrightarrow{N\to\infty} 1\,.$$

(3.6)

Since the argument of the points of C is $\pm\pi/3$, we have that

$$\int_{C+\epsilon}\left|\exp\left\{\frac{w^3}{3} - xw\right\}\right||dw| < \infty\,,$$

hence by dominated convergence

$$\tilde{J}_N^{in}(x) \xrightarrow{N\to\infty} \int_{C+\epsilon} \exp\left\{\frac{w^3}{3} - xw\right\}dw = \text{Ai}(x)\,.$$

Observe that, varying ϵ, we have different integral representations of the Airy function, which are all equivalent thanks to (3.1).

To conclude the proof, it remains to show that

$$\tilde{J}_N^{ex}(x) := \frac{\sqrt[3]{2N}}{2\pi i\gamma} \int_{T_a\cap\{|z|\ge\gamma N^{-\alpha}\}} \exp\left\{F\left(z + \frac{\epsilon\gamma}{\sqrt[3]{2N}}\right)N - x\left(z + \frac{\epsilon\gamma}{\sqrt[3]{2N}}\right)\frac{\sqrt[3]{2N}}{\gamma}\right\}dz \xrightarrow{N\to\infty} 0.$$

We may decompose the integration domain as

$$T_a \cap \{|z| \ge \gamma N^{-\alpha}\} = \mathcal{V} \cup \mathcal{O}_N \cup \overline{\mathcal{V}} \cup \overline{\mathcal{O}}_N\,,$$

where $\mathcal{V}$ and $\mathcal{O}_N$ are the vertical and oblique segments respectively given by

$$\mathcal{V} := \left\{\Re(z) = a,\ 0 \le \arg(z) \le \frac{\pi}{3}\right\}, \quad \mathcal{O}_N := \left\{\gamma N^{-\alpha} \le |z| \le 2a,\ \arg(z) = \frac{\pi}{3}\right\},$$

and $\overline{\mathcal{V}}$ and $\overline{\mathcal{O}}_N$ are their complex conjugates. We thus estimate

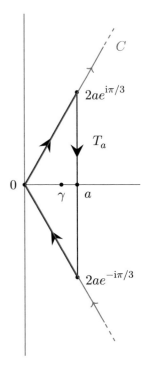

Fig. 2 The red path C is involved in the integral representation of the Airy function. The black contour T_a refers to the steepest descent analysis in the proof of Proposition 3.1

$$\left| \tilde{J}_N^{\mathrm{ex}}(x) \right| \leq \frac{\mathcal{L}(T_a) \sqrt[3]{2N}}{2\pi\gamma} \exp\left\{ \max\left[G_N\left(z + \frac{\epsilon\gamma}{\sqrt[3]{2N}} \right) : z \in \mathcal{V} \cup \mathcal{O}_N \cup \overline{\mathcal{V}} \cup \overline{\mathcal{O}}_N \right] \right\},$$
(3.7)

where $\mathcal{L}(\cdot)$ denotes the length of a contour and

$$G_N(z) := \Re[F(z)]N - x \, \Re(z) \frac{\sqrt[3]{2N}}{\gamma} \, .$$

Since

$$\Re[F(z)] = \log\left| \frac{\gamma + z}{\gamma - z} \right| - \frac{2}{\gamma} \Re(z) \, ,$$

it is clear that $G_N(\bar{z}) = G_N(z)$. Therefore, it suffices to bound the maximum of G_N over $\mathcal{V}$ and over $\mathcal{O}_N$. Since $\Re(z) = a$ and $a \leq |z| \leq 2a$ for $z \in \mathcal{V}$, we have that

$$\max_{z \in \mathcal{V}} G_N\left(z + \frac{\epsilon\gamma}{\sqrt[3]{2N}} \right) \leq -cN - xa \frac{\sqrt[3]{2N}}{\gamma} \, ,$$
(3.8)

where

$$c := \frac{2}{\gamma} a - \log\left(\frac{2a + \epsilon\gamma + \gamma}{a - \gamma} \right) \, .$$
(3.9)

If we fix a large enough a such that c is positive, the maximum in (3.8) is asymptotic to $-cN$ and diverges to $-\infty$. On the other hand, for all z such that $\arg(z) = \pi/3$, we have that

$$G_N(z) = \left[\frac{1}{2} \log \frac{\gamma^2 + \gamma|z| + |z|^2}{\gamma^2 - \gamma|z| + |z|^2} - \frac{|z|}{\gamma} \right] N - x \frac{|z|\sqrt[3]{2N}}{2} \frac{1}{\gamma},$$

whose derivative w.r.t. the modulus is

$$\frac{d}{d|z|} G_N(z) = -\frac{|z|^2(2\gamma^2 + |z|^2)}{\gamma(\gamma^4 + \gamma^2|z|^2 + |z|^4)} N - \frac{x\sqrt[3]{2N}}{2\gamma}.$$

A trivial estimate then gives

$$\frac{d}{d|z|} G_N(z) \le -\frac{\gamma N^{1-2\alpha}(2\gamma^2 + \gamma^2 N^{-2\alpha})}{(\gamma^4 + \gamma^2(2a)^2 + (2a)^4)} - \frac{x\sqrt[3]{2N}}{2\gamma} \qquad \text{for } z \in \mathcal{O}_N.$$

Since $\alpha < 1/3$, no matter the sign of x, the above derivative is negative for N large enough, so $G_N(z)$ is decreasing w.r.t. $|z|$ in $\mathcal{O}_N$. By continuity, for N large, $G_N(z + \epsilon\gamma/\sqrt[3]{2N})$ is also decreasing w.r.t. $|z|$ in $\mathcal{O}_N$, hence

$$\max_{z \in \mathcal{O}_N} G_N\left(z + \frac{\epsilon\gamma}{\sqrt[3]{2N}} \right) = G_N\left(\gamma N^{-\alpha} e^{i\pi/3} + \frac{\epsilon\gamma}{\sqrt[3]{2N}} \right)$$

$$= \left[\log \left| \frac{1 + N^{-\alpha} e^{i\pi/3} + \epsilon(2N)^{-1/3}}{1 - N^{-\alpha} e^{i\pi/3} - \epsilon(2N)^{-1/3}} \right| - N^{-\alpha} - \frac{2\epsilon}{\sqrt[3]{2N}} \right] N - x \left(\frac{N^{-\alpha}}{2} + \frac{\epsilon}{\sqrt[3]{2N}} \right) \sqrt[3]{2N}.$$

After a tedious computation, which uses the third order Taylor expansion of $\log(1 + \delta)$ as $\delta \to 0$, we obtain that

$$\max_{z \in \mathcal{O}_N} G_N\left(z + \frac{\epsilon\gamma}{\sqrt[3]{2N}} \right) = -\frac{2}{3} N^{1-3\alpha} + o(N^{1-3\alpha}) - x(2^{-2/3} N^{1/3-\alpha} + \epsilon). \quad (3.10)$$

Since $\alpha < 1/3$, the latter expression is asymptotic to $-(2/3)N^{1-3\alpha}$ and diverges to $-\infty$. Thanks to estimates (3.7), (3.8) and (3.10), we thus conclude that $\tilde{J}_N^{\text{ex}}(x)$ vanishes (at least choosing a large enough a). $\qquad\square$

The proof of Proposition 3.1 directly provides a uniform bound on $\tilde{J}_N$, which will also turn out to be useful in the next sections.

Corollary 3.2 *Let $\tilde{J}_N(x)$ be defined as in (3.3) and $s \in \mathbb{R}$. Then, there exist two positive constants c_1 and c_2 such that*

$$\sup_{N \in \mathbb{N}} \left| \tilde{J}_N(x) \right| \le c_1 e^{-c_2 x} \qquad \forall x \in [s, \infty).$$

Proof Since by continuity the converging sequence $\tilde{J}_N(x)$ is bounded uniformly in N on any compact set, it suffices to prove the claim for $s = 0$. The proof is then a straightforward consequence of the estimates obtained in the proof of Proposition 3.1. Using the notation adopted there, we will show that the uniform exponential bound is valid for both $\tilde{J}_N^{\mathrm{in}}$ and $\tilde{J}_N^{\mathrm{ex}}$, i.e. the contributions near and away from the critical point respectively. From (3.5) and (3.6), it follows that for all $x \in [0, \infty)$

$$\sup_{N \in \mathbb{N}} \left| \tilde{J}_N^{\mathrm{in}}(x) \right| \lesssim e^{-\epsilon x} \int_{C+\epsilon} e^{\Re(w^3)/3} \left| dw \right|,$$

with ϵ chosen to be strictly positive. By definition of the contour C, the above integral converges, providing the desired exponential bound for $\tilde{J}_N^{\mathrm{in}}$. On the other hand, estimates (3.7), (3.8) and (3.10) show that for all $N \in \mathbb{N}$ and $x \in [0, \infty)$

$$\left| \tilde{J}_N^{\mathrm{ex}}(x) \right| \leq \left[\frac{\mathcal{L}(T_a)\sqrt[3]{2N}}{2\pi\gamma} e^{-\min\{cN,(2/3)N^{1-3\alpha}+o(N^{1-3\alpha})\}} \right] e^{-x \min\{a\sqrt[3]{2N}/\gamma, 2^{-2/3}N^{1/3-\alpha}+\epsilon\}}$$

$$\leq c' e^{-x \min\{a\sqrt[3]{2}/\gamma, 2^{-2/3}+\epsilon\}},$$

where c is the constant (positive if a is chosen large enough) defined in (3.9), and c' is an upper bound for the vanishing sequence inside the square bracket above. This provides the desired exponential bound for $\tilde{J}_N^{\mathrm{ex}}$. $\qquad\square$

4 Point-to-Line Last Passage Percolation and GOE

We will now specialize the results of Sects. 2 and 3 to the models described in Sect. 1. We first analyze the point-to-line dLPP, writing its distribution function as a Fredholm determinant. The starting point is the next theorem, which is a result of (1.10) and (1.11), combined with the Cauchy-Binet identity.

Theorem 4.1 ([6]) *The distribution function of the point-to-line directed last passage percolation $\tau_{2N}^{\mathrm{flat}}$ with exponential waiting times as in (1.3) is a ratio of $N \times N$ determinants:*

$$\mathbb{P}\left(\tau_{2N}^{\mathrm{flat}} \leq u \right) = \frac{\det \left(H_u^{\mathrm{flat}}(\alpha_i, \beta_j) \right)_{1 \leq i, j \leq N}}{\det(C(\alpha_i, \beta_j))_{1 \leq i, j \leq N}} \tag{4.1}$$

for $u > 0$, where $C(z, w) := (z + w)^{-1}$ and

$$H_u^{\mathrm{flat}}(z, w) := e^{-u(z+w)} \int_0^u (e^{zx} - e^{-zx})(e^{wx} - e^{-wx}) \, dx .$$

We next obtain the Fredholm determinant.

Theorem 4.2 *The distribution of τ_{2N}^{flat} with exponential waiting times as in (1.3) is given by*

$$\mathbb{P}\big(\tau_{2N}^{\text{flat}} \le u\big) = \det(I - K_{N,u}^{\text{flat}})_{L^2(\mathbb{R}_+)} , \tag{4.2}$$

where $K_{N,u}^{\text{flat}} : L^2(\mathbb{R}_+) \to L^2(\mathbb{R}_+)$ is the operator defined through the kernel

$$K_{N,u}^{\text{flat}}(\lambda, \xi) = \frac{1}{(2\pi i)^2} \int_{\Gamma_1} dz \int_{\Gamma_2} dw \; e^{-\lambda z - \xi w} \; \widehat{H}_u^{\text{flat}}(z, w) \prod_{m=1}^{N} \frac{(z + \beta_m)(w + \alpha_m)}{(z - \alpha_m)(w - \beta_m)} . \tag{4.3}$$

Here, $\Gamma_1, \Gamma_2 \subset \mathbb{C}_+$ are any positively oriented simple closed contours such that Γ_1 encloses $\alpha_1, \ldots, \alpha_N$, Γ_2 encloses $\beta_1, \ldots, \beta_N$ as well as the whole Γ_1, and

$$\widehat{H}_u^{\text{flat}}(z, w) = \frac{e^{-2uz}}{w - z} + \frac{e^{-2uw}}{z - w} + \frac{e^{-2u(z+w)}}{z + w} . \tag{4.4}$$

Proof The claim is an immediate consequence of Theorem 4.1 (see formula (4.1)) and Theorem 2.1. According to (2.2), function $\widehat{H}_u^{\text{flat}}$ in (4.3) is defined through the relation $H_u^{\text{flat}} = C - \widehat{H}_u^{\text{flat}}$ (using the notation of Theorem 4.1 for H_u^{flat}), i.e.

$$\widehat{H}_u^{\text{flat}}(z, w) = \frac{1}{z + w} - e^{-u(z+w)} \int_0^u (e^{zx} - e^{-zx})(e^{wx} - e^{-wx}) \, dx .$$

If we assume that Γ_2 encloses Γ_1 (so that $z \ne w$ for all z, w), integrating the above expression yields (4.4). By symmetry, the other inclusion would lead to similar results. □

Proof of Theorem 1.1 The starting point is the Fredholm determinant formula of Theorem 4.2. We will first show the pointwise convergence of the kernel after suitable rescaling, and next sketch the (standard) argument for the convergence of the Fredholm determinant. Setting $\alpha_m = \beta_m = \gamma$ for all m, so that the waiting times are all exponential with rate 2γ (see parametrization (1.3)), kernel (4.3) reads as

$$K_{N,u}^{\text{flat}}(\lambda, \xi) = \frac{1}{(2\pi i)^2} \int_{\Gamma_1} dz \int_{\Gamma_2} dw \; e^{-\lambda z - \xi w} \; \widehat{H}_u^{\text{flat}}(z, w) \left[\frac{(z + \gamma)(w + \gamma)}{(z - \gamma)(w - \gamma)} \right]^N ,$$

where $\widehat{H}_u^{\text{flat}}$ is given by formula (4.4). Our kernel is thus a sum of three double contour integrals, each corresponding to one of the addends in (4.4). In the second one *only*, we swap the two contours taking into account the residue at the pole $w = z$. We can then readily write the kernel as the sum of four terms:

$$K_{N,u}^{\text{flat}} = K_{N,u}^{\text{flat},1} + K_{N,u}^{\text{flat},2} + K_{N,u}^{\text{flat},3} + K_{N,u}^{\text{flat},4} ,$$

where the first one corresponds to the above mentioned residue:

$$K_{N,u}^{\text{flat},1}(\lambda,\xi) := -\frac{1}{2\pi i}\int_{\Gamma_1} dz\ e^{-(2u+\lambda+\xi)z}\left[\frac{\gamma+z}{\gamma-z}\right]^{2N}, \tag{4.5}$$

and the other three terms are

$$K_{N,u}^{\text{flat},2}(\lambda,\xi) := \frac{1}{(2\pi i)^2}\int_{\Gamma_1} dz \int_{\Gamma_2} dw\ e^{-\lambda z-\xi w}\frac{e^{-2uz}}{w-z}\left[\frac{(z+\gamma)(w+\gamma)}{(z-\gamma)(w-\gamma)}\right]^N, \tag{4.6}$$

$$K_{N,u}^{\text{flat},3}(\lambda,\xi) := K_{N,u}^{\text{flat},2}(\xi,\lambda), \tag{4.7}$$

$$K_{N,u}^{\text{flat},4}(\lambda,\xi) := \frac{1}{(2\pi i)^2}\int_{\Gamma_1} dz \int_{\Gamma_2} dw\ e^{-\lambda z-\xi w}\frac{e^{-2u(z+w)}}{z+w}\left[\frac{(z+\gamma)(w+\gamma)}{(z-\gamma)(w-\gamma)}\right]^N. \tag{4.8}$$

Step 1: Main contribution in the kernel. The Airy kernel emerges from a rescaling of $K_{N,u}^{\text{flat},1}$ through Proposition 3.1, whereas the other terms turn out to be negligible under the same rescaling, as we will see. From now on, fixing $r\in\mathbb{R}$ once for all, we will take u to be $u_N := 2N/\gamma + rN^{1/3}$, as in (1.6). Moreover, we denote by $\tilde{\Psi}$ the rescaling of any function $\Psi(\lambda,\xi)$ by the factor $\sqrt[3]{2N}/\gamma$:

$$\tilde{\Psi}(\lambda,\xi) := \frac{\sqrt[3]{2N}}{\gamma}\Psi\left(\frac{\sqrt[3]{2N}}{\gamma}\lambda, \frac{\sqrt[3]{2N}}{\gamma}\xi\right). \tag{4.9}$$

By Proposition 3.1, $\tilde{K}_{N,u_N}^{\text{flat},1}$ has a non-trivial limit:

$$\tilde{K}_{N,u_N}^{\text{flat},1}(\lambda,\xi) = -\frac{\sqrt[3]{4N}}{\sqrt[3]{2}\gamma}\frac{1}{2\pi i}\int_{\Gamma_1} dz\ \exp\left\{-z\left[\frac{2}{\gamma}2N + \left(\frac{\lambda+\xi}{\sqrt[3]{2}} + 2^{1/3}\gamma r\right)\frac{\sqrt[3]{4N}}{\gamma}\right]\right\}\left[\frac{\gamma+z}{\gamma-z}\right]^{2N}$$
$$\xrightarrow{N\to\infty} 2^{-1/3}\operatorname{Ai}\left(2^{-1/3}(\lambda+\xi) + 2^{1/3}\gamma r\right).$$

We thus need to replace our whole kernel with its rescaling by the factor $\sqrt[3]{2N}/\gamma$:

$$\tilde{K}_{N,u_N}^{\text{flat}} = \tilde{K}_{N,u_N}^{\text{flat},1} + \tilde{K}_{N,u_N}^{\text{flat},2} + \tilde{K}_{N,u_N}^{\text{flat},3} + \tilde{K}_{N,u_N}^{\text{flat},4}.$$

This does not affect formula (4.2), as it just amounts to a change of variables in the multiple integrals defining the Fredholm determinant expansion (see (2.1)), so that:

$$\mathbb{P}\left(\tau_{2N}^{\text{flat}} \le u_N\right) = \det(I - \tilde{K}_{N,u_N}^{\text{flat}})_{L^2(\mathbb{R}_+)}.$$

Step 2: Vanishing terms in the kernel. We will now show that all the remaining terms $\tilde{K}_{N,u_N}^{\text{flat},i}(\lambda,\xi)$ for $i=2,3,4$ vanish, starting from $\tilde{K}_{N,u_N}^{\text{flat},2}(\lambda,\xi)$. For this purpose, we specify the contours appropriately. We choose Γ_1 to be a circle of radius ρ_1 around γ, where $0 < \rho_1 < \gamma$. Next, we choose Γ_2 to be a semicircle of radius ρ_2 centered at δ, where $0 < \delta < \gamma - \rho_1$, composed by concatenating the segment $\delta + i[-\rho_2, \rho_2]$ and the arc parametrized by $\delta + \rho_2 e^{i\theta}$ for $\theta \in [-\pi/2, \pi/2]$. It is clear that both contours lie in the right half-plane and, for ρ_2 large enough, Γ_2 encloses Γ_1. Rescaling (4.6), setting $u := u_N$, and using the fact that $\lambda, \xi \geq 0$ and $\delta \leq \Re(z), \Re(w) \leq \delta + \rho_2$ for $z \in \Gamma_1$ and $w \in \Gamma_2$, we estimate

$$
\begin{aligned}
\left|\tilde{K}^{\mathrm{flat},2}_{N,u_N}(\lambda,\xi)\right| &\leq \frac{(2N)^{1/3}\mathcal{L}(\Gamma_1)\mathcal{L}(\Gamma_2)}{\gamma(2\pi)^2\mathrm{dist}(\Gamma_1,\Gamma_2)}\,\mathrm{e}^{-(\lambda+\xi)\delta\sqrt[3]{2N}/\gamma}\,\mathrm{e}^{(m_1+m_2)N+2|r|(\delta+\rho_2)N^{1/3}} \\
&\leq c\,\mathrm{e}^{-(\lambda+\xi)\delta\sqrt[3]{2}/\gamma}\exp\left\{(m_1+m_2)N+2|r|(\delta+\rho_2)N^{1/3}+\frac{1}{3}\log N\right\}.
\end{aligned}
\tag{4.10}
$$

In the first inequality, we have denoted by $\mathcal{L}(\cdot)$ the length of a curve and by $\mathrm{dist}(\cdot,\cdot)$ the Euclidean distance in $\mathbb{C}$. In the second inequality, c is a constant depending on the parameters γ, δ, ρ_1 and ρ_2 only, whereas m_1 and m_2 are defined by

$$
m_1 := \max_{z\in\Gamma_1}\left\{-\frac{4}{\gamma}\Re(z)+\log\left|\frac{z+\gamma}{z-\gamma}\right|\right\}, \qquad m_2 := \max_{w\in\Gamma_2}\log\left|\frac{w+\gamma}{w-\gamma}\right|.
$$

A trivial estimate yields

$$
m_1 \leq -4\left(1-\frac{\rho_1}{\gamma}\right)+\log\left(1+2\frac{\gamma}{\rho_1}\right).
$$

Now, the function

$$
g(t) := -4(1-t)+\log\left(1+\frac{2}{t}\right)
$$

attains its minimum for $t \in (0,1)$ at $t_0 := \sqrt{3/2}-1$, with $g(t_0) < 0$; hence, choosing $\rho_1 := t_0\gamma$, we have that $m_1 < 0$. On the other hand, we estimate

$$
m_2 \leq \max_{\Re(w)=\delta}\log\left|\frac{w+\gamma}{w-\gamma}\right|+\max_{|w-\delta|=\rho_2}\log\left|\frac{w+\gamma}{w-\gamma}\right| \leq \log\frac{\gamma+\delta}{\gamma-\delta}+\log\frac{\rho_2+\delta+\gamma}{\rho_2+\delta-\gamma}.
$$

We can now choose $\delta > 0$ small enough and ρ_2 big enough such that $m_2 < -m_1$. It thus follows that, for certain values of ρ_1, ρ_2 and δ, the quantity after the last inequality in (4.10) decays exponentially with rate N, allowing us to conclude that $\tilde{K}^{\mathrm{flat},2}_{N,u_N}(\lambda,\xi)$ vanishes as $N \to \infty$, for all $\lambda, \xi \in \mathbb{R}_+$. Note that, in (4.10), the exponential containing variables λ and ξ does not play any role here, but will provide a useful estimate in the next step.

Because of (4.7), we have that estimate (4.10) is exactly valid for $\tilde{K}^{\mathrm{flat},3}_{N,u_N}(\lambda,\xi)$ too, so that this term also vanishes.

Finally, an estimate similar to (4.10) holds for $\tilde{K}^{\mathrm{flat},4}_{N,u_N}(\lambda,\xi)$: To see this, we make the same contour choice as we made in that case with the aim of showing that $\tilde{K}^{\mathrm{flat},2}_{N,u_N}(\lambda,\xi)$ vanishes. Rescaling (4.8) and setting $u := u_N$, we then obtain

$$\left| \tilde{K}_{N,u_N}^{\text{flat},4}(\lambda, \xi) \right| \leq \frac{(2N)^{1/3} \mathcal{L}(\Gamma_1) \mathcal{L}(\Gamma_2)}{\gamma (2\pi)^2 \text{dist}(\Gamma_1, -\Gamma_2)} \, e^{-(\lambda+\xi)\delta \sqrt[3]{2N}/\gamma} \, e^{(m_1+m_2)N + 4|r|(\delta+\rho_2)N^{1/3}}$$

$$\leq c' \, e^{-(\lambda+\xi)\delta \sqrt[3]{2}/\gamma} \exp\left\{ (m_1+m_2)N + 4|r|(\delta + \rho_2)N^{1/3} + \frac{1}{3}\log N \right\},$$

$$(4.11)$$

where the constant c' depends on γ, δ, ρ_1 and ρ_2 only. We have already proved that $m_1 + m_2 < 0$ for a certain choice of ρ_1, ρ_2 and δ, hence $\tilde{K}_{N,u_N}^{\text{flat},4}(\lambda, \xi)$ also vanishes.

Step 3: Convergence of Fredholm determinants. In the first two steps, we have proven that

$$\lim_{N\to\infty} \tilde{K}_{N,u_N}^{\text{flat}}(\lambda, \xi) = 2^{-1/3} \, \text{Ai}\left(2^{-1/3}(\lambda + \xi) + 2^{1/3}\gamma r \right) \qquad (4.12)$$

for all $\lambda, \xi \in \mathbb{R}_+$. We now need to show the convergence of the corresponding Fredholm determinants on $L^2(\mathbb{R}_+)$. The argument is standard, and based on the series expansion (2.1) of the Fredholm determinant. Notice first that there exist two positive constants c_1 and c_2 such that

$$\sup_{N\in\mathbb{N}} \left| \tilde{K}_{N,u_N}^{\text{flat}}(\lambda, \xi) \right| \leq c_1 \, e^{-c_2\lambda}$$

for all $\lambda, \xi \in \mathbb{R}_+$. The exponential bound for $\tilde{K}_{N,u_N}^{\text{flat},1}$ comes from Corollary 3.2, whereas the estimates for the remaining terms directly follow from (4.10) and (4.11). Hadamard's bound then implies that

$$\left| \det(\tilde{K}_{N,u_N}^{\text{flat}}(\lambda_i, \lambda_j))_{i,j=1}^n \right| \leq n^{n/2} \prod_{i=1}^n c_1 \, e^{-c_2\lambda_i} \ .$$

It follows that

$$\left| \det(I - \tilde{K}_{N,u_N}^{\text{flat}})_{L^2(\mathbb{R}_+)} \right| \leq 1 + \sum_{n=1}^\infty \frac{1}{n!} \int_0^\infty \cdots \int_0^\infty \left| \det(K_N(\lambda_i, \lambda_j))_{i,j=1}^n \right| d\lambda_1 \cdots \lambda_n$$

$$\leq 1 + \sum_{n=1}^\infty \frac{n^{n/2}}{n!} \left(\int_0^\infty c_1 \, e^{-c_2\lambda} \, d\lambda \right)^n < \infty.$$

These inequalities, apart from providing a further proof that the Fredholm determinants of our kernels are well-defined, allow us to conclude, by dominated convergence, that limit (4.12) still holds when passing to the corresponding Fredholm determinants on $L^2(\mathbb{R}_+)$. After a rescaling of the limiting kernel by a factor $2^{-2/3}$, one can see that the operator on $L^2(\mathbb{R}_+)$ defined through the kernel $(\lambda, \xi) \mapsto 2^{-1/3} \, \text{Ai}\left(2^{-1/3}(\lambda + \xi) + s \right)$ has the same Fredholm determinant as the operator $\mathcal{K}_1$ on $L^2([s, \infty))$ defining the Tracy-Widom GOE distribution $F_1(s)$ as in (1.4). This concludes the proof. $\qquad\qquad\Box$

5   Point-to-Half-Line Last Passage Percolation and Airy$_{2\to1}$

In this section, we analyze the point-to-half-line dLPP, using again the general results of Sects. 2 and 3. The starting point is the next theorem, which is a result of (1.9), (1.10) and (1.12), combined with the Cauchy-Binet identity.

Theorem 5.1 ([6]) *The distribution function of the point-to-half-line directed last passage percolation $\tau_{2N}^{\text{h-flat}}$ with exponential waiting times as in (1.3) is a ratio of $N \times N$ determinants:*

$$\mathbb{P}\big(\tau_{2N}^{\text{h-flat}} \leq u\big) = \frac{\det\big(H_u^{\text{h-flat}}(\alpha_i, \beta_j)\big)_{1 \leq i,j \leq N}}{\det(C(\alpha_i, \beta_j))_{1 \leq i,j \leq N}}, \tag{5.1}$$

for $u > 0$, where $C(z, w) := (z + w)^{-1}$ and

$$H_u^{\text{h-flat}}(z, w) := e^{-u(z+w)} \int_0^u (e^{zx} - e^{-zx}) e^{wx} \, dx \, .$$

We now write the distribution function as a Fredholm determinant.

Theorem 5.2 *The distribution of $\tau_{2N}^{\text{h-flat}}$ with exponential waiting times as in (1.3) is given by*

$$\mathbb{P}\big(\tau_{2N}^{\text{h-flat}} \leq u\big) = \det(I - K_{N,u}^{\text{h-flat}})_{L^2(\mathbb{R}_+)} \, , \tag{5.2}$$

where $K_{N,u}^{\text{h-flat}} : L^2(\mathbb{R}_+) \to L^2(\mathbb{R}_+)$ is the operator defined through the kernel

$$K_{N,u}^{\text{h-flat}}(\lambda, \xi) = \frac{1}{(2\pi i)^2} \int_{\Gamma_1} dz \int_{\Gamma_2} dw \, e^{-\lambda z - \xi w} \, \widehat{H}_u^{\text{h-flat}}(z, w) \prod_{m=1}^N \frac{(z + \beta_m)(w + \alpha_m)}{(z - \alpha_m)(w - \beta_m)} \, . \tag{5.3}$$

Here, $\Gamma_1, \Gamma_2 \subset \mathbb{C}_+$ are any positively oriented simple closed contours such that Γ_1 encloses $\alpha_1, \ldots, \alpha_N$, Γ_2 encloses $\beta_1, \ldots, \beta_N$ as well as the whole Γ_1, and

$$\widehat{H}_u^{\text{h-flat}}(z, w) = \frac{e^{-u(z+w)}}{z + w} + \frac{e^{-u(z+w)}}{z - w} + \frac{e^{-2uz}}{w - z} \, . \tag{5.4}$$

Proof The claim follows from formula (5.1) and Theorem 2.1. Function $\widehat{H}_u^{\text{h-flat}}$ in (5.3) is defined through the relation $H_u^{\text{h-flat}} = C - \widehat{H}_u^{\text{h-flat}}$ (using the notation of Theorem 5.1 for $H_u^{\text{h-flat}}$), i.e.

$$\widehat{H}_u^{\text{h-flat}}(z, w) = \frac{1}{z + w} - e^{-u(z+w)} \int_0^u (e^{zx} - e^{-zx}) e^{wx} \, dx \, .$$

If we assume that Γ_2 encloses Γ_1 (so that $z \neq w$ for all z, w), integrating the above expression yields (5.4).

Proof of Theorem 1.2 In order to perform the asymptotics of formula (5.2) in the i.i.d. case, we set $\alpha_m = \beta_m = \gamma$ for all m in parametrization (1.3). Our kernel (5.3) thus becomes

$$K_{N,u}^{\text{h-flat}} = K_{N,u}^{\text{h-flat},1} + K_{N,u}^{\text{h-flat},2} + K_{N,u}^{\text{h-flat},3},$$

where

$$K_{N,u}^{\text{h-flat},1}(\lambda, \xi) = \frac{1}{(2\pi i)^2} \int_{\Gamma_1} dz \int_{\Gamma_2} dw \, e^{-\lambda z - \xi w} \, \frac{e^{-u(z+w)}}{z+w} \left[\frac{(z+\gamma)(w+\gamma)}{(z-\gamma)(w-\gamma)} \right]^N,$$

$$K_{N,u}^{\text{h-flat},2}(\lambda, \xi) = \frac{1}{(2\pi i)^2} \int_{\Gamma_1} dz \int_{\Gamma_2} dw \, e^{-\lambda z - \xi w} \, \frac{e^{-u(z+w)}}{z-w} \left[\frac{(z+\gamma)(w+\gamma)}{(z-\gamma)(w-\gamma)} \right]^N,$$

$$K_{N,u}^{\text{h-flat},3}(\lambda, \xi) = \frac{1}{(2\pi i)^2} \int_{\Gamma_1} dz \int_{\Gamma_2} dw \, e^{-\lambda z - \xi w} \, \frac{e^{-2uz}}{w-z} \left[\frac{(z+\gamma)(w+\gamma)}{(z-\gamma)(w-\gamma)} \right]^N.$$

For the steepest descent analysis of the first two terms, we are going to adapt the proof of Proposition 3.1, taking into account that we now have double contour integrals instead of single ones. Noticing that $K_{N,u}^{\text{h-flat},1}$ and $K_{N,u}^{\text{h-flat},2}$ only differ for the sign in $(z \pm w)^{-1}$, we study both at the same time, denoting by $K_{N,u}^{\pm}$ either of them:

$$K_{N,u}^{\pm}(\lambda, \xi) := \frac{1}{(2\pi i)^2} \int_{\Gamma_1} dz \int_{\Gamma_2} dw \, e^{-\lambda z - \xi w} \, \frac{e^{-u(z+w)}}{z \pm w} \left[\frac{(z+\gamma)(w+\gamma)}{(z-\gamma)(w-\gamma)} \right]^N.$$

We replace the contour Γ_1 with $T_{R_1} + 2\epsilon\gamma/\sqrt[3]{2N}$ and the contour Γ_2 with $T_{R_2} + \epsilon\gamma/\sqrt[3]{2N}$, for some $\gamma < R_1 < R_2$ and $\epsilon > 0$; Here, as in the proof of Proposition 3.1, T_R is the negatively oriented triangular path with vertices 0, $2R\,e^{i\pi/3}$ and $2R\,e^{-i\pi/3}$. Notice that changing the orientation of both paths does not yield any change of sign in the double contour integral; moreover, the first contour is still enclosed by the second one, and the singularities at $(z \pm w)^{-1}$ are not crossed by the deformed contours (the infinitesimal shifts of T_{R_1} and T_{R_2} are also done for this sake). Set now $u = u_N := 2N/\gamma + rN^{1/3}$ and denote by $\tilde{\Psi}$ the rescaling of any function $\Psi(\lambda, \xi)$ by the factor $\sqrt[3]{2N}/\gamma$, as in (4.9). We can thus write

$$\tilde{K}_{N,u_N}^{\pm}(\lambda, \xi) = \frac{\sqrt[3]{2N}}{\gamma(2\pi i)^2} \int_{T_{R_1} + \frac{2\epsilon\gamma}{\sqrt[3]{2N}}} dz \int_{T_{R_2} + \frac{\epsilon\gamma}{\sqrt[3]{2N}}} dw \, \frac{e^{NF(z) - \lambda_r z \sqrt[3]{2N}/\gamma} \, e^{NF(w) - \xi_r w \sqrt[3]{2N}/\gamma}}{z \pm w},$$

where $\lambda_r := \lambda + 2^{-1/3}\gamma r$, $\xi_r := \xi + 2^{-1/3}\gamma r$, and $F(z) := \log(\gamma + z) - \log(\gamma - z) - 2z/\gamma$. Since the main contribution in the integral is expected to come from $z = w = 0$, which is the critical point of F, we split the above integral into the following sum:

$$\tilde{K}_{N,u_N}^{\pm} = \tilde{K}_{N,u_N}^{\pm,\text{in},\text{in}} + \tilde{K}_{N,u_N}^{\pm,\text{in},\text{ex}} + \tilde{K}_{N,u_N}^{\pm,\text{ex},\text{in}} + \tilde{K}_{N,u_N}^{\pm,\text{ex},\text{ex}}.$$

Here, the first superscript "in" ("ex") indicates that the integration w.r.t. z is performed only in the interior (exterior, respectively) of the ball $\{|z - 2\epsilon\gamma/\sqrt[3]{2N}| \le \gamma N^{-\alpha}\}$ for some exponent $\alpha > 0$ to be specified later on, while the second superscript "in" ("ex") indicates that the integration w.r.t. w is performed only in the interior (exterior, respectively) of the ball $\{|w - \epsilon\gamma/\sqrt[3]{2N}| \le \gamma N^{-\alpha}\}$. In $\tilde{K}_{N,u_N}^{\pm,\text{in},\text{in}}$, after the changes of variables $\tilde{z} := z\sqrt[3]{2N}/\gamma$ and $\tilde{w} := w\sqrt[3]{2N}/\gamma$, we obtain

$$\tilde{K}_{N,u_N}^{\pm,\text{in},\text{in}}(\lambda, \xi) = \frac{1}{(2\pi i)^2} \int_{C+2\epsilon} d\tilde{z}\, \exp\left\{\frac{\tilde{z}^3}{3} - \lambda_r \tilde{z} + R\left(\frac{\gamma}{\sqrt[3]{2N}}\tilde{z}\right)N\right\} \mathbb{1}_{\{|\tilde{z}-2\epsilon| \le \sqrt[3]{2}N^{1/3-\alpha}\}}$$

$$\times \int_{C+\epsilon} d\tilde{w}\, \exp\left\{\frac{\tilde{w}^3}{3} - \xi_r \tilde{w} + R\left(\frac{\gamma}{\sqrt[3]{2N}}\tilde{w}\right)N\right\} \mathbb{1}_{\{|\tilde{w}-\epsilon| \le \sqrt[3]{2}N^{1/3-\alpha}\}} \frac{1}{\tilde{z} \pm \tilde{w}},$$

where C is the piecewise linear path going from $e^{-i\pi/3}\infty$ to the origin to $e^{i\pi/3}\infty$, and R is the remainder defined by (3.4). The indicator functions clearly converge to 1 if $\alpha < 1/3$. As in the proof of Proposition 3.1, one can also show that the remainders, even when multiplied by N, vanish uniformly for $\tilde{z}, \tilde{w}$ in the support of the integrand, if we choose $1/4 < \alpha < 1/3$. Applying dominated convergence, one can see that

$$\lim_{N\to\infty} \tilde{K}_{N,u_N}^{\pm,\text{in},\text{in}}(\lambda, \xi) = \frac{1}{(2\pi i)^2} \int_{C+2\epsilon} d\tilde{z} \int_{C+\epsilon} d\tilde{w}\, \frac{1}{\tilde{z} \pm \tilde{w}} \exp\left\{\frac{\tilde{z}^3}{3} - \lambda_r \tilde{z} + \frac{\tilde{w}^3}{3} - \xi_r \tilde{w}\right\}.$$

Using similar arguments as in the proof of Proposition 3.1, together with the bound $|z \pm w|^{-1} \le 2\sqrt[3]{2N}/(\sqrt{3}\epsilon\gamma)$, one can see that the other terms $\tilde{K}_{N,u_N}^{\pm,\text{in},\text{ex}}$, $\tilde{K}_{N,u_N}^{\pm,\text{ex},\text{in}}$, and $\tilde{K}_{N,u_N}^{\pm,\text{ex},\text{ex}}$ vanish exponentially fast in the limit $N \to \infty$. We thus have:

$$\lim_{N\to\infty} \tilde{K}_{N,u_N}^{\text{h-flat},1}(\lambda, \xi) = \frac{1}{(2\pi i)^2} \int_{C+2\epsilon} d\tilde{z} \int_{C+\epsilon} d\tilde{w}\, \frac{1}{\tilde{z} + \tilde{w}}\, e^{\tilde{z}^3/3 - \lambda_r \tilde{z}}\, e^{\tilde{w}^3/3 - \xi_r \tilde{w}},$$

$$\lim_{N\to\infty} \tilde{K}_{N,u_N}^{\text{h-flat},2}(\lambda, \xi) = \frac{1}{(2\pi i)^2} \int_{C+2\epsilon} d\tilde{z} \int_{C+\epsilon} d\tilde{w}\, \frac{1}{\tilde{z} - \tilde{w}}\, e^{\tilde{z}^3/3 - \lambda_r \tilde{z}}\, e^{\tilde{w}^3/3 - \xi_r \tilde{w}}.$$

We will now rewrite these expressions as integrals of Airy functions. In the first one, since $\Re(\tilde{z} + \tilde{w}) > 0$ for all $\tilde{z}$ and $\tilde{w}$, we can make the substitution $(\tilde{z} + \tilde{w})^{-1} = \int_0^\infty e^{-(\tilde{z}+\tilde{w})x}\, dx$. The resulting $d\tilde{z}\, d\tilde{w}\, dx$ integral is absolutely convergent, hence Fubini's Theorem can be applied to obtain:

$$\lim_{N\to\infty} \tilde{K}_{N,u_N}^{\text{h-flat},1}(\lambda, \xi) = \int_0^\infty \left[\frac{1}{2\pi i} \int_{C+2\epsilon} e^{\tilde{z}^3/3 - (\lambda_r+x)\tilde{z}}\, d\tilde{z}\right]\left[\frac{1}{2\pi i} \int_{C+\epsilon} e^{\tilde{w}^3/3 - (\xi_r+x)\tilde{w}}\, d\tilde{w}\right] dx$$

$$= \int_0^\infty \text{Ai}(\lambda_r + x)\, \text{Ai}(\xi_r + x)\, dx,$$

according to definition (3.1) of Airy function. In $\tilde{K}_{N,u_N}^{\text{h-flat},2}$, we deform the contour $C + \epsilon$ into the straight line l_ϵ going from $\epsilon - i\infty$ to $\epsilon + i\infty$; since now $\Re(\tilde{z} - \tilde{w}) \ge \epsilon > 0$ for all $\tilde{z}$ and $\tilde{w}$, we can make the substitution $(\tilde{z} - \tilde{w})^{-1} = \int_0^\infty e^{-(\tilde{z}-\tilde{w})x}\, dx$. The resulting $d\tilde{z}\, d\tilde{w}\, dx$ integral is absolutely convergent, hence Fubini's Theorem can be applied to obtain:

$$\lim_{N \to \infty} \tilde{K}_{N,u_N}^{\text{h-flat},2}(\lambda, \xi) = \int_0^\infty \left[\frac{1}{2\pi i} \int_{C+2\epsilon} e^{\tilde{z}^3/3 - (\lambda_r + x)\tilde{z}} \, d\tilde{z} \right] \left[\frac{1}{2\pi i} \int_{l_\epsilon} e^{\tilde{w}^3/3 - (\xi_r - x)\tilde{w}} \, d\tilde{w} \right] dx$$

$$= \int_0^\infty \text{Ai}(\lambda_r + x) \, \text{Ai}(\xi_r - x) \, dx \, .$$

We remark that the second square bracket above is an Airy function as well, since the path l_ϵ can be deformed back to a contour, like $C + \epsilon$, whose arguments at ∞ are $\pm \pi/3$.

We finally notice that $K_{N,u_N}^{\text{h-flat},3}(\lambda, \xi)$ equals exactly the term $K_{N,u_N}^{\text{flat},2}(\lambda, \xi)$ defined in the proof of Theorem 1.1. Therefore, as we have already proved there, the rescaled version $\tilde{K}_{N,u_N}^{\text{h-flat},3}(\lambda, \xi)$ vanishes as $N \to \infty$.

We conclude that, as a whole, our rescaled kernel has the following limit:

$$\lim_{N \to \infty} \tilde{K}_{N,u_N}^{\text{h-flat}}(\lambda, \xi) = \mathcal{K}_{2 \to 1}(\lambda_r, \xi_r) = \mathcal{K}_{2 \to 1}(\lambda + 2^{-1/3}\gamma r, \xi + 2^{-1/3}\gamma r) \, ,$$

where $\mathcal{K}_{2 \to 1}$ is defined in (1.8). Using the key fact that all contours are chosen to have positive distance from the imaginary axis (as in the analogous estimates obtained in Corollary 3.2 and in the proof of Theorem 1.1), one can show that there exist two positive constants c_1 and c_2 such that, for all $\lambda, \xi \in \mathbb{R}_+$,

$$\sup_{N \in \mathbb{N}} \left| \tilde{K}_{N,u_N}^{\text{h-flat}}(\lambda, \xi) \right| \le c_1 \, e^{-c_2 \lambda} \, .$$

The latter bound provides, as in the third step of the proof of Theorem 1.1, the right estimates for the series expansion of $\det(I - \tilde{K}_{N,u_N}^{\text{h-flat}})_{L^2(\mathbb{R}_+)}$ in order to justify its convergence. The claim thus follows from the Fredholm determinant representation (1.7) of $F_{2 \to 1}$. □

Acknowledgements This article is dedicated to Raghu Varadhan on the occasion of his 75th birthday. Exact solvability is arguably not within Raghu's signature style. However, the second author learned about random polymer models and the $t^{1/3}$ law from Raghu, as being a challenging problem, while he was working under his direction on a somewhat different PhD topic. Since that time he has always had the desire to make some contributions in this direction and would like to thank him for, among other things, having provided this stimulus.

The work of EB was supported by EPSRC via grant EP/M506679/1. The work of NZ was supported by EPSRC via grant EP/L012154/1.

References

1. Baik, J., Rains, E.M.: The asymptotics of monotone subsequences of involutions. Duke Math. J. **109**(2), 205–281 (2001)
2. Borodin, A., Ferrari, P.L., Sasamoto, T.: Transition between Airy1 and Airy2 processes and TASEP fluctuations. Comm. Pure Appl. Math. **61**, 1603–1629 (2008)
3. Borodin, A., Vadim, G.: Lectures on integrable probability. In: Probability and Statistical Physics in St. Petersburg, Proceedings of Symposia in Pure Mathematics, vol. 91 (2012)

4. Bisi, E.: Random polymers via orthogonal Whittaker and symplectic Schur functions. PhD thesis (2018). arXiv:1810.03734
5. Bisi, E., Zygouras, N.: Transition between characters of classical groups, decompositions of Gelfand-Tsetlin patterns, and last passage percolation (In preparation)
6. Bisi, E., Zygouras, N.: Point-to-line polymers and orthogonal Whittaker functions. Trans. Am. Math. Soc. (to appear). arXiv:1703.07337
7. Corwin, I., O'Connell, N., Seppäläinen, T., Zygouras, N.: Tropical combinatorics and Whittaker functions. Duke Math. J. **163**(3), 513–563 (2014)
8. Ferrari, P.L.: Polynuclear growth on a flat substrate and edge scaling of GOE eigenvalues. Comm. Math. Phys. **252**, 77–109 (2004)
9. Ferrari, P.L., Spohn, H.: A determinantal formula for the GOE Tracy-Widom distribution. J. Phys. A Math. Gen. **38**, L557–L561 (2005)
10. Fulton, W., Harris, J.: Representation Theory—A First Course. Graduate Texts in Mathematics, vol. 129. Springer (1991)
11. Forrester, P.J.: Log-Gases and Random Matrices (LMS-34). Princeton University Press (2010)
12. Johansson, K.: Shape fluctuations and random matrices. Commun. Math. Phys. **209**(2), 437–476 (2000)
13. Johansson, K.: Random growth and random matrices. In: European Congress of Mathematics, pp. 445–456 (2001)
14. Kirillov, A.N.: Introduction to tropical combinatorics. In: Proceedings of Nagoya 2000 2nd International Workshop on Physics and Combinatorics, pp. 82–150. World Scientific, Singapore (2001)
15. König, W.: Orthogonal polynomial ensembles in probability theory. Probab. Surv. **2**, 385–447 (2005)
16. Le Doussal, P., Calabrese, P.: The KPZ equation with flat initial condition and the directed polymer with one free end. J. Stat. Mech. Theory Exper. **06**, P06001 (2012)
17. Nguyen, V.-L., Zygouras, N.: Variants of geometric RSK, geometric PNG and the multipoint distribution of the log-Gamma polymer. Int. Math. Res. Not. **15**, 4732–4795 (2017)
18. O'Connell, N., Seppäläinen, T., Zygouras, N.: Geometric RSK correspondence, Whittaker functions and symmetrized random polymers. Invent. Math. **197**(2), 361–416 (2014)
19. Okounkov, A.: Infinite wedge and random partitions. Selecta Math. **7**(1), 57–81 (2001). New Series
20. Sasamoto, T.: Spatial correlations of the 1D KPZ surface on a flat substrate. J. Phys. A Math. Gen. **38**, L549 (2005)
21. Seppäläinen, T.: Scaling for a one-dimensional directed polymer with boundary conditions. Ann. Probab. **40**(1), 19–73 (2012)
22. Schütz, G.M.: Exact solution of the master equation for the asymmetric exclusion process. J. Stat. Phys. **88**(1), 427–445 (1997)
23. Tracy, C.A., Widom, H.: On orthogonal and symplectic matrix ensembles. Comm. Math. Phys. **177**(3), 727–754 (1996)
24. Tracy, C.A., Widom, H.: Correlation functions, cluster functions, and spacing distributions for random matrices. J. Stat. Phys. **92**(5), 809–835 (1998)
25. Tracy, C.A., Widom, H.: Matrix kernels for the Gaussian orthogonal and symplectic ensembles. Annales de l'Institut Fourier **55**(6), 2197–2207 (2005)

A Large Deviations Principle for the Polar Empirical Measure in the Two-Dimensional Symmetric Simple Exclusion Process

Claudio Landim, Chih-Chung Chang and Tzong-Yow Lee

Abstract We prove an energy estimate for the polar empirical measure of the two-dimensional symmetric simple exclusion process. We deduce from this estimate and from results in (Chang et al. in Ann Probab 32:661–691, (2004) [2]) large deviations principles for the polar empirical measure and for the occupation time of the origin.

Keywords Exclusion process · Hydrodynamic limit · Large deviations · Occupation time

MSC Codes 82C22 · 60K35 · 60F10

1 Introduction

We presented in [2] a large deviations principle for the occupation time of the origin in the two-dimensional symmetric simple exclusion process. The proof relies on a large deviations principle for the "polar" empirical measure. After the paper was published and after T-Y Lee passed away, A. Asselah pointed to us that there was a flaw in the argument. The proofs of the lower and upper bound of the large deviations principle for the polar measure were correct, but the bounds did not match.

We correct this inaccuracy in this article by showing that we may restrict the upper bound to measures with finite energy, that is, to absolutely continuous measures $\mu(dr) = m(r)dr$ whose density m has generalized derivatives, denoted by m', such that $\int_{\mathbb{R}_+} [m'(r)]^2/\sigma(m(r)) \, dr < \infty$, where $\sigma(a) = a(1-a)$.

C. Landim (✉) · T.-Y. Lee
IMPA, Estrada Dona Castorina 110, CEP 22460 Rio de Janeiro,
Brasil and CNRS UPRES-A 6085, Université de Rouen, 76128 Mont Saint Aignan, France
e-mail: landim@impa.br

C.-C. Chang
Department of Mathematics, National Taiwan University, Taipei, Taiwan, ROC
e-mail: ccchang@math.ntu.edu.tw

© Springer Nature Switzerland AG 2019
P. Friz et al. (eds.), *Probability and Analysis in Interacting Physical Systems*, Springer Proceedings in Mathematics & Statistics 283,
https://doi.org/10.1007/978-3-030-15338-0_8

215

The large deviations principle of the occupation time of the origin is correct as stated in [2], and follows, through a contraction principle, from the amended version of the large deviations principle for the polar measure presented here.

There are many reasons to examine the large deviations of the occupation time in dimension 2. On the one hand, the unusual large deviations decay rate $t/\log t$, with a logarithmic correction which appears in critical dimensions. This is the case, for example, of the survival probability of a random walk among a Poisson system of moving traps on $\mathbb{Z}^2$, which can also be interpreted as the solution of a parabolic Anderson model with a random time-dependent potential [3] [note that the large deviations decay rate in this model coincides in all dimensions with the one of the occupation time in the symmetric simple exclusion process].

Another reason to study occupation time large deviations in dimension 2 is the unexpected possibility to derive an explicit formula (cf. Eq. (2.6)) for the large deviations rate function. Finally, the method by itself may be of interest in other contexts. It has been shown [5] that in dimension 1 the occupation time large deviations, whose decay rate is $\sqrt{t}$, is related to the large deviations of the empirical measure. Here, in dimension 2, it is shown to be connected to the large deviations of the polar measure. It is conceivable that in higher dimensions, where the decay rate is t, the large deviations are associated to some other type of empirical measure.

We refer to [2] for further references and for an historical background of this problem. We wish to thank A. Asselah for pointing to us the flaw in [2], K. Mallick and K. Tsunoda for stimulating discussion on occupation time large deviations and drawing our attention to the recent papers [9, 10]. These exchanges encouraged us to try to fill the gap left in [2].

2 Notation and Results

The speeded-up, symmetric simple exclusion process on $\mathbb{Z}^2$ is the continuous-time Markov process on $\{0, 1\}^{\mathbb{Z}^2}$ whose generator, denoted by L_T, acts on functions $f : \{0, 1\}^{\mathbb{Z}^2} \to \mathbb{R}$ which depends only on a finite number of coordinates as

$$(L_T f)(\eta) = \frac{T}{2} \sum_{j=1}^{2} \sum_{x \in \mathbb{Z}^2} \{f(\sigma^{x,x+e_j}\eta) - f(\eta)\}.$$

In this formula, $\{e_1, e_2\}$ is the canonical basis of $\mathbb{R}^2$, and $\sigma^{x,y}\eta$ is the configuration obtained from η by exchanging the occupation variables $\eta(x)$ and $\eta(y)$:

$$(\sigma^{x,y}\eta)(z) = \begin{cases} \eta(z) & \text{if } z \neq x, y, \\ \eta(x) & \text{if } z = y, \\ \eta(y) & \text{if } z = x. \end{cases}$$

Denote by $\nu_\alpha, 0 \leq \alpha \leq 1$, the Bernoulli product measure on $\{0, 1\}^{\mathbb{Z}^2}$ with marginals given by

$$\nu_\alpha\{\eta, \ \eta(x) = 1\} \ = \ \alpha, \quad \text{for } x \in \mathbb{Z}^2.$$

A simple computation shows that $\{\nu_\alpha, \ 0 \le \alpha \le 1\}$ is a one-parameter family of reversible invariant measures.

Denote by $D(\mathbb{R}_+, \{0, 1\}^{\mathbb{Z}^2})$ the space of right continuous functions $x : \mathbb{R}_+ \to \{0, 1\}^{\mathbb{Z}^2}$ with left limits, endowed with the Skorohod topology. The elements of $D(\mathbb{R}_+, \{0, 1\}^{\mathbb{Z}^2})$ are represented by η_s, $s \ge 0$. Let $\mathbb{P}_\alpha = \mathbb{P}_{T,\alpha}$, $0 \le \alpha \le 1$, be the probability measure on $D(\mathbb{R}_+, \{0, 1\}^{\mathbb{Z}^2})$ induced by Markov process whose generator is L_T starting from ν_α. Expectation with respect to $\mathbb{P}_\alpha$ is denoted by $\mathbb{E}_\alpha$.

Denote by $\mathcal{M}$ the space of locally finite, nonnegative measures on $[0, \infty)$. Let $\sigma_T : \mathbb{Z}_*^2 := \mathbb{Z}^2 \setminus \{0\} \to \mathbb{R}_+$ be given by

$$\sigma_T(x) \ = \ \frac{\log |x|}{\log T},$$

where $|x|$ represents the Euclidean norm of x, $|x|^2 = x_1^2 + x_2^2$. Denote by $\mu^{1,T} : \{0, 1\}^{\mathbb{Z}^2} \to \mathcal{M}$ the "polar" empirical measure on $\mathbb{R}_+$ induced by a configuration η:

$$\mu^{1,T}(\eta) \ = \ \frac{1}{2\pi \log T} \sum_{x \in \mathbb{Z}_*^2} \eta(x) \frac{1}{|x|^2} \delta_{\sigma_T(x)}.$$

Here, δ_v is the Dirac measure concentrated on $v \in \mathbb{R}_+$. Notice the factor 2π on the denominator to avoid unpleasant coefficients in the limit. Denote by $\bar{\mu}^T : D(\mathbb{R}_+, \{0, 1\}^{\mathbb{Z}^2}) \to \mathcal{M}$ the measure on $\mathbb{R}_+$ obtained as the time integral of the measures $\mu^{1,T}$:

$$\bar{\mu}^T \ = \ \int_0^1 \mu^{1,T}(\eta_s) \, ds. \tag{2.1}$$

The main result of this article establishes a large deviations principle for the measure $\bar{\mu}^T$ under $\mathbb{P}_\alpha$.

Denote by $\mathbb{1}$ the configuration in which all sites are occupied, $\mathbb{1}(x) = 1$ for all $x \in \mathbb{Z}^2$. The measures $\mu^{1,T}, \bar{\mu}^T$ are nonnegative and bounded above by the measure $\lambda_T = \mu^{1,T}(\mathbb{1})$: for all nonnegative, continuous function $H : (0, \infty) \to \mathbb{R}$ with compact support, and all elements of $\{0, 1\}^{\mathbb{Z}^2}$, $D(\mathbb{R}_+, \{0, 1\}^{\mathbb{Z}^2})$,

$$0 \le \int_{\mathbb{R}_+} H \, d\mu^{1,T} \le \int_{\mathbb{R}_+} H \, d\lambda_T \quad \text{and} \quad 0 \le \int_{\mathbb{R}_+} H \, d\bar{\mu}^T \le \int_{\mathbb{R}_+} H \, d\lambda_T.$$

On the other hand, an elementary computation shows that there exists a finite universal constant C_0 such that

$$\lambda_T([a, b]) \ \le \ (b - a) + \frac{C_0}{\log T} \tag{2.2}$$

for all $0 < a < b < \infty$, $T > 1$. It is therefore natural to introduce the space $\mathcal{M}_c$, $c > 0$, of nonnegative, locally finite measures μ defined on the Borel sets of $(0, \infty)$ and such that $\mu([a, b]) \leq (b - a) + c$ for every $0 < a < b < \infty$:

$$\mathcal{M}_c = \left\{ \mu \in \mathcal{M} : \mu([a, b]) \leq (b - a) + c \text{ for } 0 < a < b < \infty \right\}.$$

The uniform bound on the measure of the intervals makes the set $\mathcal{M}_c$ endowed with the vague topology a compact, separable metric space. Let $\mathcal{M}_0$ be the subspace of $\mathcal{M}_c$ of all measures which are absolutely continuous with respect to the Lebesgue measure and whose density is bounded by 1. The subspace $\mathcal{M}_0$ is closed (and thus compact).

Remark 2.1 The vague topology is the natural one in the investigation of the law of large numbers and the large deviations of the empirical measure in infinite volume, the test functions being the continuous ones with compact support, cf. [6, 8]. Actually, in the problem considered here, Lemma 3.4 asserts that the analysis can be confined to the interval $[0, 1/2]$, as $\bar{\mu}^T$ is super-exponentially close to $\alpha \, dr$ on $(1/2, \infty)$.

Let $C_K((0, \infty))$ be the space of continuous functions $G : (0, \infty) \to \mathbb{R}$ with a compact support, and let $C_K^n((0, \infty))$, $n \geq 1$, be the space of compactly supported functions $F : (0, \infty) \to \mathbb{R}$ whose n-th derivative is continuous. Denote by $\mathcal{Q} : \mathcal{M}_0 \to \mathbb{R}_+$ the energy functional given by

$$\mathcal{Q}(m(r)\,dr) = \sup_{G \in C_K^1((0, \infty))} \left\{ -\int_0^\infty G'(r)\, m(r)\, dr - \int_0^\infty \sigma(m(r))\, G(r)^2\, dr \right\},$$
$$(2.3)$$

where $\sigma(a) = a(1 - a)$ stands for the mobility of the exclusion process. By [1, Lemma 4.1], the functional $\mathcal{Q}$ is convex and lower-semicontinuous. Moreover, if $\mathcal{Q}(m(r)\,dr)$ is finite, m has a generalized derivative, denoted by m', and

$$\mathcal{Q}(m(r)\,dr) = \frac{1}{4} \int_0^\infty \frac{[m'(r)]^2}{\sigma(m(r))}\, dr.$$

Fix $0 < \alpha < 1$, and let $\mathcal{M}_{0,\alpha}$ the space of measures in $\mathcal{M}_0$ whose densities are equal to α on $(1/2, \infty)$: $\mathcal{M}_{0,\alpha} = \{\mu(dr) = m(r)dr \in \mathcal{M}_0 : m(r) = \alpha \text{ a.s. in } (1/2, \infty)\}$. Denote by $I_{\mathcal{Q},\alpha} : \mathcal{M}_c \to \mathbb{R}_+$ the functional given by

$$I_{\mathcal{Q},\alpha}(\mu) = \begin{cases} \pi\,\mathcal{Q}(\mu) & \text{if } \mu \in \mathcal{M}_{0,\alpha}, \\ +\infty & \text{otherwise.} \end{cases} \qquad (2.4)$$

Since the set $\mathcal{M}_{0,\alpha}$ is convex and closed, the functional $I_{\mathcal{Q},\alpha}$ inherits from $\mathcal{Q}$ the convexity and the lower-semicontinuity. Furthermore, as $\mathcal{M}_c$ is compact and $I_{\mathcal{Q},\alpha}$ lower semi-continuous, the level sets of $I_{\mathcal{Q},\alpha}$ are compact. Next assertion is the main result of the article.

Theorem 2.2 *For every closed subset F of $\mathcal{M}_c$ and every open subset G of $\mathcal{M}_c$,*

$$\limsup_{T \to \infty} \frac{\log T}{T} \log \mathbb{P}_\alpha \left[\bar{\mu}^T \in F \right] \leq - \inf_{\mu \in F} I_{Q,\alpha}(\mu),$$

$$\liminf_{T \to \infty} \frac{\log T}{T} \log \mathbb{P}_\alpha \left[\bar{\mu}^T \in G \right] \geq - \inf_{\mu \in G} I_{Q,\alpha}(\mu).$$

Moreover, the rate functional $I_{Q,\alpha} : \mathcal{M}_c \to \mathbb{R}_+$ is convex, lower semi-continuous and has compact level sets.

Remark 2.3 We explain in this remark the flaw in [2]. Denote by $\hat{I}_\alpha : \mathcal{M}_0 \to \mathbb{R}_+$ the functional given by

$$\hat{I}_\alpha (m(r) \, dr) \;=\; \sup_{G \in C_k^1((0,1/2))} \left\{ - \int_0^{1/2} G'(r) \, m(r) \, dr \;-\; \int_0^{1/2} \sigma(m(r)) \, G(r)^2 \, dr \right\}.$$

Note that the supremum is carried over functions whose support is now contained in $(0, 1/2)$. Let $I_\alpha : \mathcal{M}_c \to \mathbb{R}_+$ be given by

$$I_\alpha(\mu) \;=\; \begin{cases} \hat{I}_\alpha(\mu) & \text{if } \mu \in \mathcal{M}_{0,\alpha}, \\ +\infty & \text{otherwise.} \end{cases} \tag{2.5}$$

Section 5 of [2] shows that I_α is an upper bound for the large deviations principle. This upper bound is not sharp. Consider, for instance, the measure $\mu_\beta(dr) = m_\beta(r) dr$, $\beta \neq \alpha$, where $m_\beta(r) = \beta$ for $0 \leq r < 1/2$ and $m_\beta(r) = \alpha$ for $r \geq 1/2$. By (2.5), $I_\alpha(\mu_\beta) = 0$, which is clearly not sharp.

The problem lies in the proof of Lemma 6.3, at the end of page 686. It is claimed there that if $I_\alpha(\mu) < \infty$ for an absolutely continuous measure $\mu(dr) = m(r) dr$, there exists a sequence of smooth functions m_n such that $m_n(r) = \alpha$ for $r \geq 1/2$, $\mu_n(dr) = m_n(r) dr \to \mu$ in the vague topology, and $I_\alpha(m_n(r) dr) \to I_\alpha(\mu)$. This is not true for the measure μ_β introduced in the previous paragraph.

Remark 2.4 For a measure μ in $\mathcal{M}_{0,\alpha}$ with finite energy, $\mathcal{Q}(\mu) < \infty$,

$$I_{Q,\alpha}(\mu) \;=\; \frac{\pi}{4} \int_0^{1/2} \frac{[m'(r)]^2}{\sigma(m(r))} \, dr \;=\; I_\alpha(\mu).$$

However, for measures in $\mathcal{M}_{0,\alpha}$ with infinite energy, $I_{Q,\alpha}(\mu) = \infty$, while $I_\alpha(\mu)$ might be finite. For example, $I_{Q,\alpha}(\mu_\beta) = \infty$ and $I_\alpha(\mu_\beta) = 0$, where μ_β is the measure introduced in the previous remark.

This remark shows that what is missing in the proof of the large deviations principle in [2] is the derivation of the property that measures with infinite energy in $\mathbb{R}_+$ have infinite cost. Note that for measures $\mu(dr) = m(r) \, dr$ in $\mathcal{M}_{0,\alpha}$ and such that $I_\alpha(\mu) < \infty$, the finiteness of the energy is a property of the measure in a vicinity of $1/2$ because $m(r) = \alpha$ for $r > 1/2$ and the energy of μ on the interval $(0, 1/2)$ is finite by definition of I_α.

Corollary 4.3 asserts that measures with infinite energy in $\mathbb{R}_+$ have infinite cost. Its proof relies on Proposition 3.6, a new result which states that a superexponential two-blocks estimate for the cylinder function $[\eta(0) - \eta(e_j)]^2$ holds on the entire space $\mathbb{Z}^2$, and not only on $\{x \in \mathbb{Z}^2 : |x| < T^{1/2-\delta}\}$. Proposition 3.6 is restricted to the local function $[\eta(0) - \eta(e_j)]^2$ because the concavity of the map $\beta \mapsto E_{\nu_\beta}[\{\eta(0) - \eta(e_j)\}^2]$ is used. We refer to Remark 3.7 for further comments on this result.

Remark 6.2 explains why it is possible to prove an energy estimate in the whole space $\mathbb{Z}^2$, but it is not possible to handle, in the proof of the large deviations upper bound, perturbations defined in the entire space. Actually, in the upper bound, the dynamics is perturbed only in a ball $\{x \in \mathbb{Z}^2 : |x| < T^{1/2-\delta}\}$ because, as anticipated in Remark 2.1, the measure $\bar{\mu}^T$ is super-exponentially close to $\alpha\, dr$ on $(1/2, \infty)$.

The energy estimate for the empirical measure $\bar{\mu}^T$ requires some care because the measure is defined on $\mathbb{R}_+$, a one-dimensional space, and the system evolves the two-dimensional space $\mathbb{Z}^2$. This difficulty is surmounted through formula (4.5) and Lemma 4.1.

Let $\mathbf{1}\{[a, b]\}$ be the indicator function of the interval $[a, b]$. A large deviations principle for the occupation time of the origin follows from Theorem 2.2, the fact that the occupation time of the origin is super-exponentially close to $\varepsilon^{-1} \int_{\mathbb{R}_+} \mathbf{1}\{[0, \varepsilon]\}\, \bar{\mu}^T(dr)$, and a contraction principle. The proof of this result is presented in Sect. 7 of [2]. We recall it here since it is one of the main motivations for Theorem 2.2.

Theorem 2.5 *For every closed subset F of $[0, 1]$ and every open subset G of $[0, 1]$,*

$$\limsup_{T \to \infty} \frac{\log T}{T} \log \mathbb{P}_\alpha\Big[\int_0^1 \eta_s(0)\, ds \in F \Big] \le - \inf_{\beta \in F} \Upsilon_\alpha(\beta),$$

$$\liminf_{T \to \infty} \frac{\log T}{T} \log \mathbb{P}_\alpha\Big[\int_0^1 \eta_s(0)\, ds \in G \Big] \ge - \inf_{\beta \in G} \Upsilon_\alpha(\beta),$$

where $\Upsilon_\alpha : [0, 1] \to \mathbb{R}_+$ is the rate function given by

$$\Upsilon_\alpha(\beta) = \frac{\pi}{2}\Big\{ \sin^{-1}(2\beta - 1) - \sin^{-1}(2\alpha - 1) \Big\}^2.$$

Actually the rate function Υ_α is derived through the variational problem [which comes from the contraction principle]

$$\Upsilon_\alpha(\beta) = \inf_m \frac{\pi}{4} \int_0^{1/2} \frac{m'(r)^2}{\sigma(m(r))}\, dr, \tag{2.6}$$

where the infimum is carried over all smooth functions $m : [0, 1/2] \to \mathbb{R}$ such that $m(0) = \beta, m(1/2) = \alpha$.

The article is organized as follows. In Sect. 3, we state the superexponential estimates and in the following one the energy estimate. In Sect. 5, we present an alternative formula for the large deviations rate functional and derive some of its properties. In Sects. 6 and 7 we prove the upper bound and the lower bounds of the large deviations principle.

3   Superexponential Estimate

We present in this section some superexponential estimates needed in the proof of the large deviations principle. We start with an elementary estimate. Denote by ϕ the approximation of the identity given by $\phi(r) = (1/2)\mathbf{1}\{[-1, 1]\}$, where $\mathbf{1}\{[-1, 1]\}$ represents the indicator of the interval $[-1, 1]$. For $r, \delta > 0$, let $\phi_{r,\delta}$ be the family of approximations induced by ϕ: $\phi_\delta(s) = \delta^{-1}\phi(s/\delta)$, $\phi_{r,\delta}(s) = \phi_\delta(s - r)$.

It will be simpler to work with a continuous family of approximations of the identity. Let ψ_δ be a nonnegative, continuous function, bounded by $1/(2\delta)$, which coincides with ϕ_δ on $[-(\delta - \delta^2), (\delta - \delta^2)]$ and whose support is contained in $[-(\delta + \delta^2), (\delta + \delta^2)]$. Set $\psi_{r,\delta}(s) = \psi_\delta(s - r)$.

Denote by $\mu(f)$ the integral of a continuous and compactly supported function $f : (0, \infty) \to \mathbb{R}$ with respect a the measure $\mu \in \mathcal{M}_c$:

$$\mu(f) = \int_{\mathbb{R}_+} f(r) \, \mu(dr).$$

By construction, for all $b > 0$, there exists a finite constant $C = C(b)$ such that

$$\limsup_{T \to \infty} \sup_{2\delta \leq r \leq b} \sup_\eta \left| \mu^{1,T}(\phi_{r,\delta}) - \mu^{1,T}(\psi_{r,\delta}) \right| \leq C\delta. \tag{3.1}$$

A similar estimate holds if $\mu^{1,T}$ is replaced by $\bar{\mu}^T$. Note that for each measure $\mu \in \mathcal{M}_c$, $\mu(\psi_{r,\delta})$ is a continuous function of the parameter r because ψ_δ is a bounded, continuous function.

The next comparison between a Riemann sum with its integral counterpart will also be used repeatedly.

Lemma 3.1 *Let $H : \mathbb{R}_+ \to \mathbb{R}$ be a Lipschitz-continuous function with compact support in (a, b), $0 < a < b < \infty$. Then, there exists a finite constant C_0 depending only on $\|H\|_\infty = \sup_{r \in \mathbb{R}_+} |H(r)|$ and on the Lipschitz constant of H such that*

$$\left| \frac{1}{\log T} \sum_{x \in \mathbb{Z}_*^2} H(\sigma_T(x)) \frac{1}{|x|^2} - 2\pi \int_{\mathbb{R}_+} H(r) \, dr \right| \leq \frac{C_0}{T^a}.$$

Proof Let $\square_x = [x_1, x_1 + 1) \times [x_2, x_2 + 1)$. The proof consists in comparing

$$H(\sigma_T(x)) \frac{1}{|x|^2} \quad with \quad \int_{\square_x} H(\sigma_T(z)) \frac{1}{|z|^2} \, dz,$$

and then the sum over x of the second term in this formula with the $2\pi \int_{\mathbb{R}_+} H(r)\, dr$. Details are left to the reader. $\qquad\square$

Remark 3.2 Let J be a function in $C^1_K((0, \infty))$. We will apply the previous result to $H(r) = J(r)\, \mu^{1,T}(\psi_{r,\delta})$ in Lemma 3.3 and to $H(r) = J(r)^2\, \sigma(\mu^{1,T}(\psi_{r,\delta}))$ in Corollary 4.2. The proof of Lemma 3.1 relies on the finiteness of $\|H\|_\infty$ and on the Lipschitz property of H. Both conditions are fulfilled by the map $r \to \mu^{1,T}(\psi_{r,\delta})$ on compact intervals of $(0, \infty)$. On the one hand, for all $0 < 2\delta < r$, $0 \le \mu^{1,T}(\psi_{r,\delta}) \le 1$. On the other hand, as ψ_δ is a Lipschitz-continuous function, by definition of $\psi_{r,\delta}$, for each $\delta > 0$, there exists a finite constant $C(\delta)$ such that $|\mu^{1,T}(\psi_{r,\delta}) - \mu^{1,T}(\psi_{s,\delta})| \le C(\delta)|r - s|$ for all r, $s \ge 2\delta$. Thus, Lemma 3.1 holds for these functions with a contant C_0 which also depends on δ.

The next estimate will be used to introduce space averages through the regularity of the test function and a summation by parts. Denote by $\mathbb{A}_{T,\delta}(x) \subset \mathbb{Z}^2$, $\delta > 0$, $|x| > T^{2\delta}$, the annulus

$$\mathbb{A}_{T,\delta}(x) = \left\{ y \in \mathbb{Z}^2 : |x|\, T^{-\delta} \le |y| \le |x|\, T^\delta \right\}.$$

Let $J \in C_K((0, \infty))$ be a Lipschitz continuous function whose support is contained in $[a, b]$. There exists a finite constant $C(J)$, depending only on J, such that for all $0 < \delta \le a/2$,

$$\limsup_{T \to \infty} \sup_{x \in \mathbb{Z}^2} \left| J(\sigma_T(x)) - \frac{1}{4\pi\delta \log T} \sum_{y \in \mathbb{A}_{T,\delta}(x)} \frac{1}{|y|^2}\, J(\sigma_T(y)) \right| \le C(J)\,\delta. \quad (3.2)$$

Note that we may restrict the supremum to the points x such that $|x| \ge T^{a-\delta}$.

Lemma 3.3 *Let $J \in C^1_K((0, \infty))$. There exists a finite constant C_0, depending only on J, such that*

$$\limsup_{T \to \infty} \sup_\eta \left| \int_{\mathbb{R}_+} J(r)\, \mu^{1,T}(dr) - \int_{\mathbb{R}_+} J(r)\, \mu^{1,T}(\psi_{r,\delta})\, dr \right| \le C_0\,\delta,$$

A similar result is in force with $\mu^{1,T}$ replaced by $\bar{\mu}^T$.

Proof This result is a simple consequence of (3.2), a summation by parts, the bound (3.1) and Remark 3.2. $\qquad\square$

We continue with two lemmata whose proofs are similar to the one of Lemma 5.1 in [2].

Lemma 3.4 *For every $\varrho > 0$ and continuous function $H : [1/2, \infty) \to \mathbb{R}$ with compact support,*

$$\limsup_{T \to \infty} \frac{\log T}{T} \log \mathbb{P}_\alpha \left[\left| \int_{\mathbb{R}_+} H(r)\, \bar{\mu}^T(dr) - \alpha \int_{1/2}^\infty H(r)\, dr \right| > \varrho \right] = -\infty.$$

Let A be a finite subset of $\mathbb{Z}^2$, and denote by η_A the local function $\eta_A = \prod_{x \in A} \eta(x)$. For a continuous function $H : (0, \infty) \to \mathbb{R}$ with compact support and $j = 1, 2$, let

$$W_{T,j}^{H,A}(\eta) = \frac{1}{\log T} \sum_{x \in \mathbb{Z}_*^2} \frac{1}{|x|^2} \frac{x_j^2}{|x|^2} H(\sigma_T(x)) \{\eta_{A+x} - \alpha^{|A|}\},$$

where $A + x = \{y + x : y \in A\}$.

Lemma 3.5 *For every finite subset A of $\mathbb{Z}^2$, $\varrho > 0$, $j = 1, 2$, and continuous function $H : [1/2, \infty) \to \mathbb{R}$ with compact support,*

$$\limsup_{T \to \infty} \frac{\log T}{T} \log \mathbb{P}_\alpha \left[\left| \int_0^1 W_{T,j}^{H,A}(\eta_s) \, ds \right| > \varrho \right] = -\infty.$$

Since any local function can be expressed as a linear combination of function of type η_A this result extends to all local functions.

Consider a continuous, non-negative function $J : \mathbb{R}_+ \to \mathbb{R}$ with compact support in $(0, \infty)$. Let $W_T^{J,\delta}$ be the local function defined as

$$W_T^{J,\delta}(\eta) = \frac{1}{\log T} \sum_{j=1}^2 \sum_{x \in \mathbb{Z}} J(\sigma_T(x)) \frac{x_j^2}{|x|^4} \left\{ [\eta(x + e_j) - \eta(x)]^2 - 2\sigma(m_{\delta,T}(x, \eta)) \right\},$$

$$(3.3)$$

where

$$m_{\delta,T}(x, \eta) = \begin{cases} \mu^{1,T}(\psi_{\sigma_T(x),\delta}) & \text{if } \sigma_T(x) < 1/2, \\ \alpha & \text{if } \sigma_T(x) \geq 1/2. \end{cases}$$

Proposition 3.6 *Let $J : \mathbb{R}_+ \to \mathbb{R}$ be a non-negative function of class C^1 with compact support in $(0, \infty)$. For every $\varrho > 0$,*

$$\limsup_{\delta \to 0} \limsup_{T \to \infty} \frac{\log T}{T} \log \mathbb{P}_\alpha \left[\int_0^1 W_T^{J,\delta}(\eta_s) \, ds > \varrho \right] = -\infty.$$

Remark 3.7 The concavity of the mobility σ plays an important role in the proof of this proposition. We are not able to prove the so-called superexponential two-blocks estimate, but only a mesoscopic superexponential estimate. The concavity of σ permits to insert inside σ macroscopic averages through Jensen's inequality. This argument provides an inequality which, fortunately, goes in the right direction.

For the same reasons, we are not able to prove this proposition for the absolute value of the time integral.

The proof of this proposition is divided in several steps. Denote by $\mathbb{R}_{T,\kappa}^{(1)}$, $\mathbb{R}_{T,\kappa}^{(2)}$, $\mathbb{R}_T^{(3)}$, $0 < \kappa < 1/2$, the subsets of $\mathbb{Z}^2$ defined by

$$\mathbb{R}_{T,\kappa}^{(1)} = \{x \in \mathbb{Z}^2 : \sigma_T(x) < 1/2 - \kappa\}, \quad \mathbb{R}_T^{(3)} = \{x \in \mathbb{Z}^2 : \sigma_T(x) > 1/2\},$$
$$\mathbb{R}_{T,\kappa}^{(2)} = \{x \in \mathbb{Z}^2 : (1/2) - \kappa \le \sigma_T(x) \le 1/2\}.$$

A. The region $\mathbb{R}_T^{(3)}$. On the region $\mathbb{R}_T^{(3)}$, $m_{\delta,T}(x, \eta) = \alpha$. Let $W_T^{(3),J}$ be the local function defined as

$$W_T^{(3),J}(\eta) = \frac{1}{\log T} \sum_{j=1}^{2} \sum_{x \in \mathbb{R}_T^{(3)}} J(\sigma_T(x)) \frac{x_j^2}{|x|^4} \left\{ \left[\eta(x + e_j) - \eta(x)\right]^2 - 2\sigma(\alpha) \right\}.$$

By Lemma 3.5, for every $\varrho > 0$,

$$\limsup_{T \to \infty} \frac{\log T}{T} \log \mathbb{P}_\alpha \left[\left| \int_0^1 W_T^{(3),J}(\eta_s) \, ds \right| > \varrho \right] = -\infty. \tag{3.4}$$

B. The region $\mathbb{R}_T^{(2)}$. Let $\kappa = \kappa(J, c)$, $c > 0$, be such that

$$\frac{1}{\log T} \sum_{x \in \mathbb{R}_{T,\kappa}^{(2)}} \frac{1}{|x|^2} J(\sigma_T(x)) \le c \cdot \tag{3.5}$$

Taking $\kappa = \kappa(J, \varrho/3)$ in the definition of the regions $\mathbb{R}_T^{(i)}$, the contribution of the region $\mathbb{R}_T^{(2)}$ to the sum defining $W_{T,j}^{J,\delta}(\eta_s)$ is bounded in absolute value by $\varrho/3$ because the absolute value of the expression inside braces in (3.3) is bounded by 1.

C. The region $\mathbb{R}_T^{(1)}$. Denote by $\mathbb{T}_\pi$ the one-dimensional torus $[-\pi, \pi)$, and by $\Theta(x)$ the angle of $x \in \mathbb{R}^2 \setminus \{0\}$ so that $x = (|x| \cos \Theta(x), |x| \sin \Theta(x))$.

Fix a positive function $q : (0, 1] \to (0, 1]$ decreasing to 0 slower than the identity: $\lim_{\varepsilon \to 0} q(\varepsilon)/\varepsilon = \infty$. For $0 < \varepsilon < r_0$, let

$$\iota_+ = \iota_+(\varepsilon, r_0, T) = \frac{1}{\log T} \log \frac{T^{r_0} + T^\varepsilon}{T^{r_0}},$$

$$\iota_- = \iota_-(\varepsilon, r_0, T) = \frac{1}{\log T} \log \frac{T^{r_0}}{T^{r_0} - T^\varepsilon} \cdot$$

Denote by $M_{T,\varepsilon}^{r,\theta}(\eta)$, $\theta \in \mathbb{T}_\pi$, the weighted average of particles in the polar cube $[r - \iota_-, r + \iota_+] \times [\theta - q(\varepsilon), \theta + q(\varepsilon)]$ for a configuration η:

$$M_{T,\varepsilon}^{r,\theta}(\eta) = \frac{1}{2(\iota_+ + \iota_-)q(\varepsilon) \log T} \sum_{\substack{T^r - T^\varepsilon \le |z| \le T^r + T^\varepsilon \\ \theta - q(\varepsilon) \le \Theta(z) \le \theta + q(\varepsilon)}} \frac{\eta(z)}{|z|^2} \cdot \tag{3.6}$$

Note that this average is performed over a mesoscopic polar square.

Let $W_T^{J,\kappa,\varepsilon}$ be the local function defined as

$$W_T^{J,\kappa,\varepsilon}(\eta) = \frac{1}{\log T} \sum_{x \in \mathbb{R}_{T,\kappa}^{(1)}} \sum_{j=1}^{2} J(\sigma_T(x)) \frac{x_j^2}{|x|^4} \Psi_{T,j,\varepsilon}(x,\eta),$$

where $\Psi_{T,j,\varepsilon}(x,\eta)$ is given by

$$\Psi_{T,j,\varepsilon}(x,\eta) = \left[\eta(x+e_j) - \eta(x)\right]^2 - 2\sigma\left(M_{T,\varepsilon}^{\sigma_T(x),\Theta(x)}\right).$$

Next lemma is the superexponential estimate presented in Lemma 4.1 of [2].

Lemma 3.8 *For any function J satisfying the assumptions of Proposition 3.6, $0 < \kappa < 1/2$, and $\varrho > 0$,*

$$\limsup_{\varepsilon \to 0} \limsup_{T \to \infty} \frac{\log T}{T} \log \mathbb{P}_\alpha\left[\left|\int_0^1 W_T^{J,\kappa,\varepsilon}(\eta_s)\, ds\right| > \varrho\right] = -\infty.$$

To replace the average $M_{T,\varepsilon}^{\sigma_T(x),\Theta(x)}$ over a mesoscopic square by a macroscopic object we use the concavity of σ. Fix a non-negative, Lipschitz-continuous function $J : \mathbb{R}_+ \to \mathbb{R}$ whose support is contained in (a,b) for $0 < a < b < \infty$. We claim that there exists a contant C_0, depending only on J, such that for all $0 < \varepsilon \leq \delta < \kappa/2$, $\kappa \leq \min\{1/2, a\}$,

$$\frac{2}{\log T} \sum_{j=1}^{2} \sum_{x \in \mathbb{R}_{T,\kappa}^{(1)}} J(\sigma_T(x)) \frac{x_j^2}{|x|^4} \sigma\left(M_{T,\varepsilon}^{\sigma_T(x),\Theta(x)}\right)$$

$$\leq \frac{2}{\log T} \sum_{x \in \mathbb{R}_{T,\kappa}^{(1)}} J(\sigma_T(x)) \frac{1}{|x|^2} \sigma\left(m_\delta(\sigma_T(x),\eta)\right) + C_0\delta + o_T(1), \tag{3.7}$$

where $o_T(1) \to 0$ as $T \to \infty$, uniformly over η.

To prove this assertion, on the left-hand side, sum in j to replace $\sum_j x_j^2$ by $|x|^2$. Add and subtract an average of J over the set $\mathbb{A}_{T,\delta}(x)$ introduced just above (3.2). By (3.2), performing a summation by parts, we conclude that for $\delta \leq \kappa/2$, the left-hand side of (3.7) is bounded above by

$$\frac{2}{\log T} \sum_{y \in \mathbb{R}_{T,\kappa-\delta}^{(1)}} \frac{1}{|y|^2} J(\sigma_T(y)) \frac{1}{4\pi\delta \log T} \sum_{x \in \mathbb{A}_{T,\delta}(y)} \frac{1}{|x|^2} \sigma\left(M_{T,\varepsilon}^{\sigma_T(x),\Theta(x)}\right) + C(J)\delta,$$

where the sum over x is also restricted to the set $\mathbb{R}_{T,\kappa}^{(1)}$.

The sum for y such that $T^{(1/2)-\kappa-\delta} \leq |y| \leq T^{(1/2)-\kappa+\delta}$ is bounded by $C(J)\delta$. We may thus remove these terms from the sum by paying this price. For y such that $|y| \leq T^{(1/2)-\kappa-\delta}$ we may remove in the second sum the restriction that $x \in \mathbb{R}_{T,\kappa}^{(1)}$.

After removing this restriction we may insert in the first sum the term y such that $T^{(1/2)-\kappa-\delta} \leq |y| \leq T^{(1/2)-\kappa}$ by paying an extra error bounded by $C(J)\delta$. This shows that the previous sum is less than or equal to

$$\frac{2}{\log T} \sum_{y \in \mathbb{R}_{T,\kappa}^{(1)}} \frac{1}{|y|^2} J(\sigma_T(y)) \frac{1}{4\pi\delta \log T} \sum_{x \in \mathbb{A}_{T,\delta}(y)} \frac{1}{|x|^2} \sigma\left(M_{T,\varepsilon}^{\sigma_T(x),\Theta(x)}\right) + C(J)\delta.$$

Substituting x by y, as σ is concave, the previous expression is bounded above by

$$\frac{2}{\log T} \sum_{x \in \mathbb{R}_{T,\kappa}^{(1)}} \frac{1}{|x|^2} J(\sigma_T(x)) \sigma\left(\frac{1}{4\pi\delta \log T} \sum_{y \in \mathbb{A}_{T,\delta}(x)} \frac{1}{|y|^2} M_{T,\varepsilon}^{\sigma_T(y),\Theta(y)}\right) + C(J)\delta.$$

Recall the definition (3.6) of $M_{T,\delta}^{\sigma_T(y),\Theta(y)}$ and to sum by parts inside σ to bound the previous expression by.

$$\frac{2}{\log T} \sum_{x \in \mathbb{R}_{T,\kappa}^{(1)}} \frac{1}{|x|^2} J(\sigma_T(x)) \sigma\left(\mu^{1,T}\left(\phi_{\sigma_T(x),\delta}\right)\right) + C(J)\delta.$$

To complete the proof of (3.7), it remains to recall (3.1) to replace $\phi_{\sigma_T(x),\delta}$ by $\psi_{\sigma_T(x),\delta}$ inside σ.

We summarize in Lemma 3.9 the estimate obtained in the region $\mathbb{R}_{T,\kappa}^{(1)}$. The statement requires some notation. For a continuous function $J : \mathbb{R}_+ \to \mathbb{R}$ with compact support in $(0, \infty)$, $\delta > 0$, $\kappa > 0$, let

$$W_{T,\kappa}^{J,\delta}(\eta) = \frac{1}{\log T} \sum_{j=1}^{2} \sum_{x \in \mathbb{R}_{T,\kappa}^{(1)}} J(\sigma_T(x)) \frac{x_j^2}{|x|^4} \Psi_{\delta,T}(j, x, \eta), \tag{3.8}$$

where $\Psi_{\delta,T}(j, x, \eta) = \left[\eta(x + e_j) - \eta(x)\right]^2 - 2\sigma\left(m_{\delta,T}(x, \eta)\right),$

and $m_{\delta,T}(x, \eta)$ is defined below (3.3). Next lemma follows from Lemma 3.8 and (3.7) by taking $\delta = \varepsilon$.

Lemma 3.9 *Let $J : \mathbb{R}_+ \to \mathbb{R}$ be a non-negative function of class C^1 with compact support in $(0, \infty)$. For every $\varrho > 0$, $0 < \kappa < 1/2$,*

$$\limsup_{\delta \to 0} \limsup_{T \to \infty} \frac{\log T}{T} \log \mathbb{P}_\alpha\left[\int_0^1 W_{T,\kappa}^{J,\delta}(\eta_s)\, ds > \varrho\right] = -\infty.$$

Proof (Proof of Proposition 3.6) Fix a function J and $\varrho > 0$ and recall the definition of the regions $\mathbb{R}_{T,\kappa}^{(1)}, \mathbb{R}_{T,\kappa}^{(2)}, \mathbb{R}_{T,\kappa}^{(3)}$, introduced just below the statement of the proposition. Let $0 < \kappa = \kappa(J, \varrho/3)$ for (3.5) to hold with $c = \varrho/3$. Fix this κ and decompose the sum over x in (3.3) according to these 3 regions.

By definition of κ, the sum over the region $\mathbb{R}^{(2)}_{T,\kappa}$ is bounded by $\varrho/3$. Assertion (3.4) takes care of the region $\mathbb{R}^{(3)}_{T,\kappa}$ and Lemma 3.9 of the region $\mathbb{R}^{(1)}_{T,\kappa}$. $\qquad\square$

4   Energy Estimate

We prove in this section a microscopic energy estimate. It follows from this result that measures with infinite energy have infinite cost in the large deviations principle. This crucial point in the proof of the large deviations dates back to [7].

The following elementary observation will repeatedly be used in the sequel. For any sequence $M_T \to \infty$, and positive sequences a_T, b_T,

$$\limsup_{T\to\infty} \frac{1}{M_T} \log(a_T + b_T) = \max\left\{ \limsup_{T\to\infty} \frac{1}{M_T} \log a_T \,,\, \limsup_{T\to\infty} \frac{1}{M_T} \log b_T \right\}.$$
(4.1)

The Dirichlet form of a function also plays a role in this section. For a local function f, denote by $\mathcal{E}_T(f) = \mathcal{E}_{T,\alpha}(f)$ the Dirichlet form of f:

$$\mathcal{E}_T(f) \;=\; \langle f,\, (-L_T)f \rangle_{\nu_\alpha},$$

where $\langle f,\, g \rangle_{\nu_\alpha}$ represents the scalar product in $L^2(\nu_\alpha)$. An elementary computation provides an explicit formula for the Dirichlet form:

$$\mathcal{E}_T(f) \;=\; \frac{T}{4} \sum_{j=1}^{2} \sum_{x\in\mathbb{Z}^2} \int \left\{ f(\sigma^{x,x+e_j}\eta) - f(\eta) \right\}^2 \nu_\alpha(d\eta).$$
(4.2)

Lemma 4.1 is the main estimate of this section. For a continuous function $H : \mathbb{R}_+ \to \mathbb{R}$ with compact support in $(0, \infty)$, let $V_T(\eta) = V_T^H(\eta)$ be given by

$$V_T(\eta) \;=\; \sum_{j=1}^{2} \sum_{x\in\mathbb{Z}^2} \frac{x_j}{|x|^2} H\left(\frac{\log|x|}{\log T}\right) \left[\eta(x + e_j) - \eta(x)\right]$$

$$-\; \frac{1}{\log T} \sum_{j=1}^{2} \sum_{x\in\mathbb{Z}^2} \frac{x_j^2}{|x|^4} H\left(\frac{\log|x|}{\log T}\right)^2 \left[\eta(x + e_j) - \eta(x)\right]^2.$$

Most of the time we omit the superscript H of $V_T^H(\eta)$.

Lemma 4.1 Let $H : \mathbb{R}_+ \to \mathbb{R}$ be a continuous function with compact support in $(0, \infty)$. Then, for all $\ell \geq 1$,

$$\limsup_{T\to\infty} \frac{\log T}{T} \log \mathbb{P}_\alpha\left[\int_0^1 V_T^H(\eta_s)\, ds \geq \ell \right] \;\leq\; -\ell.$$

Proof By Chebyshev's exponential inequality, it is enough to prove that

$$\limsup_{T\to\infty} \frac{\log T}{T} \log \mathbb{E}_\alpha\left[\exp\left\{\frac{T}{\log T}\int_0^1 V_T(\eta_s)\,ds\right\}\right] \leq 0. \qquad (4.3)$$

By Feynman-Kac' formula (cf. [4, Sect. A.1.7]), the left hand side is bounded by

$$\limsup_{T\to\infty}\sup_f \left\{\int V_T(\eta)\,f(\eta)\,\nu_\alpha(d\eta) - (\log T)\,D_T(f)\right\}, \qquad (4.4)$$

where the supremum is carried over all densities f and $D_T(f)$ represents the Dirichlet form of $\sqrt{f}$ defined in (4.2): $f \geq 0$, $\int f\,d\nu_\alpha = 1$, $D_T(f) = \mathcal{E}_T(\sqrt{f})$.

Consider the linear (in H) term of $V_T(\eta)$. Performing a change of variables $\eta' = \sigma^{x,x+e_j}$ we obtain that

$$\sum_{j=1}^2 \sum_{x\in\mathbb{Z}^2} \frac{x_j}{|x|^2} H\left(\frac{\log|x|}{\log T}\right) \int [\eta(x+e_j) - \eta(x)]\,f(\eta)\,\nu_\alpha(d\eta)$$

$$= -\frac{1}{2}\sum_{j=1}^2 \sum_{x\in\mathbb{Z}^2} \frac{x_j}{|x|^2} H\left(\frac{\log|x|}{\log T}\right) \int [\eta(x+e_j) - \eta(x)]\left[f(\sigma^{x,x+e_j}\eta) - f(\eta)\right]\nu_\alpha(d\eta)$$

Write the difference $f(\sigma^{x,x+e_j}\eta) - f(\eta)$ as $(b-a) = (\sqrt{b} - \sqrt{a})(\sqrt{b} + \sqrt{a})$ and apply Young's inequality to bound the previous expression by

$$\frac{1}{4\log T}\sum_{j=1}^2 \sum_{x\in\mathbb{Z}^2} \frac{x_j^2}{|x|^4} H\left(\frac{\log|x|}{\log T}\right)^2 \int [\eta(x+e_j) - \eta(x)]^2 \left[\sqrt{f(\sigma^{x,x+e_j}\eta)} + \sqrt{f(\eta)}\right]^2 \nu_\alpha(d\eta)$$

$$+ \frac{\log T}{4}\sum_{j=1}^2 \sum_{x\in\mathbb{Z}^2} \int \left[\sqrt{f(\sigma^{x,x+e_j}\eta)} - \sqrt{f(\eta)}\right]^2 \nu_\alpha(d\eta),$$

By (4.2), the second line is $(\log T)\,D_T(f)$ and cancels with the second term in (4.4). On the other hand, since $(\sqrt{b} + \sqrt{a})^2 \leq 2(a + b)$, a change of variables $\eta' = \sigma^{x,x+e_j}\eta$ yields that the first line is equal to

$$\frac{1}{\log T}\sum_{j=1}^2 \sum_{x\in\mathbb{Z}^2} \frac{x_j^2}{|x|^4} H\left(\frac{\log|x|}{\log T}\right)^2 \int [\eta(x+e_j) - \eta(x)]^2\,f(\eta)\,\nu_\alpha(d\eta),$$

which is exactly the quadratic (in H) term in $V_T(\eta)$. This proves (4.3), and therefore the lemma.

For the proof of the large deviations principle, we need to restate Lemma 4.1 in terms of the polar measure $\bar{\mu}^T$. For the piece which is linear in H this is just a summation by parts. For the one which is quadratic in H, it relies on the superexponential estimates presented in the previous section.

Fix a smooth function $H : \mathbb{R}_+ \to \mathbb{R}$ with compact support in $[a, b]$, where $0 < a < b$. We claim that

$$- 2\pi \int H'(r), \mu^{1,T}(dr) = \sum_{j=1}^{2} \sum_{x \in \mathbb{Z}^2} \frac{x_j}{|x|^2} H\left(\frac{\log |x|}{\log T}\right) [\eta(x + e_j) - \eta(x)] + R_T,$$

(4.5)

where the absolute value of R_T is bounded by $C_0 T^{-a}$ for some finite constant C_0 which depends only on H. This result follows from a summation by parts on the right-hand side. The derivative of H provides the term on the left-hand side. The divergence of $(x_1/|x|^2, x_2/|x|^2)$ vanishes because $\log |x|$ is harmonic and $x_j/|x|^2 = \partial_{x_j} \log |x|$.

For a continuous function $H : \mathbb{R}_+ \to \mathbb{R}$ with compact support in $(0, \infty)$, let $V_{T,\delta}(\eta) = V_{T,\delta}^H(\eta)$ be given by

$$V_{T,\delta}(\eta) = - 2\pi \int_{\mathbb{R}_+} H'(r) \mu^{1,T}(\psi_{r,\delta}) dr - 4\pi \int_{\mathbb{R}_+} H(r)^2 \sigma\big(m_{\delta,T}(r)\big) dr,$$

where

$$m_{\delta,T}(r) = \begin{cases} \mu^{1,T}(\psi_{r,\delta}) & \text{if } r < 1/2, \\ \alpha & \text{if } r \geq 1/2. \end{cases}$$

Corollary 4.2 *Let $H : \mathbb{R}_+ \to \mathbb{R}$ be a function in $C_K^1((0, \infty))$. Then, for all $\ell \geq 1$,*

$$\limsup_{\delta \to 0} \limsup_{T \to \infty} \frac{\log T}{T} \log \mathbb{P}_\alpha\left[\int_0^1 V_{T,\delta}^H(\eta_s) ds \geq \ell \right] \leq -(\ell - 1).$$

Proof By Lemma 4.1 and (4.1), it is enough to show that

$$\limsup_{\delta \to 0} \limsup_{T \to \infty} \frac{\log T}{T} \log \mathbb{P}_\alpha\left[\int_0^1 \left\{ V_{T,\delta}^H(\eta_s) - V_T^H(\eta_s) \right\} ds \geq 1 \right] \leq -\infty.$$

(4.6)

The sums $V_{T,\delta}^H$, V_T^H are expressed as a difference between a linear term in H and a quadratic term in H. We compare separately the linear and the quadratic terms. By (4.5) and Lemma 3.3, the absolute value of the difference between the linear terms is uniformly bounded by $1/2$ for δ small enough.

We turn to the quadratic terms. Apply Remark 3.2 to replace the integral $\int H(r)^2 \sigma\big(m_{\delta,T}(r)\big) dr$ by a Riemannian sum. After this step, the difference of the quadratic terms is seen to be equal to $W_T^{H^2,\delta}(\eta)$ introduced in (3.3). Assertion (4.6) for the quadratic piece follows therefore from Proposition 3.6.

The previous result rephrases Lemma 4.1 in terms of the polar measure $\mu^{1,T}$. We go one step further integrating in time to express the estimate in terms of $\bar{\mu}^T$. For a continuous function $H : \mathbb{R}_+ \to \mathbb{R}$ with compact support in $(0, \infty)$, let $W_{T,\delta}^H$ be given by

$$W_{T,\delta}^H = 2\pi \left\{ - \int_{\mathbb{R}_+} H'(r)\,\bar{\mu}^T(\psi_{r,\delta})\,dr - 2 \int_{\mathbb{R}_+} H(r)^2\,\sigma\big(\overline{m}_{\delta,T}(r)\big)\,dr \right\},$$

where

$$\overline{m}_{\delta,T}(r) = \begin{cases} \bar{\mu}^T(\psi_{r,\delta}) & \text{if } r < 1/2, \\ \alpha & \text{if } r \geq 1/2. \end{cases}$$

The next result follows from the previous corollary and from the concavity of σ.

Corollary 4.3 *Let* $H : \mathbb{R}_+ \to \mathbb{R}$ *be a function in* $C_K^1((0,\infty))$. *Then, for all* $\ell \geq 1$,

$$\limsup_{\delta \to 0} \limsup_{T \to \infty} \frac{\log T}{T} \log \mathbb{P}_\alpha\Big[W_{T,\delta}^H \geq \ell \Big] \leq -(\ell - 1).$$

One recognizes in $W_{T,\delta}^H$ the germ of an energy functional. For a function G in $C_K^1((0,\infty))$, let $\mathcal{Q}_{\alpha,G} : \mathcal{M}_0 \to \mathbb{R}$ be given by

$$\mathcal{Q}_{\alpha,G}(m(r)\,dr) = - \int_0^\infty G'(r)\,m(r)\,dr - 2 \int_0^\infty \sigma(m_\alpha(r))\,G(r)^2\,dr, \quad (4.7)$$

where

$$m_\alpha(r) = \begin{cases} m(r) & \text{if } r < 1/2, \\ \alpha & \text{if } r \geq 1/2. \end{cases} \quad (4.8)$$

With this notation, Corollary 4.3 can be restated as follows. Let $H : \mathbb{R}_+ \to \mathbb{R}$ be a function in $C_K^1((0,\infty))$. Then, for all $\ell \geq 1$,

$$\limsup_{\delta \to 0} \limsup_{T \to \infty} \frac{\log T}{T} \log \mathbb{P}_\alpha\Big[2\pi\,\mathcal{Q}_{\alpha,H}(\bar{\mu}_\delta^T) \geq \ell \Big] \leq -(\ell - 1). \quad (4.9)$$

where, for a measure $\mu \in \mathcal{M}_c$, $\mu_\delta \in \mathcal{M}_0$ represents the absolutely continuous measure whose density m_δ is given by $\mu(\psi_{r,\delta})$: for all functions $G \in C_K((0,\infty))$,

$$\int_0^\infty G(r)\,\mu_\delta(dr) = \int_0^\infty G(r)\,\mu(\psi_{r,\delta})\,dr. \quad (4.10)$$

5 Energy and Rate Function I_α

We present in this section some properties of the large deviations rate functional.
    Fix $0 < \alpha < 1$. Denote by $\mathcal{Q}_\alpha : \mathcal{M}_0 \to \mathbb{R}_+$ the energy functional given by

$$\mathcal{Q}_\alpha(\mu) = \sup_{G \in C_K^1((0,\infty))} \mathcal{Q}_{\alpha,G}(\mu), \quad (5.1)$$

where $\mathcal{Q}_{\alpha,G}$ is defined by (4.7). Next result is Lemma 4.1 in [1].

Lemma 5.1 *The functional $\mathcal{Q}_\alpha : \mathcal{M}_0 \to \mathbb{R}_+$ is convex and lower-semicontinuous. Moreover, if $\mathcal{Q}_\alpha(m(r)dr) < \infty$, then $m(r)$ has a generalized derivative, denoted by $m'(r)$, and*

$$\mathcal{Q}_\alpha(m(r)dr) = \frac{1}{8} \int_0^\infty \frac{m'(r)^2}{\sigma(m_\alpha(r))} dr. \tag{5.2}$$

Let $C^2(\mathbb{R}_+, \alpha)$ be the space of twice continuously differentiable functions $\gamma : [0, \infty) \to (0, 1)$ such that γ' has a compact support in $(0, 1/2)$ and such that $\gamma(r) = \alpha$ for r sufficiently large. There exists therefore $0 < \beta < 1$ and $0 < \varepsilon < 1/4$ such that $\gamma(r) = \beta$ for $r \le \varepsilon$, and $\gamma(r) = \alpha$ for $r \ge 1/2 - \varepsilon$, $\varepsilon \le \gamma(s) \le 1 - \varepsilon$ for all $s \ge 0$. For each γ in $C^2(\mathbb{R}_+, \alpha)$, let $\Gamma = \Gamma_{\gamma,\alpha} : \mathbb{R}_+ \to \mathbb{R}$ be given by

$$\Gamma(u) = \frac{1}{2} \left\{ \log \frac{\gamma(u)}{1 - \gamma(u)} - \log \frac{\alpha}{1 - \alpha} \right\}. \tag{5.3}$$

Note that the space $\{\Gamma'_{\gamma,\alpha}, \gamma \in C^2(\mathbb{R}_+, \alpha)\}$ corresponds to the space $C_K^1((0, 1/2))$. Fix $\gamma \in C^2(\mathbb{R}_+, \alpha)$, and let $J_\gamma : \mathcal{M}_0 \to \mathbb{R}$ be the rate-functional given by

$$J_\gamma(m(r)\,dr) = -\pi \int_0^\infty \Gamma''(r)\, m(r)\, dr - \pi \int_0^\infty \sigma(m(r))\, \Gamma'(r)^2\, dr. \tag{5.4}$$

Recall from (2.4) that we denote by $\mathcal{M}_{0,\alpha}$ the space of absolutely continous measures whose density is equal to α on $(1/2, \infty)$. Denote by $\mathcal{M}_{0,\alpha}^{\mathcal{Q}}$ the set of measures in $\mathcal{M}_{0,\alpha}$ with finite energy:

$$\mathcal{M}_{0,\alpha}^{\mathcal{Q}} = \{\mu(dr) = m(r)\, dr \in \mathcal{M}_{0,\alpha} : \mathcal{Q}_\alpha(\mu) < \infty\}. \tag{5.5}$$

Let $J_\gamma^{\mathcal{Q}} : \mathcal{M}_c \to \mathbb{R} \cup \{+\infty\}$ the functional given by

$$J_\gamma^{\mathcal{Q}}(\mu) = \begin{cases} J_\gamma(\mu) & \text{if } \mu \in \mathcal{M}_{0,\alpha}^{\mathcal{Q}}, \\ +\infty & \text{otherwise,} \end{cases} \tag{5.6}$$

Note that $J_\gamma^{\mathcal{Q}}$ is defined on $\mathcal{M}_c$, while J_γ is only defined on $\mathcal{M}_0$.

Let $J^{\mathcal{Q}} : \mathcal{M}_c \to \mathbb{R}_+$ be given by

$$J^{\mathcal{Q}}(\mu) = \sup_{\gamma \in C^2(\mathbb{R}_+, \alpha)} J_\gamma^{\mathcal{Q}}(\mu). \tag{5.7}$$

Since the set $\{\Gamma'_{\gamma,\alpha} : \gamma \in C^2(\mathbb{R}_+, \alpha)\}$ coincides with the set $C_K^1((0, 1/2))$, on the set $\mathcal{M}_{0,\alpha}^{\mathcal{Q}}$ the functional $J^{\mathcal{Q}}$ can be rewritten as

$$J^{\mathcal{Q}}(m(r)dr) = \pi \sup_{H \in C_K^1((0,1/2))} \left\{ -\int_0^\infty H'(r)\, m(r)\, dr - \int_0^\infty \sigma(m(r))\, H(r)^2\, dr \right\}.$$

The proof of Lemma 5.1 yields that if $J^{\mathcal{Q}}(m(r)dr) < \infty$, then m has a generalized derivative in $(0, 1/2)$, denoted by m', and

$$J^{\mathcal{Q}}(m(r)dr) = \frac{\pi}{4} \int_0^{1/2} \frac{[m'(r)]^2}{\sigma(m(r))} dr. \tag{5.8}$$

The next results asserts that in the definition of the rate function $J^{\mathcal{Q}}$, we can replace the set $C_K^1((0, 1/2))$ by the larger one $C_K^1((0, \infty))$.

Lemma 5.2 *For $\mu \in \mathcal{M}_{0,\alpha}^{\mathcal{Q}}$,*

$$J^{\mathcal{Q}}(m(r)dr) = \pi \sup_{H \in C_K^1((0,\infty))} \left\{ -\int_0^\infty H'(r) m(r) dr - \int_0^\infty \sigma(m(r)) H(r)^2 dr \right\}. \tag{5.9}$$

In particular, $J^{\mathcal{Q}} = I_{\mathcal{Q},\alpha}$ on $\mathcal{M}_c$.

Proof Denote by $Q(m(r)dr)$ the right hand side of (5.9). It is clear that $J^{\mathcal{Q}}(\mu) \leq Q(\mu)$ for all $\mu \in \mathcal{M}_0$. We prove the reverse inequality for measures in $\mathcal{M}_{0,\alpha}^{\mathcal{Q}}$.

Fix $\mu(dr) = m(r)dr \in \mathcal{M}_{0,\alpha}^{\mathcal{Q}}$. We claim that $Q(\mu) < \infty$. Indeed, recall the definition of $\mathcal{Q}_\alpha$ introduced in (5.1). In formula (4.7), take $G/2$ in place of G, to obtain that

$$-\int_0^\infty G'(r) m(r) dr - \int_0^\infty \sigma(m_\alpha(r)) G(r)^2 dr \leq 2 \mathcal{Q}_{\alpha, G/2}(\mu) \leq 2 \mathcal{Q}_\alpha(\mu)$$

for all $G \in C_K^1((0, \infty))$. Since $m(r) = \alpha$ for $r \geq 1/2$, in the variational formula which defines Q, we may replace $\sigma(m(r))$ by $\sigma(m_\alpha(r))$. After this replacement, optimizing over G yields that $Q(\mu) \leq 2 \mathcal{Q}_\alpha(\mu)$. As μ belongs to $\mathcal{M}_{0,\alpha}^{\mathcal{Q}}$, $\mathcal{Q}_\alpha(\mu) < \infty$ so that $Q(\mu) < \infty$, as claimed.

By Lemma 5.1, since $Q(\mu) < \infty$, m has a generalized derivative, denoted by m', and

$$Q(\mu) = = \frac{\pi}{4} \int_0^\infty \frac{[m'(r)]^2}{\sigma(m(r))} dr.$$

Since $m(r) = \alpha$ for $r \geq 1/2$, $m'(r) = 0$ a.s. on $[1/2, \infty)$, and the range of the previous integral can be reduced to $[0, 1/2]$, which proves that $Q(\mu) = J^{\mathcal{Q}}(\mu)$ in view of (5.8).

To prove the second assertion of the lemma, observe first that both functionals coincide on the set $\mathcal{M}_{0,\alpha}^{\mathcal{Q}}$. Indeed, the right hand side of (5.9) is just $\pi \mathcal{Q}$, where $\mathcal{Q}$ has been introduced in (2.3), and $I_{\mathcal{Q},\alpha}$ is equal to $\pi \mathcal{Q}$ on $\mathcal{M}_{0,\alpha}^{\mathcal{Q}}$. It remains to show that $I_{\mathcal{Q},\alpha} = J^{\mathcal{Q}} = \infty$ on $[\mathcal{M}_{0,\alpha}^{\mathcal{Q}}]^c$. For $J^{\mathcal{Q}}$ this follows by definition. For $I_{\mathcal{Q},\alpha}$ the identity holds on $[\mathcal{M}_{0,\alpha}]^c$ by definition. On the set $\mathcal{M}_{0,\alpha} \setminus \mathcal{M}_{0,\alpha}^{\mathcal{Q}}$, $I_{\mathcal{Q},\alpha}(\mu) = \pi \mathcal{Q}(\mu) = \infty$.   □

6 The Upper Bound

The proof of the upper bound is similar to the one presented in [2], but relies on the energy estimate proved in the previous section to restrict the set of measures to the ones with finite energy on $\mathbb{R}_+$.

We follow [1] with a minor improvement. Instead of considering $\bar{\mu}^T$ as a density function on $\mathbb{R}_+$ we defined here $\bar{\mu}^T$ as a measure with mass points. This is more natural, but creates an extra minor difficulty, as we have to show that at the level of the large deviations, we may exclude measures which are not absolutely continuous.

The proof of the large deviations principle is based on the following perturbations of the dynamics. Fix $\gamma \in C^2(\mathbb{R}_+, \alpha)$ and recall from (5.3) the definition of the function $\Gamma : \mathbb{R}_+ \to \mathbb{R}$ introduced in. Let $\Gamma_T(x)$, $T > 0$, $x \in \mathbb{Z}^2$, be given by

$$\Gamma_T(x) = \Gamma(\sigma_T(x)).$$

Denote by $L_{T,\gamma}$ the generator of the inhomogeneous exclusion process in which a particle jumps from x to y at rate $\exp\{\Gamma_T(y) - \Gamma_T(x)\}$:

$$(L_{T,\gamma} f)(\eta) = \frac{T}{2} \sum_{x \in \mathbb{Z}^2} \sum_{y:|x-y|=1} \eta(x)\{1 - \eta(y)\} e^{\Gamma_T(y) - \Gamma_T(x)} [f(\sigma^{x,y}\eta) - f(\eta)].$$

Denote by $\nu_{T,\gamma}$ the product measure on $\{0, 1\}^{\mathbb{Z}^2}$, with marginals given by

$$\nu_{T,\gamma}\{\eta, \eta(x) = 1\} = \gamma(\sigma_T(x)). \tag{6.1}$$

The measure $\nu_{T,\gamma}$ coincides with ν_α outside a ball of radius $T^{1/2-\varepsilon}$ centered at the origin, for some $\varepsilon > 0$. Moreover, a simple computation shows that $\nu_{T,\gamma}$ is an invariant reversible measure for the Markov process with generator $L_{T,\gamma}$. Denote by $\mathbb{P}_{T,\gamma}$ the probability measure on $D(\mathbb{R}_+, \{0, 1\}^{\mathbb{Z}^2})$ induced by Markov process whose generator is $L_{T,\gamma}$ and which starts from $\nu_{T,\gamma}$.

This section is organized as follows. We first define four subsets of measures whose complements have superexponentially small probabilities. Then, we show that on these sets a family of martingales can be expressed in terms of the polar measure $\bar{\mu}^T$. These explicit formulae and a min-max argument due to Varadhan permit to conclude the proof of the upper bound.

A. Polar measure at $[1/2, \infty)$. Let $\{G_m : m \geq 1\}$ be a sequence of functions in $C_K((1/2, \infty))$ which is dense with respect to the supremum norm. For $\kappa > 0$ and $n \geq 1$, denote by $A_{n,\kappa}$ the closed subspace of $\mathcal{M}_c$ defined by

$$A_{n,\kappa} = \left\{\mu \in \mathcal{M}_c : \left|\int G_m(r)\,\mu(dr) - \alpha \int G_m(r)\,dr\right| \leq \kappa \text{ for } 1 \leq m \leq n\right\}. \tag{6.2}$$

By Lemma 3.4 and (4.1), for every $\kappa > 0$ and $n \geq 1$,

$$\limsup_{T \to \infty} \frac{\log T}{T} \log \mathbb{P}_\alpha \big[\bar{\mu}^T \notin A_{n,\kappa} \big] = -\infty. \tag{6.3}$$

B. Energy Functionals. Recall the definition of the functionals $\mathcal{Q}_{\alpha,H}$ defined by (4.7). Fix a sequence $\{H_p : p \geq 1\}$ of smooth functions, $H_p \in C_K^2(\mathbb{R}_+)$, dense in $C_K^1(\mathbb{R}_+)$. For $q \geq 1, \ell > 0$, let $B_{q,\ell}$ be the set of paths with truncated energy bounded by ℓ:

$$B_{q,\ell} = \big\{ \mu \in \mathcal{M}_0 : 2\pi \max_{1 \leq p \leq q} \mathcal{Q}_{\alpha,H_p}(\mu) \leq \ell \big\}. \tag{6.4}$$

By (4.1) and (4.9), for any $q \geq 1$ and $\ell > 0$

$$\limsup_{\delta \to 0} \limsup_{T \to \infty} \frac{\log T}{T} \log \mathbb{P}_\alpha \big[\bar{\mu}_\delta^T \notin B_{q,\ell} \big] \leq -(\ell - 1). \tag{6.5}$$

C. Absolutely Continuous Measures. Let $\{F_i : i \geq 1\}$ be a sequence of nonnegative functions in $C_K((0, 1/2))$ which is dense with respect to the supremum norm in the space of nonnegative functions in $C_K((0, 1/2))$. For $\kappa > 0$ and $m \geq 1$, denote by $C_{m,\kappa}$ the closed subspace of $\mathcal{M}_c$ defined by

$$C_{m,\kappa} = \Big\{ \mu \in \mathcal{M}_c : \int F_i(r) \, \mu(dr) \leq \int F_i(r) \, dr + \kappa \text{ for } 1 \leq i \leq m \Big\}. \tag{6.6}$$

By (2.2), for every $\kappa > 0$ and $m \geq 1$,

$$\limsup_{T \to \infty} \frac{\log T}{T} \log \mathbb{P}_\alpha \big[\bar{\mu}^T \notin C_{m,\kappa} \big] = -\infty. \tag{6.7}$$

D. Ergodic Bounds. Fix $\gamma \in C^2(\mathbb{R}_+, \alpha)$, $\delta > 0$ and $\varrho > 0$. Recall from (3.8) the definition of the local function $\Psi_{\delta,T}(j, x, \eta)$. Let $B_{T,\gamma}^{\delta,\varrho}$ be the set defined by

$$B_{T,\gamma}^{\delta,\varrho} = \Big\{ \eta : \int_0^1 W_{T,\gamma}^\delta(\eta_s) \, ds \leq \varrho \Big\}, \tag{6.8}$$

where

$$W_{T,\gamma}^\delta(\eta) = \frac{1}{2 \log T} \sum_{x \in \mathbb{Z}_*^2} \sum_{j=1}^2 \frac{x_j^2}{|x|^4} \Gamma'(\sigma_T(x))^2 \, \Psi_{\delta,T}(j, x, \eta).$$

As the support of Γ' is contained in $(0, 1/2)$, by Lemma 3.9, for every $\gamma \in C^2(\mathbb{R}_+, \alpha)$, $\varrho > 0$

$$\limsup_{\delta \to 0} \limsup_{T \to \infty} \frac{\log T}{T} \log \mathbb{P}_\alpha \big[(B_{T,\gamma}^{\delta,\varrho})^c \big] = -\infty. \tag{6.9}$$

E. Radon-Nikodym Derivatives. Fix $\gamma \in C^2(\mathbb{R}_+, \alpha)$. Recall from the paragraph below (6.1) the definition of the measure $\mathbb{P}_{T,\gamma}$, and denote by $d\mathbb{P}_\alpha/d\mathbb{P}_{T,\gamma}$ the Radon-Nikodym derivative of the measure $\mathbb{P}_\alpha$ with respect to the measure $\mathbb{P}_{T,\gamma}$ restricted to the σ-algebra generated by η_s, $0 \le s \le 1$.

The Radon-Nikodym derivative $d\mathbb{P}_{T,\gamma}/d\mathbb{P}_\alpha$ can be written as the product of three exponentials:

$$\frac{d\mathbb{P}_{T,\gamma}}{d\mathbb{P}_\alpha} = \Psi_{\text{stat}} \, \Psi_{\text{pot}} \, \Psi_{\text{dyn}}. \tag{6.10}$$

The first exponential corresponds to the Radon-Nikodym derivative of the initial states: $d\nu_{T,\gamma}/d\nu_\alpha$:

$$\Psi_{\text{stat}} = \exp \sum_{x \in \mathbb{Z}_*^2} \left\{ \eta_0(x) \log \left(\frac{\gamma(\sigma_T(x))}{\alpha} \right) + [1 - \eta_0(x)] \log \left(\frac{1 - \gamma(\sigma_T(x))}{1 - \alpha} \right) \right\}.$$

The second one is associated to the potential $V(\eta) = \sum_{x \in \mathbb{Z}_*^2} \Gamma_T(x) \eta(x)$:

$$\Psi_{\text{pot}} = \exp \left\{ \sum_{x \in \mathbb{Z}_*^2} \Gamma_T(x) \{\eta_1(x) - \eta_0(x)\} \right\}.$$

The last one is the exponential corrector which turns $e^{V(\eta_s)}$ a martingale: $\Psi_{\text{dyn}} = \exp\{-\int_0^1 e^{-V(\eta_s)} L_T e^{V(\eta_s)} \, ds\}$, so that

$$\Psi_{\text{dyn}} = \exp \left\{ -\frac{T}{2} \int_0^t \sum_{x \in \mathbb{Z}^2} \sum_{y:|y-x|=1} \eta_s(x) [1 - \eta_s(y)] \{e^{\Gamma_T(y) - \Gamma_T(x)} - 1\} \, ds \right\}.$$

where $\Gamma_T(z)$ has been defined above (6.1).

Assume that the support of γ' is contained in $[a, b] \subset (0, 1/2)$. In this case, $|\log \Psi_{\text{stat}}|$ and $|\log \Psi_{\text{pot}}|$ are bounded by $C_0 T^{2b} \ll T$. On the other hand, by a Taylor's expansion and the harmonicity of $\log |x|$ in $\mathbb{R}^2$,

$$\frac{\log T}{T} \log \Psi_{\text{dyn}} = -\pi \int \Gamma''(r) \bar{\mu}^T (dr) - \int_0^1 W_\gamma(\eta_s) \, ds + o_T(1), \tag{6.11}$$

where

$$W_\gamma(\eta) = \frac{1}{4 \log T} \sum_{j=1}^2 \sum_{x \in \mathbb{Z}} [\Gamma'(\sigma_T(x))]^2 \frac{x_j^2}{|x|^4} [\eta(x + e_j) - \eta(x)]^2, \tag{6.12}$$

and $\lim_T o_T(1) = 0$.

It follows from the previous estimates that there exists a finite constant C_0 depending only on γ such that

$$\left| \log \frac{d\mathbb{P}_{T,\gamma}}{d\mathbb{P}_\alpha} \right| \le C_0 \frac{T}{\log T} . \tag{6.13}$$

Recall from (5.4) the definition of the functional J_γ, and from (4.10) the definition of the measure μ_δ. Next result follows from the estimates of Ψ_{stat}, Ψ_{pot}, (6.11) and (3.1) to replace $\phi_{r,\delta}$ by $\psi_{r,\delta}$.

Lemma 6.1 *Fix* $\gamma \in C^2(\mathbb{R}_+, \alpha)$, $0 < \delta \le \varrho$. *There exists a finite constant* C_0, *depending only on* γ, *such that on the set* $B_{T,\gamma}^{\delta,\varrho}$ *introduced in* (6.8),

$$\log \frac{d\mathbb{P}_\alpha}{d\mathbb{P}_{T,\gamma}} \le -\frac{T}{\log T} J_\gamma(\bar{\mu}_\delta^T) + C_0 \varrho \frac{T}{\log T} .$$

Remark 6.2 In the proof of the large deviations upper bound, the pieces Ψ_{stat} and Ψ_{pot} of the Radon-Nikodym derivative $d\mathbb{P}_{T,\gamma}/d\mathbb{P}_\alpha$ are the ones which forbid perturbations γ which are not constant outside a compact subset of $(0, 1/2)$. Indeed, if the support of γ' has a nonempty intersection with $(1/2, \infty)$ Ψ_{stat} and Ψ_{pot} are of an order much larger than $\exp\{T/\log T\}$ because of the volume of the region $\{x \in \mathbb{Z}^2 : T^a \le |x| \le T^b\}$ for $a \ge 1/2$.

This is not the case of Ψ_{dyn} due to the presence of the factor $|x|^{-2}$. Indeed, as shown in the proof of Proposition 3.6, to estimate Ψ_{dyn} in the case of a perturbation γ which is not constant outside a compact subset of $(0, 1/2)$, we may divide $\mathbb{Z}^2$ in three regions $\mathbb{R}_{T,\kappa}^{(1)}$, $\mathbb{R}_{T,\kappa}^{(2)}$ and $\mathbb{R}_{T,\kappa}^{(3)}$. All terms in the first region belong to the set $\{x \in \mathbb{Z}^2 : |x| \le T^{1/2-\kappa}\}$ and can be handled as in [2]. The sum over $\mathbb{R}_{T,\kappa}^{(2)}$ is negligible if κ is small (cf. Eq. (3.5)), while the sum over $\mathbb{R}_{T,\kappa}^{(3)}$ is fixed, as proved in Lemma 3.5.

This explains why we are able to prove an energy estimate on $\mathbb{R}_+$ and not just on $(0, 1/2)$: the expression which appears in the proof of the energy estimate stated in Lemma 4.1 is similar to Ψ_{dyn} and there are no terms corresponding to Ψ_{stat} and Ψ_{pot}.

F. Proof of the Upper Bound. We are now in a position to prove the upper bound. Fix $\gamma \in C^2(\mathbb{R}_+, \alpha)$, and let Γ be the function associated to γ by (5.3). Fix $\varrho > 0$, $\delta > 0$, $\varepsilon > 0$, $\kappa_1 > 0$, $\kappa_2 > 0$, $q \ge 1$, $n \ge 1$, $m \ge 1$, $\ell \ge 1$, and recall the definition of the sets $A_{n,\kappa}$, $B_{q,\ell}$, $C_{m,\kappa}$, $B_{T,\gamma}^{\delta,\varrho}$ introduced in (6.2), (6.4), (6.6) and (6.8). Let $B_{q,\ell}^{\varepsilon,T} = \{\bar{\mu}_\varepsilon^T \in B_{q,\ell}\}$. It follows from (4.1), (6.3), (6.5), (6.7) and (6.9) that for any subset A of $\mathcal{M}_c$,

$$\limsup_{T \to \infty} \frac{\log T}{T} \log \mathbb{P}_\alpha[\bar{\mu}^T \in A]$$

$$\le \max \left\{ \limsup_{T \to \infty} \frac{\log T}{T} \log \mathbb{P}_\alpha\left[\bar{\mu}^T \in A \cap A_{n,\kappa_1} \cap C_{m,\kappa_2}, B_{T,\gamma}^{\delta,\varrho} \cap B_{q,\ell}^{\varepsilon,T}\right], R_\gamma \right\},$$

where $R_\gamma = \max\{C_\gamma(\delta, \varrho), C(\varepsilon, q, \ell)\}$ and

$$\limsup_{\delta \to 0} C_\gamma(\delta, \varrho) = -\infty \quad \text{and} \quad \limsup_{\varepsilon \to 0} C(\varepsilon, q, \ell) \le -(\ell - 1) \tag{6.14}$$

for all $\varrho > 0$, $q \geq 1$, $\ell \geq 1$, $\gamma \in C^2(\mathbb{R}_+, \alpha)$.

To estimate the right hand side of the penultimate formula, observe that

$$\mathbb{P}_\alpha[\bar{\mu}^T \in D, \, B_{T,\gamma}^{\delta,\varrho} \cap B_{q,\ell}^{\varepsilon,T}] \;=\; \mathbb{E}_{T,\gamma}\Big[\frac{d\mathbb{P}_\alpha}{d\mathbb{P}_{T,\gamma}} \mathbf{1}\{\bar{\mu}^T \in D, \, B_{T,\gamma}^{\delta,\varrho} \cap B_{q,\ell}^{\varepsilon,T}\}\Big],$$

where $\mathbb{E}_{T,\gamma}$ represents the expectation with respect to $\mathbb{P}_{T,\gamma}$, and $D = A \cap A_{n,\kappa_1} \cap C_{m,\kappa_2}$. By Lemma 6.1, on the set $B_{T,\gamma}^{\delta,\varrho}$, if $\delta \leq \varrho$,

$$\log \frac{d\mathbb{P}_\alpha}{d\mathbb{P}_{T,\gamma}} \;\leq\; -\frac{T}{\log T} J_{\gamma,\delta}(\bar{\mu}^T) + C(\gamma)\varrho \frac{T}{\log T} \cdot$$

where $J_{\gamma,\delta} : \mathcal{M}_c \to \mathbb{R}$ is the functional given by

$$J_{\gamma,\delta}(\mu) \;=\; J_\gamma(\mu_\delta).$$

On the set $\{\bar{\mu}^T \in A_{n,\kappa_1} \cap C_{m,\kappa_2}\} \cap B_{q,\ell}^{\varepsilon,T}$, we may replace the functional $J_{\gamma,\delta}(\bar{\mu}^T)$ by $J_{\gamma,\delta}^{\varepsilon,q,\ell,n,\kappa_1,m,\kappa_2}(\bar{\mu}^T)$, where

$$J_{\gamma,\delta}^{\varepsilon,q,\ell,n,\kappa_1,m,\kappa_2}(\mu) \;=\; \begin{cases} J_{\gamma,\delta}(\mu) & \text{if } \mu \in A_{n,\kappa_1} \cap C_{m,\kappa_2} \text{ and } \mu_\varepsilon \in B_{q,\ell}\,, \\ +\infty & \text{otherwise.} \end{cases}$$

To avoid long formulas, write $J_{\gamma,\delta}^{\varepsilon,q,\ell,n,\kappa_1,m,\kappa_2}$ as $J_{\gamma,\delta}^\star$. Note that $J_{\gamma,\delta}^\star$ is lower semi-continuous because the set $A_{n,\kappa_1} \cap C_{m,\kappa_2} \cap \{\mu : \mu_\varepsilon \in B_{q,\ell}\}$ is closed.

Up to this point, we proved that for all $\delta \leq \varrho$,

$$\limsup_{T \to \infty} \frac{\log T}{T} \log \mathbb{P}_\alpha[\bar{\mu}^T \in A]$$

$$\leq \max\Big\{ \sup_{\mu \in A} -J_{\gamma,\delta}^\star(\mu) + C(\gamma)\varrho\,, \, C_\gamma(\delta, \varrho)\,, \, C(\varepsilon, q, \ell)\Big\}.$$

where $C(\gamma)$ is a finite constant which depends only on γ, while the other terms satisfy (6.14).

Optimize the previous inequality with respect to all parameters and assume that the set A is closed (and therefore compact because so is $\mathcal{M}_c$). Since, for each fixed set of parameters, the functional $J_{\gamma,\delta}^\star$ is lower semi-continuous, we may apply the arguments presented in [4, Lemma A2.3.3] to exchange the supremum with the infimum. In this way we obtain that the last expression is bounded above by

$$\sup_{\mu \in A} \;\inf_{\substack{\varepsilon,q,\ell,n,\kappa_1,m,\kappa_2 \\ \gamma,\delta \leq \varrho}} \; \max\Big\{ -J_{\gamma,\delta}^\star(\mu) + C(\gamma)\varrho\,, \, C_\gamma(\delta, \varrho)\,, \, C(\varepsilon, q, \ell)\Big\}.$$

Fix $\mu \in \mathcal{M}_c$, and let $n \uparrow \infty$, and $\kappa_1 \downarrow 0$, and then $m \uparrow \infty$, and $\kappa_2 \downarrow 0$ in $J_{\gamma,\delta}^\star(\mu)$. Keep in mind that μ is fixed as well as $J_{\gamma,\delta}(\mu)$, the only object which is changing with the variables n, κ_1, m and κ_2 is the set at which $J_{\gamma,\delta}^\star$ takes the value $+\infty$. Use the closeness of the sets $A_{n,\kappa_1}, C_{m,\kappa_2}$ to conclude that the previous expression is bounded by

$$\sup_{\mu \in A} \inf_{\substack{\varepsilon,q,\ell \\ \gamma,\delta \leq \varrho}} \max \left\{ -J_{\gamma,\delta}^{\varepsilon,q,\ell}(\mu) + C(\gamma)\varrho, \, C_\gamma(\delta,\varrho), \, C(\varepsilon,q,\ell) \right\},$$

where

$$J_{\gamma,\delta}^{\varepsilon,q,\ell}(\mu) = \begin{cases} J_{\gamma,\delta}(\mu) & \text{if } \mu \in \mathcal{M}_{0,\alpha} \text{ and } \mu_\varepsilon \in B_{q,\ell}, \\ +\infty & \text{otherwise,} \end{cases}$$

and $\mathcal{M}_{0,\alpha}$ is the set introduced just below (2.3).

Let now $\varepsilon \downarrow 0$. We claim that for all $\mu \in \mathcal{M}_c$,

$$J_{\gamma,\delta}^{q,\ell}(\mu) \leq \liminf_{\varepsilon \to 0} J_{\gamma,\delta}^{\varepsilon,q,\ell}(\mu), \tag{6.15}$$

where

$$J_{\gamma,\delta}^{q,\ell}(\mu) = \begin{cases} J_{\gamma,\delta}(\mu) & \text{if } \mu \in \mathcal{M}_{0,\alpha} \cap B_{q,\ell}, \\ +\infty & \text{otherwise,} \end{cases} \tag{6.16}$$

Indeed, fix $\mu \in \mathcal{M}_c$. We may assume that $\mu \in \mathcal{M}_{0,\alpha}$, otherwise $J_{\gamma,\delta}^{\varepsilon,q,\ell}(\mu) = J_{\gamma,\delta}^{q,\ell}(\mu) = \infty$ for all $\varepsilon > 0$. Note that $\mu_\varepsilon \to \mu$ as $\varepsilon \downarrow 0$. Since $B_{q,\ell}$ is a closed set, if $\mu \notin B_{q,\ell}$, $\mu_\varepsilon \notin B_{q,\ell}$ for ε small enough and both sides of (6.15) are equal to $+\infty$. It remains to consider the case $\mu \in B_{q,\ell}$. Here, by definition, $J_{\gamma,\delta}^{q,\ell}(\mu) = J_{\gamma,\delta}(\mu) \leq J_{\gamma,\delta}^{\varepsilon,q,\ell}(\mu)$ for all $\varepsilon > 0$, which proves claim (6.15).

In view of the second bound in (6.14), up to this point we proved that for all closed subset A of $\mathcal{M}_c$,

$$\limsup_{T \to \infty} \frac{\log T}{T} \log \mathbb{P}_\alpha[\bar\mu^T \in A]$$

$$\leq -\inf_{\mu \in A} \sup_{q,\ell,\gamma,\delta \leq \varrho} \min \left\{ J_{\gamma,\delta}^{q,\ell}(\mu) - C(\gamma)\varrho, \, -C_\gamma(\delta,\varrho), \, \ell - 1 \right\},$$

where $J_{\gamma,\delta}^{q,\ell}$ is given by (6.16). We claim that for all $\mu \in \mathcal{M}_c$,

$$J_{\gamma,\delta}^\ell(\mu) \leq \sup_q J_{\gamma,\delta}^{q,\ell}(\mu), \tag{6.17}$$

where

$$J_{\gamma,\delta}^\ell(\mu) = \begin{cases} J_{\gamma,\delta}(\mu) & \text{if } \mu \in \mathcal{M}_{0,\alpha} \text{ and } \mathcal{Q}_\alpha(\mu) \leq \ell, \\ +\infty & \text{otherwise.} \end{cases}$$

Indeed, suppose first that $\mathcal{Q}_\alpha(\mu) > \ell$. In this case, since H_p is a dense sequence, for all q sufficiently large, $\max_{1 \le p \le 1} \mathcal{Q}_{\alpha, H_p}(\mu) > \ell$ so that both sides of (6.17) are equal to $+\infty$. On the other hand, if $\mathcal{Q}_\alpha(\mu) \le \ell$ both sides are equal to $J_{\gamma, \delta}(\mu)$. This proves the claim.

We now assert that for all $\mu \in \mathcal{M}_c$,

$$J^{\mathcal{Q}}_{\gamma, \delta}(\mu) \le \sup_\ell \min \left\{ J^\ell_{\gamma, \delta}(\mu), \ \ell - 1 \right\}, \tag{6.18}$$

where

$$J^{\mathcal{Q}}_{\gamma, \delta}(\mu) = \begin{cases} J_{\gamma, \delta}(\mu) \text{ if } \mu \in \mathcal{M}_{0, \alpha} \text{ and } \mathcal{Q}_\alpha(\mu) < \infty, \\ +\infty \quad \text{ otherwise,} \end{cases} \tag{6.19}$$

Indeed, if $J^\ell_{\gamma, \delta}(\mu) = \infty$ for all $\ell \ge 1$, there is nothing to prove. If this is note the case, by definition of $J^\ell_{\gamma, \delta}$, $\mathcal{Q}_\alpha(\mu) \le m$ for some $m \ge 1$, and $J^\ell_{\gamma, \delta}(\mu) = J_{\gamma, \delta}(\mu) = J^{\mathcal{Q}}_{\gamma, \delta}(\mu)$. This proves (6.18). Recall from (5.5) that we denote by $\mathcal{M}^{\mathcal{Q}}_{0, \alpha}$ the set of measures μ in $\mathcal{M}_c$ such that $\mu \in \mathcal{M}_{0, \alpha}$ and $\mathcal{Q}_\alpha(\mu) < \infty$, which is the set appearing in the definition of $J^{\mathcal{Q}}_{\gamma, \delta}$.

Putting together the previous two estimates we conclude that for all closed subset A of $\mathcal{M}_c$,

$$\limsup_{T \to \infty} \frac{\log T}{T} \log \mathbb{P}_\alpha[\bar{\mu}^T \in A]$$

$$\le - \inf_{\mu \in A} \sup_{\gamma, \delta \le \varrho} \min \left\{ J^{\mathcal{Q}}_{\gamma, \delta}(\mu) - C(\gamma) \varrho, \ -C_\gamma(\delta, \varrho) \right\}, \tag{6.20}$$

where $J^{\mathcal{Q}}_{\gamma, \delta}$ is the functional given by (6.19).

It remains to let $\delta \to 0$. Since J_γ is lower semi-continuous and since $\mu_\delta \to \mu$ as $\delta \to 0$, for all $\mu \in \mathcal{M}_c$,

$$J^{\mathcal{Q}}_\gamma(\mu) \le \limsup_{\delta \to 0} J^{\mathcal{Q}}_{\gamma, \delta}(\mu),$$

where $J^{\mathcal{Q}}_\gamma$ is defined in (5.6). Hence, letting $\delta \to 0$ in (6.20) and then $\varrho \to 0$, we conclude that for all closed subsets A of $\mathcal{M}_c$,

$$\limsup_{T \to \infty} \frac{\log T}{T} \log \mathbb{P}_\alpha[\bar{\mu}^T \in A] \le - \inf_{\mu \in A} \sup_\gamma J^{\mathcal{Q}}_\gamma(\mu) = - \inf_{\mu \in A} J^{\mathcal{Q}}(\mu),$$

where $J^{\mathcal{Q}}$ is the functional given by (5.7). This is the upper bound of the large deviations principle, in view of Lemma 5.2.

7   The lower bound

We prove in this section the lower bound of the large deviations principle. Most of the results are taken from [2] and are repeated here in sake of completeness.

Consider a functional $\mathcal{J} : \mathcal{M}_c \to \mathbb{R}_+ \cup \{+\infty\}$. A subset $\mathcal{M}^*$ of $\mathcal{M}_c$ is said to be $\mathcal{J}$-dense if for each $\mu \in \mathcal{M}_c$ such that $\mathcal{J}(\mu) < \infty$, there exists a sequence $\{\mu_n \in \mathcal{M}^* : n \geq 1\}$ converging vaguely to μ and such that $\lim_{n\to\infty} \mathcal{J}(\mu_n) = \mathcal{J}(\mu)$.

Denote by $\mathcal{M}_0^*$ the subset of $\mathcal{M}_c$ formed by the measures in $\mathcal{M}_0$ whose density m is smooth, bounded away from 0 and 1, and for which m' has a compact support in $(0, \infty)$. The next result follows from the proof of [1, Lemma 4.1].

Lemma 7.1 *Recall from* (2.3) *the definition of the functional* $\mathcal{Q}$. *The set* $\mathcal{M}_0^*$ *is* $\mathcal{Q}$-*dense.*

Let $\mathcal{M}_{0,\alpha}^* = \mathcal{M}_0^* \cap \mathcal{M}_{0,\alpha}$. Fix a measure μ in $\mathcal{M}_{0,\alpha}^*$, and denote its density by m. Since m' has support contained in $(0, 1/2)$ and $m(r) = \alpha$ for $r \geq 1/2$, m belongs to $C^2(\mathbb{R}_+, \alpha)$. Hence, $\mathcal{M}_{0,\alpha}^*$ corresponds to the measures whose density belongs to $C^2(\mathbb{R}_+, \alpha)$.

Corollary 7.2 *The set* $\mathcal{M}_{0,\alpha}^*$ *is* $I_{Q,\alpha}$-*dense.*

Proof Fix μ such that $I_{Q,\alpha}(\mu) < \infty$. By definition of $I_{Q,\alpha}$, μ belongs to $\mathcal{M}_{0,\alpha}$ and $I_{Q,\alpha}(\mu) = \pi \mathcal{Q}(\mu)$. By the previous lemma, there exists a sequence $v_n \in \mathcal{M}_0^*$ such that $v_n \to \mu$ and $\mathcal{Q}(v_n) \to \mathcal{Q}(\mu)$. To prove the corollary it is therefore enough to show that for every $\mu \in \mathcal{M}_0^*$ there exists a sequence $\mu_n \in \mathcal{M}_{0,\alpha}^*$ such that $\mu_n \to \mu$ and $\mathcal{Q}(\mu_n) \to \mathcal{Q}(\mu)$.

Fix such a measure $\mu(dr) = m(r)\, dr$. Since μ belongs to $\mathcal{M}_0^*$,

$$\mathcal{Q}(\mu) = \frac{1}{4} \int_0^\infty \frac{[m'(r)]^2}{\sigma(m(r))}\, dr.$$

Fix $\delta > 0$, and let $u_\delta(r) = m(r)\, \mathbf{1}\{r \leq (1/2) - \delta\} + \alpha\, \mathbf{1}\{r > (1/2) - \delta\}$. Extend u_δ to $(-\infty, 0)$ by setting $u_\delta(r) = m(0)$ for $r \leq 0$. Let $m_\delta = u_\delta * \varphi_{\delta/2}$, where $\varphi_{\delta/2}$ is a smooth approximation of the identity whose support is contained in $[-\delta/2, \delta/2]$. Denote by μ_δ the measure on $\mathbb{R}_+$ whose density is m_δ.

It is clear that μ_δ belongs to $\mathcal{M}_{0,\alpha}^*$ for δ sufficiently small and that $\mu_\delta \to \mu$ as $\delta \to 0$. By the lower semicontinuity of $\mathcal{Q}$, $\mathcal{Q}(\mu) \leq \liminf_{\delta\to 0} \mathcal{Q}(\mu_\delta)$. On the other hand, by construction, for δ sufficiently small,

$$\mathcal{Q}(\mu_\delta) = \int_0^{1/2} \frac{[m_\delta'(r)]^2}{\sigma(m_\delta(r))}\, dr \to \int_0^{1/2} \frac{[m'(r)]^2}{\sigma(m(r))}\, dr \leq \mathcal{Q}(\mu).$$

Hence, $\limsup_{\delta\to 0} \mathcal{Q}(\mu_\delta) \leq \mathcal{Q}(\mu)$, which proves the corollary.

We are now in a position to prove the lower bound. We start with a law of large numbers for the polar empirical measure under the measure $\mathbb{P}_{T,\gamma}$. This result is

Lemma 6.1 in [2]. It follows from the stationarity of the measure $\nu_{T,\gamma}$ and from the fact that it is a product measure.

Lemma 7.3 *Fix γ in $C^2(\mathbb{R}_+, \alpha)$. As $T \uparrow \infty$, the measure $\bar{\mu}^T$ converges in $\mathbb{P}_{T,\gamma}$-probability to the measure $\gamma(r)\,dr$.*

Proof of the Lower Bound. We reproduce the proof presented in [2]. Fix an open subset G of $\mathcal{M}_c$. In view of Corollary 7.2, it is enough to show that

$$\liminf_{T\to\infty} \frac{\log T}{T} \log \mathbb{P}_\alpha[\bar{\mu}^T \in G] \geq -I_{Q,\alpha}(\mu)$$

for every μ in $\mathcal{M}_{0,\alpha}^* \cap G$. Fix such a measure μ and denote its density by γ. As observed above the statement of Lemma 7.2, γ belongs to $C^2(\mathbb{R}_+, \alpha)$. Let $\mathcal{A}_\gamma = \{\bar{\mu}^T \in G\}$ and denote by $\mathbb{P}_{T,\gamma}^A$ the probability measure $\mathbb{P}_{T,\gamma}$ conditioned on the set $\mathcal{A}_\gamma$. With this notation we may write

$$\mathbb{P}_\alpha[\bar{\mu}^T \in G] = \mathbb{E}_{T,\gamma}\left[\frac{d\mathbb{P}_\alpha}{d\mathbb{P}_{T,\gamma}} \mathbf{1}\{\mathcal{A}_\gamma\}\right] = \mathbb{E}_{T,\gamma}^A\left[\frac{d\mathbb{P}_\alpha}{d\mathbb{P}_{T,\gamma}}\right] \mathbb{P}_{T,\gamma}[\mathcal{A}_\gamma].$$

By the law of large numbers stated in Lemma 7.3, $\lim_{T\to\infty} \mathbb{P}_{T,\gamma}[\mathcal{A}_\gamma] = 1$. Hence, by Jensen inequality,

$$\liminf_{T\to\infty} \frac{\log T}{T} \log \mathbb{P}_\alpha[\bar{\mu}^T \in G] = \liminf_{T\to\infty} \frac{\log T}{T} \log \mathbb{E}_{T,\gamma}^A\left[\frac{d\mathbb{P}_\alpha}{d\mathbb{P}_{T,\gamma}}\right]$$

$$\geq \liminf_{T\to\infty} \frac{\log T}{T} \mathbb{E}_{T,\gamma}^A\left[\log \frac{d\mathbb{P}_\alpha}{d\mathbb{P}_{T,\gamma}}\right] = \liminf_{T\to\infty} \frac{\log T}{T} \mathbb{E}_{T,\gamma}\left[\log \frac{d\mathbb{P}_\alpha}{d\mathbb{P}_{T,\gamma}} \mathbf{1}\{\mathcal{A}_\gamma\}\right].$$

By the bound (6.13) for the Radon-Nikodym derivative and by Lemma 7.3, last term is equal to

$$\liminf_{T\to\infty} \frac{\log T}{T} \mathbb{E}_{T,\gamma}\left[\log \frac{d\mathbb{P}_\alpha}{d\mathbb{P}_{T,\gamma}}\right]$$

which is, up to a sign, the entropy of $\mathbb{P}_{T,\gamma}$ with respect to $\mathbb{P}_\alpha$. In view of formula (6.10) for the Radon-Nikodym derivative $d\mathbb{P}_{T,\gamma}/d\mathbb{P}_\alpha$, the previous limit is equal to

$$\liminf_{T\to\infty} \mathbb{E}_{T,\gamma}\left[\pi \int_{\mathbb{R}_+} \Gamma''(r)\,\bar{\mu}^T(dr)\right] + \liminf_{T\to\infty} \mathbb{E}_{T,\gamma}\left[\int_0^1 W_\gamma(\eta_s)\,ds\right],$$

where W_γ is defined in (6.12). Since $\nu_{T,\gamma}$ is a stationary state, these expectations are easily computed. Recall Lemma 3.1 to show that the limit is equal to

$$\pi \int_0^{1/2} \left\{\Gamma''(r)\,\gamma(r) + [\Gamma'(r)]^2\,\sigma(\gamma(r))\right\} dr = -\frac{\pi}{4} \int_0^{1/2} \frac{[\gamma']^2}{\gamma(1-\gamma)}\,dr = -I_{Q,\alpha}(\mu).$$

We were allowed to integrate by parts the first term on the right-hand side because the function $\Gamma = (1/2)\{\log \gamma/(1 - \gamma) - \log \alpha/(1 - \alpha)\}$ vanishes at the boundary. This proves the lower bound.

Acknowledgements This work has been partially supported by FAPERJ CNE E-26/201.207/2014, by CNPq Bolsa de Produtividade em Pesquisa PQ 303538/2014-7, by ANR-15-CE40-0020-01 LSD of the French National Research Agency.

References

1. Bertini, L., Landim, C., Mourragui, M.: Dynamical large deviations for the boundary driven weakly asymmetric exclusion process. Ann. Probab. **37**, 2357–2403 (2009)
2. Chang, C.C., Landim, C., Lee, T.-Y.: Occupation time large deviations of two-dimensional symmetric simple exclusion process. Ann. Probab. **32**, 661–691 (2004)
3. Drewitz A., Gärtner J., Ramírez A.F., Sun R.: Survival probability of a random walk among a poisson system of moving traps. In: Deuschel, J.D., Gentz, B., König, W., von Renesse, M., Scheutzow, M., Schmock, U. (eds) Probability in Complex Physical Systems. Springer Proceedings in Mathematics, vol. 11. Springer, Berlin, Heidelberg (2012)
4. Kipnis, C., Landim, C.: Scaling Limits of Interacting Particle Systems, Grundlheren der mathematischen Wissenschaften 320. Springer, Berlin, New York (1999)
5. Landim, C.: Occupation time large deviations for the symmetric simple exclusion process. Ann. Probab. **20**, 206–231 (1992)
6. Landim, C., Yau, H.-T.: Large deviations of interacting particle systems in infinite volume Commun. Pure Appl. Math. **48**, 339–379 (1995)
7. Quastel, J., Rezakhanlou, F., Varadhan, S.R.S.: Large deviations for the symmetric simple exclusion process in dimensions $d \geq 3$. Probab. Th. Rel. Fields **113**, 1–84 (1999)
8. Rezakhanlou, F.: Hydrodynamic limit for attractive particle systems on $\mathbb{Z}d$. Commun. Math. Phys. **140**, 417–448 (1991)
9. Shiraishi, N.: Anomalous dependence on system size of large deviation functions for empirical measure. Interdiscip. Inform. Sci. **19**, 85–92 (2013)
10. Shiraishi, N.: Anomalous system size dependence of large deviation functions for local empirical measure. J. Stat. Phys. **152**, 336–352 (2013)

On the Growth of a Superlinear Preferential Attachment Scheme

Sunder Sethuraman and Shankar C. Venkataramani

Dedicated to Professor S.R.S. VARADHAN, a constant source of inspiration, on the occasion of his 75th birthday.

Abstract We consider an evolving preferential attachment random graph model where at discrete times a new node is attached to an old node, selected with probability proportional to a superlinear function of its degree. For such schemes, it is known that the graph evolution condenses, that is a.s. in the limit graph there will be a single random node with infinite degree, while all others have finite degree. In this note, we establish a.s. law of large numbers type limits and fluctuation results, as $n \uparrow \infty$, for the counts of the number of nodes with degree $k \geq 1$ at time $n \geq 1$. These limits rigorously verify and extend a physical picture of Krapivsky et al. (Phys Rev Lett 85:4629–4632, 2000 [16]) on how the condensation arises with respect to the degree distribution.

Keywords Preferential attachment · Random graphs · Degree distribution · Growth · Fluctuations · Superlinear

2000 Mathematics Subject Classification. Primary 60G20 · Secondary 05C80 · 37H10

1 Model and Results

There has been much recent interest in preferential attachment schemes which grow a random network by adding, progressively over time, new nodes to old ones based on their connectivity (cf. books and surveys [1, 5, 9, 12, 17, 19]). Part of the interest

S. Sethuraman (✉) · S. C. Venkataramani
Department of Mathematics, University of Arizona, Tucson, AZ 85721, USA
e-mail: sethuram@math.arizona.edu

S. C. Venkataramani
e-mail: shankar@math.arizona.edu

© Springer Nature Switzerland AG 2019
P. Friz et al. (eds.), *Probability and Analysis in Interacting Physical Systems*, Springer Proceedings in Mathematics & Statistics 283,
https://doi.org/10.1007/978-3-030-15338-0_9

243

is that such schemes, which might be seen as forms of reinforcement dynamics, have interesting and complex structure, relating to some 'real world' networks.

Consider the following model. At time $n = 1$, an initial network G_1 is composed of two vertices connected by a single edge. At times $n \geq 2$, a new vertex is attached to a vertex $x \in G_{n-1}$ with probability proportional to $w(d_x(n))$, that is with chance $w(d_x(n-1))/\sum_{y \in G_{n-1}} w(d_y(n-1))$, where $d_z(j)$ is the degree at time j of vertex z and $w = w(d) : \mathbb{N} \to \mathbb{R}_+$ is a 'weight' function in the form $w(d) = d^\gamma$ for a fixed $\gamma > -\infty$. In this way, a random tree is grown.

Let now $Z_k(n)$ be the number of vertices in G_n with degree k, that is $Z_k(n) = \sum_{y \in G_{n-1}} 1(d_y(n) = k)$. Also, let $S(n) = \sum_{k=1}^{n} w(k)Z_k(n)$ be the total weight of G_n and $V(n) = \max\{k \,|\, Z_k(n) \geq 1\}$ denote the largest degree among the vertices in G_n. By construction, $V(n) \leq n$. Let also $\Phi_k(n) = \sum_{\ell \geq k} Z_\ell(n)$ be the count of vertices with degree at least k in G_n.

We observe that the total number of vertices and edges in G_n are $n + 1$ and n respectively, so that we have the 'conservation laws'

$$\sum_{k=1}^{n} Z_k(n) = n + 1, \quad \text{and} \quad \sum_{k=1}^{n} kZ_k(n) = 2n \tag{1.1}$$

In [16] (see also [15]), a trichotomy of growth behaviors was observed depending on the strength of the exponent γ.

Linear case. First, when $\gamma = 1$, the scheme is often referred to as the 'Barabasi-Albert' model. Here, the degree structure satisfies, for $k \geq 1$,

$$\lim_{n \to \infty} \frac{Z_k(n)}{n} = \frac{4}{k(k+1)(k+2)} \quad \text{a.s.}$$

This limit was first proved in mean-value in [10, 16], in probability in [4], and almost-surely via different methods in [3, 6, 18, 21].

Sublinear case. Next, when $\gamma < 1$, it was shown that

$$\lim_{n \uparrow \infty} \frac{Z_k(n)}{n} = q(k) \quad \text{a.s.}$$

Although q is not a power law, it is in form where it decays faster than any polynomial [7, 21]: For $k \geq 1$,

$$q(k) = \frac{\bar{s}}{k^\gamma} \prod_{j=1}^{k} \frac{j^\gamma}{\bar{s} + j^\gamma}, \quad \text{and } \bar{s} \text{ is determined by } 1 = \sum_{k=1}^{\infty} \prod_{j=1}^{k} \frac{j^\gamma}{\bar{s} + j^\gamma}.$$

Asymptotically, as $k \uparrow \infty$, when $0 < \gamma < 1$, $\log q(k) \sim -(\bar{s}/(1-\gamma))k^{1-\gamma}$ is in 'stretched exponential' form; when $\gamma < 0$, $\log q(k) \sim \gamma k \log k$; when $\gamma = 0$, the case of uniform attachment when an old vertex is selected uniformly, $\bar{s} = 1$ and q is geometric: $q(k) = 2^{-k}$ for $k \geq 1$.

Superlinear case. Finally, when $\gamma > 1$, 'explosion' or a sort of 'condensation' happens in that in the limiting graph a random single vertex dominates in accumulating connections. Let $A = \lfloor \gamma/(\gamma - 1) \rfloor$ and let $\mathcal{T}_j$ be the collection of rooted trees with j or less nodes.

In [20], the limiting graph is shown to be a tree with the following structure a.s.

> There is a single (random) vertex v with an infinite number of children.
>
> To this vertex, v infinite copies of $\mathcal{T}$ are glued for each $\mathcal{T} \in \mathcal{T}_A$. $\qquad$ (1.2)
>
> The remaining nodes in the tree form a finite collection $\mathbb{T}$
>
> of bounded (but arbitrary) degree vertices.

We remark, by the definition of $\mathcal{T}_A$, in the limit tree, there are an infinite number of nodes of degree k for $1 \leq k \leq A$.

One may ask how the 'condensation' effect arises in the dynamics with respect to the degree distribution. In [16] (see also [15]), non-rigorous rate formulation derivations of the mean orders of growth of $\{Z_k(n)\}$ give that

$$\lim_{n \uparrow \infty} \frac{1}{n^{k-(k-1)\gamma}} E[Z_k(n)] = a_k \quad \text{when } 1 \leq k < \frac{\gamma}{\gamma - 1}$$

$$\lim_{n \uparrow \infty} \frac{1}{\log(n)} E[Z_k(n)] = b_k \quad \text{when } 2 \leq k = \frac{\gamma}{\gamma - 1} \text{ is an integer.}$$

$$\lim_{n \uparrow \infty} E[\Phi_k(n)] < \infty \quad \text{when } k > \frac{\gamma}{\gamma - 1} \qquad (1.3)$$

where

$$a_k = \prod_{j=2}^{k} \frac{w(j-1)}{j - (j-1)\gamma} \quad \text{and} \quad b_k = w(k-1)a_{k-1}. \qquad (1.4)$$

See Sect. 4.2, pp. 92–94 in [12], which discusses [16], and the roles that the ansatz and the limit,

$$E\left[\frac{Z_k(n)}{S(n)}\right] \sim \frac{E[Z_k(n)]}{E[S(n)]} \quad \text{and} \quad \lim_{n \uparrow \infty} \frac{E[S(n)]}{n^\gamma} = 1, \qquad (1.5)$$

assumed in [16], play in the approximations used to obtain (1.3).

We remark, however, in [2], that an a.s. form of the limit for $k = 1$, that is $\lim_{n \uparrow \infty} \frac{1}{n} Z_1(n) = 1$ a.s., was proved through branching process methods. Also, we observe, from (1.2), that one can deduce that $\lim_{n \uparrow \infty} \Phi_k(n) < \infty$ a.s. for $k > \gamma/(\gamma - 1)$.

The stratification of growth orders, or 'connectivity transitions', appears to be a phenomenon in a class of superlinear attachment schemes. See [15, 16] for more discussion. In this context, we remark the work [8] considers a preferential attachment urn model, different from the above graph model, and gives stratification results of the growth of mean orders of the counts of variously sized urns, and of the maximum sized urn (cf. Lemma 3.4, Theorem 3.5, Corollary. 3.6 in [8]).

In this context, the purpose of this note is to give, for the graph superlinear model, when $\gamma > 1$, a rigorous, self-contained derivation of a.s. versions of (1.3) in Theorem 9.1, and to elaborate a more general asymptotic picture, including lower order terms and fluctuations, in Theorem 9.2.

These derivations are based on rate equations and martingale analysis, with respect to the dynamics of the counts. We were inspired by the interesting works [8, 16, 20]. Although our arguments follow part of the outline in [8], they do not make use of (1.2) or its branching process/tree constructions in [20], or the more combinatorial estimates in [8]. We also remark that the methods given here seem robust and may be of use in other superlinear preferential schemes, beyond the 'standard' graph model that we have concentrated upon.

In the next section, we present our results and give their proofs in Sect. 2.

1.1 Results

Recall the formulas for $\{a_k\}$ and $\{b_k\}$ in (1.4).

Theorem 9.1 *The following structure for the degree sequence holds.*
(1) We have

(a) When $\frac{\gamma}{\gamma-1} > k \geq 1$,

$$\lim_{n\uparrow\infty} \frac{1}{n^{k-(k-1)\gamma}} Z_k(n) = a_k \text{ a.s..}$$

(b) When $k = \frac{\gamma}{\gamma-1} \geq 2$ is an integer,

$$\lim_{n\uparrow\infty} \frac{1}{\log(n)} Z_k(n) = b_k \text{ a.s..}$$

(c) When $k > \frac{\gamma}{\gamma-1}$,

$$\lim_{n\uparrow\infty} \Phi_k(n) < \infty \text{ a.s.}$$

(2) Also, a.s., for $n \geq n_0$, where is n_0 is a random time, $V(n)$ is achieved at a fixed (random) vertex, and

$$\lim_{n\uparrow\infty} \frac{1}{n} V(n) = 1.$$

Moreover, all other vertices, in the process $\{G_n\}$, are of bounded degree.

The previous theorem represents a different, more quantitative, but also less descriptive, in terms of connectivity, version of the quite detailed structure in (1.2).

We now elaborate further upon these growth orders.

Theorem 9.2 *Suppose $\frac{\gamma}{\gamma-1} > k \geq 2$, and let $k^* = \lfloor \frac{\gamma}{2(\gamma-1)} + \frac{k}{2} \rfloor$ denote the largest integer such that $k^* - (k^* - 1)\gamma \geq \frac{1}{2}[k - (k-1)\gamma]$. Then, with respect to coefficients $c_k^k = a_k, c_{k+1}^k, \ldots, c_{k^*}^k$, which can be found from Eq. (2.19), we have*

$$\frac{1}{n^{(k-(k-1)\gamma)/2}}\left\{ Z_k(n) - c_k^k n^{k+\gamma-k\gamma} - \cdots - c_{k^*}^k n^{k^*-(k^*-1)\gamma} \right\} \Rightarrow N(0, a_k).$$

Also, when $k = \frac{\gamma}{\gamma-1} \geq 2$ is an integer, we have, for $b_k = w(k-1)a_{k-1}$,

$$\frac{1}{(\log(n))^{1/2}}\left\{ Z_k(n) - b_k \log(n) \right\} \Rightarrow N(0, b_k).$$

In addition, when $\frac{\gamma}{\gamma-1} < 2$, we have

$$\lim_{n\uparrow\infty}\left(n - Z_1(n) \right) < \infty \text{ a.s. and } \lim_{n\uparrow\infty}\left(n - V(n) \right) < \infty \text{ a.s.}$$

We remark that the last claim, when $\frac{\gamma}{\gamma-1} < 2$, also follows from the description (1.2), although we will give here a different proof.

From Theorems 9.1 and 9.2, we see that, for $k \geq 2$, the variance of $Z_k(n)$ is on the same scale as the leading order deterministic contribution to $Z_k(n)$ as $n \to \infty$. This pattern, however, does not hold for $Z_1(n)$ or for $V(n)$. The fluctuations in these quantities are much smaller than one might naively expect. As we address in the following corollary, the fluctuations in Z_1, and in V, which are entirely determined by the fluctuations in Z_k for $k \geq 2$, are perfectly correlated (resp. anticorrelated) with the fluctuations in Z_2 in the limit $n \to \infty$. As seen in the proof, this 'leading order balance of fluctuations' is a reflection of the structure of the two linear, deterministic, 'constraints' (1.1) that are satisfied by the random counts $Z_k(n)$ in every realization.

Corollary 9.3 *Suppose $\frac{\gamma}{\gamma-1} > 2$ and recall $2^* = \lfloor \frac{\gamma}{2(\gamma-1)} \rfloor + 1$. With respect to coefficients $c_2^1, c_3^1, \ldots, c_{2^*}^1$, and $m_2, m_3, \ldots, m_{2^*}$ given by Eq. (2.20) and coefficients $c_2^2 = a_2, c_3^2, \ldots, c_{2^*}^2$, obtained from Eq. (2.19), we have*

$$\frac{1}{n^{1-\gamma/2}}\left\{ \begin{pmatrix} Z_1(n) - n \\ -Z_2(n) \\ V(n) - n \end{pmatrix} + \sum_{\ell=2}^{2^*} \begin{pmatrix} c_\ell^1 \\ c_\ell^2 \\ m_\ell \end{pmatrix} n^{\ell-(\ell-1)\gamma} \right\} \Rightarrow N(\mathbf{0}, a_2 \mathbb{1})$$

where $\mathbb{1}$ is the 3×3 matrix with all the entries equal to 1.
Also, when $\frac{\gamma}{\gamma-1} = 2$ (that is $\gamma = 2$), we have

$$\frac{1}{\sqrt{\log(n)}}\left\{ \begin{pmatrix} Z_1(n) - n \\ -Z_2(n) \\ V(n) - n \end{pmatrix} + \begin{pmatrix} b_2 \\ b_2 \\ b_2 \end{pmatrix} \log(n) \right\} \Rightarrow N(\mathbf{0}, b_2 \mathbb{1}).$$

We remark that the weak convergence statements in Theorem 9.2 and Corollary 9.3 appear to be the first fluctuation results for nonlinear preferential attachment schemes. We note, however, for the linear preferential model, a central limit theorem for the counts has been shown in [18].

We also note an example, when $\gamma = 5/4$, is worked out in Sect. 3 in [22].

2 Proofs

The proof section is organized as follows. After development of a martingale framework in Sect. 2.1, and combination of some stochastic analytic estimates, we prove Theorem 9.1 at the end of Sect. 2.2. Also, we prove Theorem 9.2 and Corollary 9.3 in Sect. 2.3.

2.1 Martingale Decompositions

Define, for $j \geq 1$, the increment $d_k(j + 1) = Z_k(j + 1) - Z_k(j)$. Given $\mathcal{F}_j = \sigma\{G_1, \ldots, G_j\}$, for $k \geq 2$, $d_k(j + 1)$ has distribution

$$d_k(j + 1) = \begin{cases} 1 & \text{with prob. } \frac{w(k-1)Z_{k-1}(j)}{S(j)} \\ -1 & \text{with prob. } \frac{w(k)Z_k(j)}{S(j)} \\ 0 & \text{with prob. } 1 - \frac{w(k-1)Z_{k-1}(j)}{S(j)} - \frac{w(k)Z_k(j)}{S(j)} \end{cases}$$

When $k = 1$,

$$d_1(j + 1) = \begin{cases} 1 & \text{with prob. } 1 - \frac{w(1)Z_1(j)}{S(j)} \\ 0 & \text{with prob. } \frac{w(1)Z_1(j)}{S(j)}. \end{cases}$$

Recall the count of the number of vertices with degree at least a certain level $k \geq 1$,

$$\Phi_k(j) = \sum_{\ell \geq k} Z_\ell(j).$$

Let $e_k(j + 1) = \Phi_k(j + 1) - \Phi_k(j)$ be the corresponding increment. Conditional on $\mathcal{F}_j$, for $k \geq 2$,

$$e_k(j + 1) = \begin{cases} 1 & \text{with prob. } \frac{w(k-1)Z_{k-1}(j)}{S(j)} \\ 0 & \text{with prob. } 1 - \frac{w(k-1)Z_{k-1}(j)}{S(j)} \end{cases}$$

We note $\Phi_k(j)$ is an increasing function of j. Of course, $\Phi_1(j) = j + 1$ as the number of vertices in G_j equals $j + 1$. The formula, for $k \geq 1$,

$$Z_k(n) = \Phi_k(n) - \Phi_{k+1}(n),$$

relates $\{Z_k(n)\}$ to $\{\Phi_k(n)\}$.

Finally, recall $V(j) = \max\{k | Z_k(j) \geq 1\}$ is the maximum over the degrees of all the vertices in the graph G_n. Define its increment $h(j+1) = V(j+1) - V(j)$. Conditional on $\mathcal{F}_j$,

$$h(j+1) = \begin{cases} 1 & \text{with prob. } \frac{w(V(j))Z_{V(j)}(j)}{S(j)} \\ 0 & \text{with prob. } 1 - \frac{w(V(j))Z_{V(j)}(j)}{S(j)} \end{cases}$$

Define the mean-zero martingales, with respect to $\{\mathcal{F}_j\}$,

$$M_k(n) = \sum_{j=1}^{n-1} d_k(j+1) - E\big[d_k(j+1)|\mathcal{F}_j\big]$$

$$Q_k(n) = \sum_{j=1}^{n-1} e_k(j+1) - E\big[e_k(j+1)|\mathcal{F}_j\big]$$

$$R(n) = \sum_{j=1}^{n-1} h(j+1) - E\big[h(j+1)|\mathcal{F}_j\big].$$

The conditional expectations are computed as follows.

$$E\big[d_k(j+1)|\mathcal{F}_j\big] = \begin{cases} \frac{w(k-1)Z_{k-1}(j)}{S(j)} - \frac{w(k)Z_k(j)}{S(j)} & \text{for } k \geq 2 \\ 1 - \frac{w(1)Z_1(j)}{S(j)} & \text{for } k = 1. \end{cases}$$

Also, for $k \geq 2$,

$$E\big[e_k(j+1)|\mathcal{F}_j\big] = \frac{w(k-1)Z_{k-1}(j)}{S(j)}.$$

This immediately yields

$$E\big[h(j+1)|\mathcal{F}_j\big] = \frac{w(V(j))Z_{V(j)}(j)}{S(j)}.$$

Putting together terms, we have the following martingale decompositions: For $k = 1$,

$$Z_1(n) - Z_1(1) = \sum_{j=1}^{n-1}\left(1 - \frac{w(1)Z_1(j)}{S(j)}\right) + M_1(n) \tag{2.1}$$

and for $k \geq 2$,

$$Z_k(n) - Z_k(1) = \sum_{j=1}^{n-1} E\big[d_k(j+1)|\mathcal{F}_j\big] + M_k(n)$$

$$= \sum_{j=1}^{n-1} \left(\frac{w(k-1)Z_{k-1}(j)}{S(j)} - \frac{w(k)Z_k(j)}{S(j)}\right) + M_k(n).$$

Also, for $k \geq 2$,

$$\Phi_k(n) - \Phi_k(1) = \sum_{j=1}^{n-1} \frac{w(k-1)Z_{k-1}(j)}{S(j)} + Q_k(n).$$

Finally, for the maximum degree $V(n)$,

$$V(n) - V(1) = \sum_{j=1}^{n-1} \frac{w(V(j))Z_{V(j)}(j)}{S(j)} + R(n)$$

In the above equations, since we start from initial graph G_1, composed of two vertices attached by an edge, $Z_1(1) = 2$, $Z_k(1) = 0$ for $k \geq 2$, and $V(1) = 1$. Also, $\Phi_k(1) = 0$ for $k \geq 2$.

We will want to estimate the order of the martingales. Their (predictable) quadratic variations can be computed:

$$\langle M_k\rangle(n) = \sum_{j=1}^{n-1} E\left[\big(d_k(j+1) - E[d_k(j+1)|\mathcal{F}_j]\big)^2|\mathcal{F}_j\right]$$

$$\langle Q_k\rangle(n) = \sum_{j=1}^{n-1} E\left[\big(e_k(j+1) - E[e_k(j+1)|\mathcal{F}_j]\big)^2|\mathcal{F}_j\right]$$

$$\langle R\rangle(n) = \sum_{j=1}^{n-1} E\left[\big(h(j+1) - E[h(j+1)|\mathcal{F}_j]\big)^2|\mathcal{F}_j\right].$$

These martingales have bounded increments $|d_k|, |e_k|, |h| \leq 1$, and so $\langle M_k\rangle(n)$, $\langle Q_k\rangle(n)$, $\langle R\rangle(n) \leq n$. However, we will need better estimates in the sequel. We have, when $k = 1$,

$$E\left[\big(d_1(j+1) - E[d_1(j+1)|\mathcal{F}_j]\big)^2|\mathcal{F}_j\right] = \frac{w(1)Z_1(j)}{S(j)}\left(1 - \frac{w(1)Z_1(j)}{S(j)}\right) \quad (2.2)$$

and, when $k \geq 2$, that

$$E\big[\big(d_k(j+1) - E[d_k(j+1)|\mathcal{F}_j]\big)^2|\mathcal{F}_j\big]$$

$$= \Big(1 - \Big(\frac{w(k-1)Z_{k-1}(j)}{S(j)} - \frac{w(k)Z_k(j)}{S(j)}\Big)\Big)^2 \frac{w(k-1)Z_{k-1}(j)}{S(j)}$$

$$+ \Big(-1 - \Big(\frac{w(k-1)Z_{k-1}(j)}{S(j)} - \frac{w(k)Z_k(j)}{S(j)}\Big)\Big)^2 \frac{w(k)Z_k(j)}{S(j)}$$

$$+ \Big(\frac{w(k-1)Z_{k-1}(j)}{S(j)} - \frac{w(k)Z_k(j)}{S(j)}\Big)^2 \Big(1 - \frac{w(k-1)Z_{k-1}(j)}{S(j)} + \frac{w(k)Z_k(j)}{S(j)}\Big).$$

With respect to $Q_k(n)$, for $k \geq 2$, we have

$$E\big[\big(e_k(j+1) - E[e_k(j+1)|\mathcal{F}_j]\big)^2|\mathcal{F}_j\big] \tag{2.3}$$

$$= \Big(1 - \frac{w(k-1)Z_{k-1}(j)}{S(j)}\Big)^2 \frac{w(k-1)Z_{k-1}(j)}{S(j)}$$

$$+ \Big(\frac{w(k-1)Z_{k-1}(j)}{S(j)}\Big)^2 \Big(1 - \frac{w(k-1)Z_{k-1}(j)}{S(j)}\Big).$$

For $R(n)$, we can compute

$$E\big[\big(h(j+1) - E[h(j+1)|\mathcal{F}_j]\big)^2|\mathcal{F}_j\big]$$

$$= \Big(1 - \frac{w(V(j))Z_{V(j)}(j)}{S(j)}\Big)\frac{w(V(j))Z_{V(j)}(j)}{S(j)}.$$

For later reference, we recall a law of the iterated logarithm, discussed in [13], for martingales $M(n)$, say with bounded increments, and whose predictable quadratic variation diverges a.s. as $n \to \infty$ (cf. Remark 3 in [13]): Almost surely,

$$0 < \limsup_{n\to\infty} \frac{|M(n)|}{(2\langle M\rangle(n)\log\log(\langle M\rangle(n)\vee e^2))^{1/2}} < \infty. \tag{2.4}$$

Also, a basic convergence result that we will use often is the following (cf. Theorem 5.3.1 in [11]): For a bounded difference martingale M_n, with respect to a probability 1 set of realizations,

$$\text{either } M_n \text{ converges} \tag{2.5}$$

$$\text{or both } \limsup_{n\uparrow\infty} M_n = \infty \text{ and } \liminf_{n\uparrow\infty} M_n = -\infty.$$

2.2 *Leading Order Asymptotics of $Z_k(n)$, $S(n)$ and $V(n)$*

In this section we prove Theorem 9.1 after a sequence of subsidiary results.

Recall that the total degree of the graph at time n is $\sum_{k=1}^{n} k Z_k(n) = 2n$ and $Z_k(n) = 0$ for $k > n$. One can bound

$$S(n) = \sum_{k=1}^{n} w(k) Z_k(n) \leq n^{\gamma-1} \sum_{k=1}^{n} k Z_k(n) = 2n^{\gamma}. \tag{2.6}$$

A sharper bound obtains from the following argument. From the conservation laws (1.1), we get $\sum_k (k-1) Z_k(n) = n - 1$. For $\alpha > 1$ and $k \geq 1$, we have $(k-1)^\alpha \geq k^\alpha - \alpha k^{\alpha-1}$. Using this, along with $Z_k(n) = 0$ for $k > V(n)$, we arrive at

$$\sum_{k=1}^{n} k^\alpha Z_k(n) \leq \sum_{k=1}^{n} (k-1)^\alpha Z_k(n) + \alpha \sum_{k=1}^{n} k^{\alpha-1} Z_k(n)$$

$$\leq (V(n) - 1)^{\alpha-1} \sum_{k=2}^{V(n)} (k-1) Z_k(n) + \alpha \sum_{k=1}^{V(n)} k^{\alpha-1} Z_k(n)$$

$$= (n-1)(V(n) - 1)^{\alpha-1} + \alpha \sum_{k=1}^{V(n)} k^{\alpha-1} Z_k(n) \tag{2.7}$$

For $1 < \alpha \leq 2$, we can use $k^{\alpha-1} \leq k$, $\sum k Z_k = 2n$, to get

$$\sum k^\alpha Z_k(n) \leq (n-1)(V(n) - 1)^{\alpha-1} + 2n\alpha \leq n(V(n)^{\alpha-1} + C_\alpha), \tag{2.8}$$

where $C_\alpha = 2\alpha$ is a constant that only depends on α. If $p = \lceil \alpha - 1 \rceil \geq 2$ so that $2 \leq p < \alpha \leq p + 1$, we iterate estimate (2.7) p times to get

$$\sum k^\alpha Z_k(n) \leq (n-1)(V(n) - 1)^{\alpha-1} + (n-1)\alpha(V(n) - 1)^{\alpha-2} + \cdots$$

$$+ (n-1)\alpha(\alpha-1) \cdots (\alpha - p + 2)(V(n) - 1)^{\alpha-p}$$

$$+ \alpha(\alpha-1) \cdots (\alpha - p + 1) \sum k^{\alpha-p} Z_k(n) \leq n(V(n) + C_\alpha)^{\alpha-1}$$

$$\tag{2.9}$$

where we use $0 < \alpha - p \leq 1$ and $k^{\alpha-p} \leq k$, noting $\sum k Z_k(n) = 2n$. We obtain the final estimate using the Binomial expansion (with remainder), noting that $V(n) \geq 1$, $\alpha > p \geq 2$ and $C > 0$ together imply, for some $\theta \in (0, 1)$,

$$(V(n) + C)^{\alpha-1} = V(n)^{\alpha-1} + (\alpha - 1) V(n)^{\alpha-2} C + \cdots$$

$$+ \frac{(\alpha-1)(\alpha-2) \cdots (\alpha - p + 1)}{(p-1)!} V(n)^{\alpha-p} C^{p-1}$$

$$+ \frac{(\alpha - 1)(\alpha - 2) \cdots (\alpha - p)}{p! \cdot (V(n) + \theta C)^{p+1-\alpha}} C^p$$

$$\geq V(n)^{\alpha-1} + (\alpha - 1)V(n)^{\alpha-2}C$$

$$+ \sum_{\ell=2}^{p-1} \frac{(\alpha - 1)(\alpha - 2) \cdots (\alpha - \ell)}{\ell!} V(n)^{\alpha-\ell-1} C^\ell.$$

We thus obtain (2.9) by noting $V(n) \geq 1$ and picking C_α sufficiently large, so that

$$(\alpha - 1)C_\alpha \geq \alpha + 2\alpha(\alpha - 1) \cdots (\alpha - p + 1)$$

$$C_\alpha^\ell \geq \frac{\alpha \cdot \ell!}{\alpha - \ell}, \quad \ell = 2, 3, \ldots, p - 1,$$

Applying the estimates (2.8) and (2.9) to the weight $S(n)$, we get

$$S(n) = \sum_{k=1}^n k^\gamma Z_k(n) \leq \begin{cases} n(V(n)^{\gamma-1} + C_\gamma) & \text{for } 1 < \gamma \leq 2 \\ n(V(n) + C_\gamma)^{\gamma-1} & \text{for } \gamma > 2 \end{cases} \tag{2.10}$$

Clearly, $V(n) \leq n$ so that $\limsup_n S(n)/n^\gamma \leq 1$. This estimate is better than the bound in (2.6) by a factor of 2. More importantly, the bound in (2.10) involves $V(n)$ and is therefore useful for our purposes. In passing, we note another bound of $S(n)$, namely $S(n) \leq n^\gamma + n$, is obtained in Eq. (4.2.3) in [12], by an optimization procedure.

In what follows, with respect to (possibly random) quantities $U(n)$ and $u(n) > 0$, we signify that

- $U(n)$ is $O(u(n))$ if $\limsup_{n \uparrow \infty} |U(n)|/u(n) < \infty$ a.s.
- $U(n)$ is $o(u(n))$ if $\limsup_{n \uparrow \infty} |U(n)|/u(n) = 0$ a.s.
- $U(n) \gtrsim u(n)$ if $\liminf U(n)/u(n) > 0$ a.s.
- $U(n) \lesssim u(n)$ if $\limsup U(n)/u(n) < \infty$ a.s.
- $U(n) \approx u(n)$ if $U(n) \gtrsim u(n)$ and $U(n) \lesssim u(n)$.

We will need a preliminary estimate on the growth of $V(n)$.

Proposition 9.4 For each $0 < \epsilon < 1$, the maximum degree $V(n)$ satisfies

$$\liminf_{n \to \infty} \frac{V(n)}{n^{1-\epsilon}} > 0 \text{ a.s.}$$

Proof Step 1. First, we observe that $V(n) \nearrow \infty$ a.s. The probability that the maximum value increases by 1 at time $n + 1$, given $\mathcal{F}_n$, is bounded below,

$$\frac{Z_{V(n)}(n)V(n)^\gamma}{S(n)} \geq \frac{C}{n},$$

using (2.10) and $Z_{V(n)}(n), V(n) \geq 1$, so that, respectively for $1 < \gamma \leq 2$ and $\gamma > 2$,

$$\left[\frac{V(n)^{\gamma-1}}{V(n)^{\gamma-1}+C_\gamma}\right] \geq (1+C_\gamma)^{-1} \quad \text{and} \quad \left[\frac{V(n)}{V(n)+C_\gamma}\right]^{\gamma-1} \geq (1+C_\gamma)^{1-\gamma}.$$

Hence, $\sum P(V(n+1) - V(n) = 1|\mathcal{F}_n) = \infty$, and it follows from a Borel–Cantelli lemma (Theorem 5.3.2 in [11]) that $V(n) \nearrow \infty$ a.s. We note a version of this argument is used for the model in [8].

Step 2. The idea now is consider $V(n)$ in a log scale. Consider the evolution of $\Delta(n) = \sum_{j=1}^{V(n)-1} j^{-1} \approx \log(V(n))$. As $V(n) \nearrow \infty$, we have that $\Delta(n) \nearrow \infty$ a.s. Note also, our initial condition on G_1 yields $\Delta(2) = 1$.

The increments δ of Δ are given, conditioned on $\mathcal{F}_n$, by

$$\delta(n+1) = \Delta(n+1) - \Delta(n) = \begin{cases} V(n)^{-1} & \text{with prob. } \frac{Z_{V(n)}V(n)^\gamma}{S(n)} \\ 0 & \text{with prob. } 1 - \frac{Z_{V(n)}V(n)^\gamma}{S(n)} \end{cases}$$

Let also ρ denote the associated martingale, with differences $\delta(j+1) - E[\delta(j+1)|\mathcal{F}_j]$ (bounded by 1), and quadratic variation

$$\langle \rho \rangle(j) = \sum_{j=1}^{n-1} E\big[(\delta(j+1) - E[\delta(j+1)|\mathcal{F}_j])^2|\mathcal{F}_j\big].$$

We may compute

$$E[(\delta(n+1) - E[\delta(n+1) \mid \mathcal{F}_n])^2|\mathcal{F}_n] \qquad (2.11)$$
$$= \frac{1}{V(j)^2}\left(1 - \frac{Z_{V(j)}(j)V(j)^\gamma}{S(j)}\right)\frac{Z_{V(j)}(j)V(j)^\gamma}{S(j)} \leq \frac{1}{V(j)^2}.$$

Step 3. From the martingale decomposition of Δ, we get

$$\Delta(n) = 1 + \sum_{j=2}^{n-1} \frac{Z_{V(j)}(j)V(j)^{\gamma-1}}{S(j)} + \rho(n).$$

From (2.10) and $Z_{V(n)}(n) \geq 1$, we obtain, for $n \geq 2$, that

$$\Delta(n+1) - \Delta(n) \geq \frac{V(n)^{\gamma-1}}{S(n)} + \rho(n+1) - \rho(n)$$
$$\geq \frac{D_\gamma(V(n))}{n} + \rho(n+1) - \rho(n), \qquad (2.12)$$

where

$$D_\gamma(V(n)) = \begin{cases} \left(1 + C_\gamma/V(n)^{\gamma-1}\right)^{-1} & \text{for } 1 < \gamma \leq 2, \\ \left(1 + C_\gamma/V(n)\right)^{1-\gamma} & \text{for } \gamma > 2, \end{cases}$$

so that $D_\gamma(V(n)) \nearrow 1$ as $V(n) \nearrow \infty$ for all $\gamma > 1$.

Step 4. Since $V(n) \nearrow \infty$ a.s., by bounded convergence, $\lim_{n \uparrow \infty} E[D_\gamma(V(n))] = 1$. Consequently, given $0 < \epsilon_0 < 1$, there is a sufficiently large deterministic time n_0 such that $E[D_\gamma(V(n))] \geq 1 - \epsilon_0$, for $n \geq n_0$. Summing (2.12) over times from n_0 to n yields

$$E\Delta(n+1) \geq E\Delta(n_0) + (1 - \epsilon_0) \sum_{j=n_0}^{n} \frac{1}{j}.$$

Then, $E\Delta(n) \gtrsim (1 - \epsilon_0) \log(n)$. By Jensen's inequality, and the bound $\log(V(n) - 1) \geq \Delta(n) - \Delta(2)$ a.s., following from $\int_1^b x^{-1} dx \geq \sum_{j=2}^b j^{-1}$, we have $\log EV(n) \geq E \log V(n) \gtrsim E\Delta(n)$ and so $EV(n) \gtrsim n^{1-\epsilon_0}$.

Hence, by Jensen's inequality, applied to $\phi(x) = x^{-2}$, and (2.11), we have

$$E\langle \rho \rangle(j+1) - E\langle \rho \rangle(j) \leq E\left[V(j)^{-2}\right] \leq \left(EV(j)\right)^{-2} \lesssim j^{-2(1-\epsilon_0)}.$$

Now note $E\langle \rho \rangle(n_0) \leq n_0$ by (2.11) and $V(j) \geq 1$. Then, with the choice $\epsilon_0 = 1/4$ say, monotone convergence yields

$$\lim E\langle \rho \rangle(n) = E \lim \langle \rho \rangle(n) \leq n_0 + C \sum_{j \geq n_0} j^{-2(1-\epsilon_0)} < \infty.$$

Step 5. As $\sup_n E\rho_n^2 = \sup_n E\langle \rho \rangle(n) < \infty$ by Step 4, standard martingale convergence results (cf. Theorem 5.4.5 [11]) imply that ρ_n converges to a limit random variable ρ_∞ a.s and also in L^2. We note, as $\rho_\infty \in L^2$, it is finite a.s.

Step 6. Given $0 < \epsilon < 1$, noting again $V(n) \nearrow \infty$ a.s., we may choose now a (random) time n_1 such that $D_\gamma(V(n)) \geq (1 - \epsilon)$ and $\rho(n) \geq \rho_\infty - 1$. Note also that $|\Delta(n) - \log(V(n))| < 1$ and $|\log(n) - \sum_{j=1}^{n-1} j^{-1}| < 1$ for all n. Summing (2.12), from n_1 to n, we thus have

$$\Delta(n) - \Delta(n_1) \geq (1 - \epsilon) \sum_{j=n_1}^{n-1} \frac{1}{j} + \rho(n) - \rho(n_1)$$

and so $\Delta(n) - (1 - \epsilon) \log n \geq \Delta(n_1) - (1 - \epsilon) \log(n_1) + \rho_\infty - 3$ is bounded below a.s. Hence, for a constant C, depending on the realization, we have a.s. that $V(n) \geq Cn^{1-\epsilon}$ for all large n.

We now address the growth of $Z_1(n)$ and $\{\Phi_k(n)\}$.

Proposition 9.5 *Let $\epsilon > 0$ be such that $\gamma' \equiv \gamma(1 - \epsilon) > 1$ and $\gamma'/(\gamma' - 1)$ is not an integer. We have a.s. that*

$$\lim_{n \uparrow \infty} \frac{M_1(n)}{n} = 0, \quad \lim_{n \uparrow \infty} \frac{Z_1(n)}{n} = 1 \quad \text{and} \quad \limsup_{n \uparrow \infty} \frac{\Phi_k(n)}{n^{k-(k-1)\gamma'} \vee 1} < \infty.$$

Proof Proposition 9.4 implies that, for any $0 < \epsilon < 1$, $S(j) \geq V(j)^\gamma \gtrsim j^{(1-\epsilon)\gamma}$. Write, noting the decomposition (2.1), that

$$\frac{1}{n}Z_1(n) = \frac{1}{n}Z_1(1) + \frac{1}{n}\sum_{j=1}^{n-1}\left(1 - \frac{w(1)Z_1(j)}{S(j)}\right) + \frac{1}{n}M_1(n). \qquad (2.13)$$

Since $Z_1(1) = 2$, the first term on the right-hand side equals $2n^{-1}$. The second term can be estimated using $Z_1(j) \leq j + 1$, so that

$$\frac{Z_1(j)}{S(j)} \leq \frac{j+1}{j^\gamma} = O(j^{1-\gamma'}).$$

Then, if $1 < \gamma' < 2$, the sum

$$\sum_{j=1}^{n-1}\frac{Z_1(j)}{S(j)} = O(n^{2-\gamma'}). \qquad (2.14)$$

If $\gamma' = 2$, this sum is $O(\log(n))$ and, if $\gamma' > 2$, this sum is convergent. Note, in particular, in all cases that the sum in (2.14) is $o(n)$ a.s.

We now analyze the quadratic variation $\langle M_1\rangle(n)$. Noting (2.2), we have $n + 1 - Z_1(n) = \sum_{j=1}^{n-1}\frac{w(1)Z_1(j)}{S(j)} + M_1(n)$ and

$$\frac{1}{2}\sum_{j=j_0}^{n}\frac{w(1)Z_1(j)}{S(j)} \leq \langle M_1\rangle(n) \leq \sum_{j=1}^{n}\frac{w(1)Z_1(j)}{S(j)} = o(n),$$

where j_0 is chosen large enough.

If the sum $\sum_{j=1}^{n-1}\frac{w(1)Z_1(j)}{S(j)}$ converges, then $M_1(n)$, being a bounded difference martingale, must converge, as otherwise $\lim\inf M_1(n) = -\infty$, which is impossible as $n - Z_1(n) \geq 0$ (cf. (2.5)). On the other hand, if the sum diverges, then, by the law of the iterated logarithm (2.4), $M_1(n)$ is at most $(o(n)\log\log(o(n)))^{1/2}$. In either case, $n^{-1}M_1(n) \to 0$ a.s. Also, from the decomposition (2.13), we conclude that $Z_1(n)/n \to 1$ a.s.

We now turn to the counts $\{\Phi_\ell\}$. Since $\Phi_1(j) = j + 1$, we have $\lim_{n\uparrow\infty}\frac{\Phi_1(n)}{n} = 1$. Assuming that the proposition statement for $\{\Phi_\ell\}$ holds for index $k - 1$, we will prove it for k.

Note, by the induction hypothesis,

$$\sum_{j=1}^{n-1}\frac{w(k-1)Z_{k-1}(j)}{S(j)} \leq \sum_{j=1}^{n-1}\frac{w(k-1)\Phi_{k-1}(j)}{j^\gamma} = O\left(n^{k-(k-1)\gamma'} \vee 1\right).$$

By the computations (2.3), the quadratic variation $\langle Q_k\rangle(n)$ satisfies

$$\frac{1}{2} \sum_{j=j_0}^{n-1} \frac{w(k-1)Z_{k-1}(j)}{S(j)} \le \langle Q_k \rangle(n) \le \sum_{j=1}^{n-1} \frac{w(k-1)Z_{k-1}(j)}{S(j)} = O\left(n^{k-(k-1)\gamma'} \vee 1\right),$$

where again j_0 is chosen large enough.

Consider the equation

$$\Phi_k(n) = \Phi_k(1) + \sum_{j=1}^{n-1} \frac{w(k-1)Z_{k-1}(j)}{S(j)} + Q_k(n).$$

If the sum $\sum_{j=1}^{n-1} \frac{w(k-1)Z_{k-1}(j)}{S(j)}$ converges, then as Φ_k is nonnegative, $\liminf Q_k(n)$ cannot equal $-\infty$; hence, $Q_k(n)$, being a bounded difference martingale, must also converge (cf. (2.5)). If, however, the sum diverges, by the law of the iterated logarithm (2.4), $Q_k(n)$ is of the order

$$Q_k(n) = O\left(\left(n^{k-(k-1)\gamma'} \log \log n^{k-(k-1)\gamma'}\right)^{1/2}\right).$$

In either case, we derive that $\Phi_k(n)/n^{k-(k-1)\gamma'} \vee 1 = O(1)$.

Recall, for each $k \ge 1$, that $\Phi_k(n)$ is nondecreasing. By the last proposition, for $0 < \epsilon < 1$ such that $\gamma(1-\epsilon) > 1$, there is a finite index B, in particular $B = \lceil \gamma(1 - \epsilon)/(\gamma(1-\epsilon) - 1) \rceil$, such that $\lim_{n \uparrow \infty} \Phi_B(n) =: \Phi_B(\infty)$ is a.s. finite. Moreover, there is a (random) finite time N_B such that $\Phi_B(n) = \Phi_B(\infty)$ for $n \ge N_B$.

Proposition 9.6 *Only one (random) node represented in the count $\Phi_B(\infty)$ has diverging degree, while all the other nodes in this count, and elsewhere in $\{G_n\}_{n \ge N_B}$, have bounded degree a.s.*

Proof Since the maximum degree $V(n)$ diverges a.s. (Proposition 9.4), we have $\Phi_B(N_B) \ge 1$, and also that vertices from the set corresponding to $\Phi_B(N_B)$ are selected infinitely many times. Denote the (random) finite number of vertices corresponding to $\Phi_B(N_B)$ by $\{x_1, \ldots, x_m\}$. There are countable possible configurations of this set.

Now, given that an infinite number of selections are made in a fixed finite set $\{y_1, \ldots, y_m\}$ at times $\{s_j\}$, we see that vertex y_i is chosen at time s_{j+1}, conditioned on the state at time s_j, with probability

$$w(d_{y_i}(s_j)) / \sum_{\ell=1}^{m} w(d_{y_\ell}(s_j)). \tag{2.15}$$

Let $s_0 = 1$ and $\{y_1(j), \ldots, y_m(j)\}_{j \ge 0}$ be the degrees of these vertices at times $\{s_j\}_{j \ge 0}$, according to this conditional update rule.

Such a 'discrete time weighted Polya urn', $\{y_1(j), \ldots, y_m(j)\}_{j \ge 0}$, can be embedded in a finite collection of independent continuous time weighted Polya urns,

$\{U_1, \ldots, U_m\}$. Let U_i be an urn process which increments by one at time t with rate $w\big(U_i(t)\big)$ and initial size given by $w(d_{y_i}(1))$. Since $w(d) = d^\gamma$ and $\gamma > 1$, so that $\sum w(d)^{-1} < \infty$, the urn U_i explodes at a finite time T_i a.s. Let τ_j be the jth time one of the urns $\{U_i\}_{i=1}^m$ is incremented. The discrete time weighted Polya urn system is recovered by evaluating the continuous time process at times τ_j for $j \geq 1$. Indeed, from properties of exponential r.v.'s, given the state of the system at time τ_j, the urn U_i rings at time τ_{j+1} with probability $w\big(U_i(\tau_j)\big)/\sum_{\ell=1}^m w\big(U_\ell(\tau_j)\big)$, the same as (2.15).

Now, one of the continuous time urns, say U_i, explodes first, meaning it is selected an infinite number of times. Another urn U_j cannot also explode at the same time $T_j = T_i$ since T_i, being a sum of independent exponential r.v.'s, is continuous and independent of T_j. Hence, a.s. after a (random) time, selections in the finite system are made only from one (random) urn or vertex.

Hence, by decomposing on the countable choices of the set $\{x_1, \ldots, x_m\}$ corresponding to $\Phi_B(N_B)$, we see that there is exactly one vertex in the set with diverging degree a.s. All the others in the set are only selected a finite number of times, and therefore have finite degree a.s.

With respect to the graphs $\{G_n\}$, vertices in the counts $\{Z_k\}_{k=1}^{B-1}$, by definition have degree bounded by $B - 1$. Hence, all vertices in the graph process have bounded degree, except for the maximum degree vertex, which diverges a.s. Similar types of argument can be found in [2, 8]. □

We now derive a.s. and L^1 limits with respect to $V(n)$ and $S(n)$. These we remark partly addresses the question in (1.5).

Proposition 9.7 *We have, a.s. and in L^1 that*

$$\lim_{n\uparrow\infty} \frac{1}{n} V(n) = 1 \text{ and } \lim_{n\uparrow\infty} \frac{1}{n^\gamma} S(n) = 1.$$

Proof By Proposition 9.5, $Z_1(n) = n + o(n)$ and $\Phi_2(n) = \sum_{k\geq 2} Z_k(n) = o(n)$ a.s. By Proposition 9.6, $V(n)$ is achieved at a fixed (random) vertex, say v, a.s. By (1.1) $\sum_{k\geq 1} k Z_k(n) = 2n$ so that, for all large n,

$$V(n) = \Big(2n - Z_1(n)\Big) - \sum_{k=2}^{V(n)-1} k Z_k(n)$$

Recall the index B before Proposition 9.6, and let $\hat{\Phi}_B(\infty)$ be the vertices represented by $\Phi_B(\infty)$. We have, by Propositions 9.5 and 9.6, for all large n, that

$$\sum_{k=2}^{V(n)-1} k Z_k(n) = \sum_{k=2}^{B-1} k Z_k(n) + \text{finite total degree of vertices in } \hat{\Phi}_B(\infty) \setminus \{v\}$$
$$= o(n).$$

Since $Z_1(n)/n \to 1$ a.s. by Proposition 9.5, we obtain therefore that $V(n)/n \to 1$ a.s.

Further, for all large n, the weight of the graph equals

$$S(n) = \text{weight of} v + Z_1(n) + \sum_{k=1}^{B-1} k^\gamma Z_k(n)$$

$$+ \text{finite total weight of vertices in} \hat{\Phi}_B(\infty) \setminus \{v\}$$

$$= (n + o(n))^\gamma + n + o(n) \text{ a.s.,}$$

from which the desired a.s. limit for $S(n)/n^\gamma$ is recovered.

The L^1 limits follow from bounded convergence, as $0 \le V(n) \le n$, and $0 \le S(n)/n^\gamma \le 2$, say by (2.6). □

We now show Theorem 9.1.

Proof of Theorem 9.1. To establish the first part, when $\frac{\gamma}{\gamma-1} > k \ge 1$, we show by induction that a.s.

$$\lim_{n\to\infty} \frac{1}{n^{k-(k-1)\gamma}} M_k(n) = 0 \text{ and } \lim_{n\to\infty} \frac{1}{n^{k-(k-1)\gamma}} Z_k(n) = a_k. \tag{2.16}$$

The base case $k = 1$ has been shown in Proposition 9.5. Suppose now that the claim holds for $M_{k-1}(n)$ and $Z_{k-1}(n)$. Write

$$\frac{1}{n^{k-(k-1)\gamma}} \Phi_k(n) = \frac{1}{n^{k-(k-1)\gamma}} \sum_{j=1}^{n-1} \frac{w(k-1)Z_{k-1}(j)}{S(j)} + Q_k(n).$$

By the asymptotics of $S(j)$ in Proposition 9.7, and the assumption, we have a.s.

$$\frac{Z_{k-1}(j)}{S(j)} \approx j^{k-1-(k-1)\gamma}.$$

Then, as $k - (k-1)\gamma > 0$, we have a.s.

$$\frac{1}{n^{k-(k-1)\gamma}} \sum_{j=1}^{n-1} \frac{w(k-1)Z_{k-1}(j)}{S(j)} \to \frac{w(k-1)a_{k-1}}{k-(k-1)\gamma}.$$

Also, by the computations (2.3), the quadratic variation $\langle Q_k \rangle(n)$ satisfies a.s.

$$\langle Q_k \rangle(n) \approx \sum_{j=1}^{n-1} \frac{w(k-1)Z_{k-1}(j)}{S(j)} \approx \frac{w(k-1)a_{k-1}}{k-(k-1)\gamma} n^{k-(k-1)\gamma}.$$

Then, by the law of the iterated logarithm (2.4), we have a.s.

$$Q_k(n) = O\left(\left(n^{k-(k-1)\gamma}\log\log n^{k-(k-1)\gamma}\right)^{1/2}\right).$$

Therefore, a.s., we obtain

$$\lim_{n\to\infty}\frac{1}{n^{k-(k-1)\gamma}}Q_k(n) = 0 \quad\text{and}\quad \lim_{n\to\infty}\frac{1}{n^{k-(k-1)\gamma}}\Phi_k(n) = \frac{w(k-1)a_{k-1}}{k-(k-1)\gamma}.$$

To update to $M_k(n)$ and $Z_k(n)$, we observe, by (2.2), that the quadratic variation of $M_k(n)$ is of order a.s.

$$\langle M_k\rangle(n) \lesssim \sum_{j=1}^{n-1}\frac{w(k-1)Z_{k-1}(j)}{S(j)} + \sum_{j=1}^{n-1}\frac{w(k)Z_k(j)}{S(j)} \tag{2.17}$$

$$\lesssim Cn^{k-(k-1)\gamma} + \sum_{j=1}^{n-1}\frac{\Phi_k(j)}{S(j)}$$

$$\lesssim Cn^{k-(k-1)\gamma} + Cn^{k+1-k\gamma}.$$

Note that $k+1-k\gamma < k-(k-1)\gamma$. Hence, by the law of the iterated logarithm (2.4), we have $M_k(n)/n^{k-(k-1)\gamma} \to 0$ a.s. Moreover,

$$Z_k(n) = \sum_{j=1}^{n-1}\frac{w(k-1)Z_{k-1}(j)}{S(j)} - \frac{w(k)Z_k(j)}{S(j)} + M_k(n)$$

$$= \sum_{j=1}^{n-1}\frac{w(k-1)Z_{k-1}(j)}{S(j)} + o(n^{k-(k-1)\gamma}).$$

Therefore, we have a.s.

$$\lim_{n\to\infty}\frac{Z_k(n)}{n^{k-(k-1)\gamma}} = \frac{w(k-1)a_{k-1}}{k-(k-1)\gamma} = a_k,$$

concluding the proof of (2.16).

To show the second item, when $\frac{\gamma}{\gamma-1} = k \geq 2$ is an integer, that is a.s.,

$$\lim_{n\to\infty}\frac{1}{\log(n)}M_k(n) = 0 \quad\text{and}\quad \lim_{n\to\infty}\frac{1}{\log(n)}Z_k(n) = b_k,$$

the argument is quite similar, following closely the steps above and is omitted.

We now consider the third item, when $k > \gamma/(\gamma-1)$. Let $r = \lfloor\gamma/(\gamma-1)\rfloor + 1$ be the smallest integer strictly larger than $\gamma/(\gamma-1)$. We now show that $\Phi_r(n)$ converges a.s. Then, we would conclude, for $k > \gamma/(\gamma-1)$, as $\Phi_k(n) \leq \Phi_r(n)$ and $\Phi_k(n)$ is nondecreasing, that $\Phi_k(n)$ converges a.s.

Note $r - 1 \leq \gamma/(\gamma - 1)$ and $\Phi_r(n) = \sum_{j=1}^{n-1} \frac{w(r-1)Z_{r-1}(j)}{S(j)} + Q_r(n)$. In particular, by the previous items, $Z_{r-1}(j)/S(j) \approx j^{(r-1)(1-\gamma)}$ is summable as $(r-1)(\gamma-1) > 1$. Hence, $Q_r(n)$, being a bounded difference martingale, also must converge by (2.5), as its liminf is not $-\infty$ since $\Phi_r(n) \geq 0$. Hence, $\Phi_r(n)$, being nondecreasing, converges a.s.

Finally, the second part (2) follows from directly from Propositions 9.6 and 9.7.
□

2.3 Refined Asymptotics and Fluctuations for $Z_k(n)$

We now prove Theorem 9.2 and Corollary 9.3, which give a description of the lower order nonrandom terms as well as the fluctuations in the counts $Z_k(n)$.

Proof of Theorem 9.2. We consider first the part when $2 \leq k < \gamma/(\gamma - 1)$. Recall that $A = \lfloor \gamma/(\gamma - 1) \rfloor$, and the initial decomposition of $Z_k(n)$,

$$Z_k(n) = \sum_{j=1}^{n-1} \frac{w(k-1)Z_{k-1}(j)}{S(j)} - \frac{w(k)Z_k(j)}{S(j)} + M_k(n). \qquad (2.18)$$

By (1.1) and by Theorem 9.1 that counts of vertices with degrees larger than A are bounded, we have $n + 1 = \sum_{k \geq 1} Z_k(n) = Z_1(n) + \cdots + Z_A(k) + O(1)$, and hence

$$Z_1(n) = n - Z_2(n) - \cdots - Z_A(n) + O(1).$$

Also, by the remarks below (2.6), by Theorem 9.1 again, and by Proposition 9.6 that the maximum is achieved at a fixed vertex, we have, for all large n, that $n - 1 = \sum_{k=2}^{n}(k-1)Z_k(n) = V(n) + \sum_{k=2}^{A}(k-1)Z_k(n) + O(1)$, and so $V(j) = j - \sum_{k=2}^{A}(k-1)Z_k(j) + O(1)$. Hence,

$$S(j) = V(j)^\gamma + \sum_{k=1}^{A} w(k)Z_k(j) + O(1)$$

$$= \left(j - \sum_{k=2}^{A}(k-1)Z_k(j) \right)^\gamma + \sum_{k=1}^{A} w(k)Z_k(j) + O(1).$$

One may now further decompose $Z_{k-1}(j)$, $Z_k(j)$ and $S(j)$, and successively decompose in turn the resulting counts. Recall that, when $\gamma/(\gamma - 1)$ is an integer, by Theorem 9.1, $Z_A(n) = O(\log(n))$. Recall the leading orders of $\{Z_k(n)\}_{k=1}^{A}$ in Theorem 9.1. Recall, also, the martingale $M_k(n)$, when $2 \leq k < \gamma/(\gamma - 1)$, is of order $\left(n^{(k-(k-1)\gamma)} \log \log n^{k-(k-1)\gamma} \right)^{1/2}$, using (2.17) and (2.4). Similarly, when $A =$

$\gamma/(\gamma - 1)$, the quadratic variation of $M_A(n)$ is of order $\log(n)$ and so $M_A(n)$ is of order $(\log(n) \log \log \log(n))^{1/2}$.

Any term, involving martingales, arising after the initial decomposition is of lesser order than $M_k(n)$ and may be omitted. Indeed, the largest term arising from decomposing $S(j)$ in the sum in (2.18) is $\sum_{j=1}^{n-1} \frac{w(k-1)Z_{k-1}(j)(\gamma j^{-1}M_2(j))}{j^\gamma}$, which is of order $n^{k-(k-1)\gamma-\gamma/2}\sqrt{\log \log n^{2-\gamma}}$. On the other hand, the largest term arising from decomposing the numerators in the sum in (2.18) is $\sum_{j=1}^{n-1} \frac{w(k-1)M_{k-1}(j)}{S(j)}$, which is of order $n^{(k-1-(k-2)\gamma)/2+1-\gamma}\sqrt{\log \log(n^{(k-1-(k-2)\gamma)/2})}$. Given that $\gamma > 1$, both orders are smaller than $n^{(k-(k-1)\gamma)/2}$.

Moreover, following these martingale calculations, any term $\eta(n)$ of order less than $n^{(k-1-(k-2)\gamma)/2}$ in the decomposition of $Z_{k-1}(n)$ will contribute a term in the computation of $Z_k(n)$ of order less than $n^{(k-(k-1)\gamma)/2}$. Also, any term $\zeta_\ell(n)$ in the decompositions of $Z_\ell(n)$, with respect to $S(n)$, of order less than $n^{(2-\gamma)/2}$ will give a term of order less than $n^{(k-(k-1)\gamma)/2}$ in computing $Z_k(n)$.

After a finite number of iterative decompositions, we may decompose $Z_k(n)$ as the sum of a finite number of nonrandom diverging terms of orders larger or equal to $n^{(k-(k-1)\gamma)/2}$, the martingale $M_k(n)$, and a finite number of other (possibly random) terms whose order is less than $n^{(k-(k-1)\gamma)/2}$.

By Theorem 9.1, as the dominant order of $Z_\ell(j)$ is $n^{\ell-(\ell-1)\gamma}$, when $\ell < \gamma/(\gamma - 1)$, and $Z_A(j) = O(\log(j))$ when $A = \gamma/(\gamma - 1)$, the nonrandom divergent orders, larger or equal to $n^{(k-(k-1)\gamma)/2}$, which appear in the decomposition of $Z_k(n)$ are all of the form $n^{m-(m-1)\gamma}$ where $k \leq m \leq k^*$, and k^* is the largest integer so that $k^* - (k^* - 1)\gamma \geq (k - (k-1)\gamma)/2$.

By these decompositions, the coefficients $\{c_{k+\ell}^k\}_{\ell=0}^{k^*-k}$ are found from matching the powers of n in the conditions

$$Y_k(n) - \sum_{j=1}^{n-1} \frac{w(k-1)Y_{k-1}(j) - w(k)Y_k(j)}{T(j)} = o(n^{(k-(k-1)\gamma)/2}) \qquad (2.19)$$

where, for $2 \leq k < \gamma/(\gamma - 1)$,

$$Y_k(n) = c_k^k n^{k-(k-1)\gamma} + \cdots + c_{k^*}^k n^{k^*-(k^*-1)\gamma},$$

$Y_1(n) = n - \sum_{k=2}^A Y_k(n)$, and

$$T(n) = (n - \sum_{k=2}^A (k-1)Y_k(n))^\gamma + \sum_{k=1}^A w(k)Y_k(n).$$

More precisely, the procedure is to sequentially compute, for each $\ell \geq 0$ fixed, the values $c_{k+\ell}^k$ as k runs through $2 \leq k < \gamma/(\gamma - 1)$, with the provision that $k + \ell \leq k^*$.

In Sect. 3 in [22], we reprise this procedure to derive the coefficients, and work through an illustrative example.

To show the central limit theorem, when $2 \leq k < \gamma/(\gamma - 1)$, we need to show that

$$\frac{M_k(n)}{n^{(k-(k-1)\gamma)/2}} \Rightarrow N(0, a_k).$$

Let $\xi_j^{(n)} = n^{-(k-(k-1)\gamma)/2}\{d_k(j+1) - E[d_k(j+1)|\mathcal{F}_j]\}$, for $1 \leq j \leq n-1$, define increments of the martingale array $\{M_k(j)/n^{(k-(k-1)\gamma)/2} : 1 \leq j \leq n-1, n \geq 2\}$ with respect to sigma-fields $\{\mathcal{F}_j^{(n)} = \mathcal{F}_j : 1 \leq j \leq n-1, n \geq 2\}$.

Note that, a.s.,

$$\max_{1 \leq j \leq n-1} |\xi_j^{(n)}| \leq n^{-(k-(k-1)\gamma)/2} \to 0.$$

Also, the conditional variance,

$$\sum_{j=1}^{n-1} E[(\xi_j^{(n)})^2|\mathcal{F}_j] = \frac{1}{n^{k-(k-1)\gamma}}\langle M_k \rangle(n).$$

Since $S(j) \approx j^\gamma$ (Proposition 9.7) and $Z_\ell(j) \approx a_\ell j^{\ell-(\ell-1)\gamma}$ when $1 \leq \ell \leq k$ (Theorem 9.1), noting the quadratic variation formulas (2.2), the conditional variance is of form $n^{-(k-(k-1)\gamma)} \sum_{j=1}^{n-1} \frac{w(k-1)Z_{k-1}(j)}{S(j)} + o(1)$, which converges a.s. to $w(k-1)a_{k-1}/(k-(k-1)\gamma) = a_k$.

Hence, when $2 \leq k < \gamma/(\gamma - 1)$, the standard Lindeberg assumptions for the martingale central limit theorem, Corollary 3.1 in [14], are clearly satisfied, and the desired Normal convergence holds.

When $2 \leq k = \gamma/(\gamma - 1)$ is an integer, the theorem statement follows from an easier version of the same argument. Indeed, after initial decompositions, noting that $j^{-\gamma}Z_{k-1}(j) = j^{(k-1)(1-\gamma)} = j^{-1}$ by Theorem 9.1,

$$Z_k(n) = \sum_{j=1}^{n-1} \frac{w(k-1)Z_{k-1}(j)}{S(j)} + M_k(n)$$

$$= b_k \log(n) + O(1) + M_k(n).$$

The quadratic variation of $(\log(n))^{-1/2}M_k(n)$, inspecting formulas (2.2), converges a.s. to b_k. The Normal convergence follows now as above.

When $\gamma/(\gamma - 1) < 2$, that is when $\gamma > 2$,

$$n - Z_1(n) = -2 + \sum_{j=1}^{n-1} \frac{w(1)Z_1(j)}{S(j)} - M_1(n).$$

The sum converges as the summand is of order $n^{1-\gamma}$. Since $n - Z_1(n) \geq 0$, the bounded difference martingale $M_1(n)$ must also converge, as its liminf cannot be $-\infty$ (cf. (2.5)). Hence, $Z_1(n) - n$ converges a.s.

Similarly, note (i) the remarks below (2.6), (ii) the maximum $V(n)$ is achieved at a fixed (random) vertex v and other vertices have bounded degree (Proposition 9.6), and (iii) $\Phi_{A+1}(n)$ converges (Theorem 9.1) and so the set of corresponding vertices stabilizes after a random time to a finite set $\hat{\Phi}_{A+1}(\infty)$. Then, for all large n, a.s.

$$n - V(n) = -1 + \sum_{k=2}^{A}(k-1)Z_k(n) + \text{weight of vertices in } \hat{\Phi}_{A+1}(\infty) \setminus \{v\}.$$

By Theorem 9.1, as $\gamma/(\gamma-1) < 2$, the sum in the above display converges a.s. as $n \uparrow \infty$. Hence, $V(n) - n$ converges a.s.   □

Proof of Corollary 9.3. As in the proof of Theorem 9.2, we recast (1.1) as

$$Z_1(n) = n - Z_2(n) - Z_3(n) - \cdots - Z_A(n) + O(1)$$
$$V(n) = n - Z_2(n) - 2Z_3(n) - \cdots - (A-1)Z_A(n) + O(1).$$

We now argue in the setting when $\gamma/(\gamma-1) > 2$, as the case $\gamma/(\gamma-1) = 2$ is similar and easier.

Since the order of the conditional variance of $M_2(n)$ is $n^{1-\gamma/2}$, and the martingales $M_k(n)$ are of lower order for $k > 2$, we have

$$\lim_{n\uparrow\infty} \frac{1}{n^{1-\gamma/2}}\left(Z_1(n) + M_2(n) - n + \sum_{k=2}^{2^*}\sum_{\ell=k}^{2^*} c_\ell^k n^{\ell-(\ell-1)\gamma}\right) = 0 \ a.s. \quad (2.20)$$

$$\lim_{n\uparrow\infty} \frac{1}{n^{1-\gamma/2}}\left(V(n) + M_2(n) - n + \sum_{k=2}^{2^*}\sum_{\ell=k}^{2^*} (k-1)c_\ell^k n^{\ell-(\ell-1)\gamma}\right) = 0 \ a.s.$$

and Corollary 9.3 follows with $c_\ell^1 = \sum_{k=2}^{2^*} c_\ell^k$ and $m_\ell = \sum_{k=2}^{2^*}(k-1)c_\ell^k$.   □

Acknowledgements S.V. acknowledges the support from the Simons Foundation through award 524875. S.S. acknowledges the support from the grant ARO W911NF-14-1-0179.

References

1. Albert, R., Barabási, A.-L.: Statistical mechanics of complex networks. Rev. Mod. Phys. **74**, 47–97 (2002)
2. Athreya, K.B.: Preferential attachment random graphs with general weight function. Internet Math. **4**, 401–418 (2007)
3. Athreya, K.B., Ghosh, A.P., Sethuraman, S.: Growth of preferential attachment random graphs via continuous-time branching processes. Proc. Indian Acad. Sci. Math. Sci. **118**, 473–494 (2008)
4. Bollobás, B., Riordan, O., Spencer, J., Tusnády, G.: The degree sequence of a scale-free random graph process. Random Struct. Algorithms **18**, 279–290 (2001)

5. Caldarelli, G.: Scale-Free Networks: Complex Webs in Nature and Technology. Oxford University Press, USA (2007)
6. Choi, J., Sethuraman, S.: Large deviations for the degree structure in preferential attachment schemes. Ann. Appl. Probab. **23**, 722–763 (2011)
7. Choi, J., Sethuraman, S., Venkataramani, S.C.: A scaling limit for the degree distribution in sublinear preferential attachment schemes. Random Struct. Algorithms **48**, 703–731 (2015)
8. Chung, F., Handjani, S., Jungreis, D.: Generalizations of Pólya's urn problem. Ann. Comb. **7**, 141–153 (2003)
9. Chung, F., Lu, L.: Complex graphs and networks. In: CBMS Regional Conference Series in Mathematics, vol. 107. Conference Board of the Mathematical Sciences, Washington, DC (2006)
10. Dorogovtsev, S.N., Mendes, J.F.F.: Evolution of Networks: From Biological Nets to the Internet and WWW. Oxford University Press, USA (2003)
11. Durrett, R.: Probability: Theory and Examples, 4th edn. Cambridge University Press, Cambridge (2010)
12. Durrett, R.: Random Graph Dynamics. Cambridge University Press, Cambridg (2007)
13. Fisher, E.: On the law of the iterated logarithm for martingales. Ann. Probab. **20**, 675–680 (1992)
14. Hall, P., Heyde, C.C.: Martingale Limit Theory and its Application. Academic Press, New York (1990)
15. Krapivsky, P., Redner, S.: Organization of growing random networks. Phys. Rev. E **63**, 066123-1–066123-14 (2001)
16. Krapivisky, P., Redner, S., Leyvraz, F.: Connectivity of growing random networks. Phys. Rev. Lett. **85**, 4629–4632 (2000)
17. Mitzenmacher, M.: A brief history of generative models for power law and lognormal distributions. Internet Math. **1**, 226–251 (2004)
18. Móri, T.F.: On random trees. Stud. Sci. Math. Hung. **39**, 143–155 (2002)
19. Newman, M.E.J.: Networks: An Introduction. Oxford University Press, USA (2010)
20. Oliveira, R., Spencer, J.: Connectivity transitions in networks with super-linear preferential attachment. Internet Math. **2**, 121–163 (2005)
21. Rudas, A., Tóth, B., Valkó, B.: Random trees and general branching processes. Random Struct. Algorithms **31**, 186–202 (2007)
22. Sethuraman, S., Venkataramani, S.V.: On the growth of a superlinear preferential attachment scheme (2017). arXiv: 1704.05568

A Natural Probabilistic Model on the Integers and Its Relation to Dickman-Type Distributions and Buchstab's Function

Ross G. Pinsky

Dedicated to Raghu Varadhan on the occasion of his 75th birthday

Abstract Let $\{p_j\}_{j=1}^{\infty}$ denote the set of prime numbers in increasing order, let $\Omega_N \subset \mathbb{N}$ denote the set of positive integers with no prime factor larger than p_N and let P_N denote the probability measure on Ω_N which gives to each $n \in \Omega_N$ a probability proportional to $\frac{1}{n}$. This measure is in fact the distribution of the random integer $I_N \in \Omega_N$ defined by $I_N = \prod_{j=1}^{N} p_j^{X_{p_j}}$, where $\{X_{p_j}\}_{j=1}^{\infty}$ are independent random variables and X_{p_j} is distributed as $\text{Geom}(1 - \frac{1}{p_j})$. We show that $\frac{\log n}{\log N}$ under P_N converges weakly to the Dickman distribution. As a corollary, we recover a classical result from multiplicative number theory—Mertens' formula. Let $D_{\text{nat}}(A)$ denote the natural density of $A \subset \mathbb{N}$, if it exists, and let $D_{\text{log-indep}}(A) = \lim_{N \to \infty} P_N(A \cap \Omega_N)$ denote the density of A arising from $\{P_N\}_{N=1}^{\infty}$, if it exists. We show that the two densities coincide on a natural algebra of subsets of $\mathbb{N}$. We also show that they do not agree on the sets of $n^{\frac{1}{s}}$-*smooth numbers* $\{n \in \mathbb{N} : p^+(n) \leq n^{\frac{1}{s}}\}$, $s > 1$, where $p^+(n)$ denotes the largest prime divisor of n. This last consideration concerns distributions involving the Dickman function. We also consider the sets of $n^{\frac{1}{s}}$-*rough numbers* $\{n \in \mathbb{N} : p^-(n) \geq n^{\frac{1}{s}}\}$, $s > 1$, where $p^-(n)$ denotes the smallest prime divisor of n. We show that the probabilities of these sets, under the uniform distribution on $[N] = \{1, \ldots, N\}$ and under the P_N-distribution on Ω_N, have the same asymptotic decay profile as functions of s, although their rates are necessarily different. This profile involves the Buchstab function. We also prove a new representation for the Buchstab function.

R. Pinsky (✉)
Department of Mathematics Technion—Israel Institute of Technology, Haifa 32000, Israel
e-mail: pinsky@math.technion.ac.il
URL: http://www.math.technion.ac.il/~pinsky/

© Springer Nature Switzerland AG 2019
P. Friz et al. (eds.), *Probability and Analysis in Interacting Physical Systems*, Springer Proceedings in Mathematics & Statistics 283,
https://doi.org/10.1007/978-3-030-15338-0_10

Keywords Dickman function · Dickman-type distribution · Buchstab function · Prime number · k-free numbers

2000 Mathematics Subject Classification. 60F05 · 11N25 · 11K65

1 Introduction and Statement of Results

For a subset $A \subset \mathbb{N}$, the natural density $D_{\mathrm{nat}}(A)$ of A is defined by $D_{\mathrm{nat}}(A) = \lim_{N \to \infty} \frac{|A \cap [N]|}{N}$, whenever this limit exists, where $[N] = \{1, \ldots, N\}$. The natural density is additive, but not σ-additive, and therefore not a measure. For each prime p and each $n \in \mathbb{N}$, define the nonnegative integer $\beta_p(n)$, the p-adic order of n, by $\beta_p(n) = m$, if $p^m \mid n$ and $p^{m+1} \nmid n$. Let $\delta_p(n) = \max(1, \beta_p(n))$ denote the indicator function of the set of positive integers divisible by p. It is clear that for each $m \in \mathbb{N}$, the natural density of the set $\{n \in \mathbb{N} : \beta_p(n) \geq m\}$ of natural numbers divisible by p^m is $(\frac{1}{p})^m$. More generally, it is easy to see that for $l \in \mathbb{N}$, $\{m_j\}_{j=1}^l \subset \mathbb{N}$ and distinct primes $\{p_j\}_{j=1}^l$, the natural density of the set $\{n \in \mathbb{N} : \beta_{p_j}(n) \geq m_j, \ j = 1, \ldots, l\}$ is $\prod_{j=1}^l (\frac{1}{p_j})^{m_j}$. That is, the distribution of the random vector $\{\delta_{p_j}\}_{j=1}^l$, defined on the probability space $[N]$ with the uniform distribution, converges weakly as $N \to \infty$ to the random vector $\{Y_{p_j}\}_{j=1}^l$ with independent components distributed according to the Bernoulli distributions $\{\mathrm{Ber}(\frac{1}{p_j})\}_{j=1}^l$, and the distribution of the random vector $\{\beta_{p_j}\}_{j=1}^l$ converges weakly as $N \to \infty$ to the random vector $\{X_{p_j}\}_{j=1}^l$ with independent components distributed according to the geometric distributions $\mathrm{Geom}(1 - \frac{1}{p_j})$ $\left(P(X_{p_j} = m) = (\frac{1}{p_j})^m (1 - \frac{1}{p_j}), \ m = 0, 1, \ldots\right)$. This fact is the starting point of probabilistic number theory.

Denote the primes in increasing order by $\{p_j\}_{j=1}^\infty$. In the sequel, we will assume that the random variables $\{X_{p_j}\}_{j=1}^\infty$, $\{Y_{p_j}\}_{j=1}^\infty$ with distributions as above are defined as independent random variables on some probability space, and we will use the generic notation P to denote probabilities corresponding to these random variables.

A real-valued function f defined on $\mathbb{N}$ is called a real arithmetic function. It is called *additive* if $f(nm) = f(n) + f(m)$, whenever $(m, n) = 1$. If in addition, $f(p^m) = f(p)$, for all primes p and all $m \geq 2$, then it is called *strongly additive*. Classical examples of additive arithmetic functions are, for example, $\log \frac{\phi(n)}{n}$, where ϕ is the Euler totient function, $\omega(n)$, the number of distinct prime divisors of n, $\Omega(n)$, the number of prime divisors of n counting multiplicities and $\log \sigma(n)$, where σ is the sum-of-divisors function. The first two of these functions are strongly additive while the last two are not.

If f is additive, then $f(1) = 0$. Writing $n \in \mathbb{N}$ as $n = \prod_{j=1}^\infty p_j^{\beta_{p_j}(n)}$, we have for f additive, $f(n) = \sum_{j=1}^\infty f(p_j^{\beta_{p_j}(n)})$, and for f strongly additive, $f(n) = \sum_{j=1}^\infty f(p_j^{\delta_{p_j}(n)}) = \sum_{j=1}^\infty f(p_j)\delta_{p_j}(n)$. Equivalently, for each $N \in \mathbb{N}$, we have for f additive,

$$f(n) = \sum_{j=1}^{N} f(p_j^{\beta_{p_j}(n)}), \ n \in [N],$$ (1.1)

and for f strongly additive,

$$f(n) = \sum_{j=1}^{N} f(p_j)\delta_{p_j}(n), \ n \in [N].$$ (1.2)

In light of the above discussion, it is natural to compare (1.1) to

$$\mathcal{X}_N \equiv \sum_{j=1}^{N} f(p_j^{X_{p_j}}),$$ (1.3)

and to compare (1.2) to

$$\mathcal{Y}_N \equiv \sum_{j=1}^{N} f(p_j)Y_{p_j}.$$ (1.4)

Now $\mathcal{Y}_N$ converges in distribution as $N \to \infty$ if and only if it converges almost surely, and the almost sure convergence of $\mathcal{Y}_N$ is characterized by the Kolmogorov three series theorem [8]. Since $EY_{p_j} = EY_{p_j}^2 = \frac{1}{p_j}$, it follows from that theorem that $\mathcal{Y}_N$ converges almost surely if and only if the following three series converge: 1. $\sum_{j:|f(p_j)|\le 1} \frac{f(p_j)}{p_j}$; 2. $\sum_{j:|f(p_j)|\le 1} \frac{f^2(p_j)}{p_j}$; 3. $\sum_{j:|f(p_j)|>1} \frac{1}{p_j}$. Since $P(X_{p_j} \ge 2) = \frac{1}{p_j^2}$, it follows from the Borel-Cantelli lemma that $\sum_{j=1}^{\infty} 1_{\{X_{p_j} \ge 2\}}$ is almost surely finite; thus the very same criterion also determines whether $\mathcal{X}_N$ converges almost surely. The Erdös-Wintner theorem [11] states that for additive f, the converges of these three series is a necessary and sufficient condition for the convergence in distribution as $N \to \infty$ of the random variable $f(n)$ in (1.1) on the probability space $[N]$ with the uniform distribution. In the same spirit, the Kac-Erdös theorem [12] states that if f is strongly additive and bounded, then a central limit theorem holds as $N \to \infty$ for $f(n)$ on the probability space $[N]$ with the uniform distribution, if the conditions of the Feller-Lindeberg central limit theorem [8] hold for $\mathcal{Y}_N$. An appropriate corresponding result can be stated for additive f and or unbounded f. There is also a weak law of large numbers result, which in the case of $f = \omega$ goes by the name of the Hardy-Ramanujan theorem [14]. It should be noted that the original proof of Hardy and Ramanujan was quite complicated and not at all probabilistic; however, the later and much simpler proof of Turan [24] has a strong probabilistic flavor. For a concise and very readable probabilistic approach to these results, see Billingsley [3]; for a more encyclopedic probabilistic approach, see Elliott [9, 10]; for a less probabilistic approach, see Tenenbaum [23].

Turan's paper with the proof of the Hardy-Ramanujan theorem, as well as the Erdös-Wintner theorem and several papers leading up to it, all appeared in the 1930's, and the Kac-Erdös theorem appeared in 1940. Large deviations for independent and non-identically distributed random variables have been readily available since the 1970's, thus this author certainly finds it quite surprising that until very recently no one extended the parallel between (1.2) and (1.4), or (1.1) and (1.3), to study the large deviations of (1.2) or (1.1)! See [16, 17].

Another density that is sometimes used in number theory is the *logarithmic density*, $D_{\log}$, which is defined by

$$D_{\log}(A) = \lim_{N \to \infty} \frac{1}{\log N} \sum_{n \in A \cap [N]} \frac{1}{n}, \tag{1.5}$$

for $A \subset \mathbb{N}$, whenever this limit exists. Using summation by parts, it is easy to show that if $D_{\text{nat}}(A)$ exists, then $D_{\log}(A)$ exists and coincides with $D_{\text{nat}}(A)$ [23]. (On the other hand, there are sets without natural density for which the logarithmic density exists. The most prominent of these are the sets $\{B_d\}_{d=1}^{9}$ associated with Benford's law, where B_d is the set of positive integers whose first digit is d. One has $D_{\log}(B_d) = \log_{10}(1 + \frac{1}{d})$.) Thus, also on the probability space $[N]$ with the probability measure which gives to each integer n a measure proportional to $\frac{1}{n}$, the distribution of the random vector $\{\beta_{p_j}\}_{j=1}^{l}$ converges weakly as $N \to \infty$ to the random vector $\{X_{p_j}\}_{j=1}^{l}$ with independent components distributed according to the geometric distributions $\text{Geom}(1 - \frac{1}{p_j})$.

Motivated by the background described above, in this paper we consider a sequence of probability measures on $\mathbb{N}$ which may be thought of as a synthesis between the logarithmic density $D_{\log}$ and the concept of approximating the natural density via a sequence of independent random variables. Let us denote by

$$\Omega_N = \{n \in \mathbb{N} : p_j \nmid n, j > N\}$$

the set of positive integers with no prime divisor larger than p_N. By the Euler product formula,

$$C_N \equiv \sum_{n \in \Omega_N} \frac{1}{n} = \prod_{j=1}^{N} (1 - \frac{1}{p_j})^{-1} < \infty. \tag{1.6}$$

Let P_N denote the probability measure on Ω_N for which the probability of n is proportional to $\frac{1}{n}$; namely,

$$P_N(\{n\}) = \frac{1}{C_N} \frac{1}{n}, \quad n \in \Omega_N. \tag{1.7}$$

The connection between P_N and the logarithmic density is clear; the connection between P_N and a sequence of independent random variables comes from the following proposition. Define a random positive integer $I_N \in \Omega_N$ by

$$I_N = \prod_{j=1}^{N} p_j^{X_{p_j}}.$$

Proposition 1.1 *The distribution of I_N is P_N; that is,*

$$P_N(\{n\}) = P(I_N = n), \ n \in \Omega_N.$$

Proof Let $n = \prod_{j=1}^{N} p_j^{a_j} \in \Omega_N$. We have

$$P(I_N = n) = \prod_{j=1}^{N} P(X_{p_j} = a_j) = \prod_{j=1}^{N} (\frac{1}{p_j})^{a_j} (1 - \frac{1}{p_j}) = \frac{1}{C_N} \frac{1}{n} = P_N(\{n\}). \qquad \square$$

Let $D_{\text{log-indep}}$ denote the asymptotic density obtained from P_N:

$$D_{\text{log-indep}}(A) = \lim_{N \to \infty} P_N(A \cap \Omega_N) = \lim_{N \to \infty} \frac{1}{C_N} \sum_{n \in A \cap \Omega_N} \frac{1}{n},$$

for $A \subset \mathbb{N}$, whenever the limit exists. Note that the weight functions used in calculating the asymptotic densities $D_{\text{log-indep}}$ and $D_{\log}$ have the same profile, but the sequences of subsets of $\mathbb{N}$ over which the limits are taken, namely $\{\Omega_N\}_{N=1}^{\infty}$ and $\{[N]\}_{N=1}^{\infty}$, are different. As already noted, when $D_{\text{nat}}(A)$ exists, so does $D_{\log}(A)$ and they coincide. We will show below in Proposition 1.3 that the densities $D_{\text{log-indep}}$ and D_{nat} coincide on many natural subsets of $\mathbb{N}$. However we will also show below in Theorem 1.2 that they disagree on certain important, fundamental subsets of $\mathbb{N}$.

For $k \geq 2$, a positive integer n is called k-free if $p^k \nmid n$, for all primes p. When $k = 2$, one uses the term *square-free*. Let S_k denote the set of all k-free positive integers. Let

$$\Omega_N^{(k)} = \Omega_N \cap S_k.$$

Note that $\Omega_N^{(k)}$ is a finite set; it has k^N elements. The measure P_N behaves nicely under conditioning on S_k. For $k \geq 2$, define the measure $P_N^{(k)}$ by

$$P_N^{(k)}(\cdot) = P_N(\ \cdot \ |S_k).$$

Let $\{X_{p_j}^{(k)}\}_{j=1}^{\infty}$ be independent random variables with $X_{p_j}^{(k)}$ distributed as X_{p_j} conditioned on $\{X_{p_j} < k\}$. (Assume that these new random variables are defined on the same space as the $\{X_{p_j}\}_{j=1}^{\infty}$ so that we can still use P for probabilities.) Let

$$I_N^{(k)} = \prod_{j=1}^{N} p_j^{X_{p_j}^{(k)}}.$$

Proposition 1.2 *The distribution of $I_N^{(k)}$ is $P_N^{(k)}$.*

Proof

$$P_N^{(k)}(\{n\}) = P_N(\{n\}|S_k) = P(I_N = n|X_{p_j} < k, \; j \in [N]) = P(I_N^{(k)} = n),$$

where the second equality follows from Proposition 1.1.

Remark 1.1 The measure $P_N^{(2)}$ was considered by Cellarosi and Sinai in [6]. See also the remark after Theorem 1.1 below.

We will prove the following result, which identifies a certain natural algebra of subsets of $\mathbb{N}$ on which $D_{\text{log-indep}}$ and D_{nat} coincide.

Proposition 1.3 *The densities $D_{\text{log-indep}}$ and D_{nat} coincide on the algebra of subsets of $\mathbb{N}$ generated by the inverse images of $\{\beta_{p_j}\}_{j=1}^{\infty}$ and the sets $\{S_k\}_{k=2}^{\infty}$.*

We will show that under the measure P_N as well as under the measure $P_N^{(k)}$, the random variable $\frac{\log n}{\log N}$, with $n \in \Omega_N$ in the case of P_N and $n \in \Omega_N^{(k)}$ in the case of $P_N^{(k)}$, converges in distribution as $N \to \infty$ to the distribution whose density is $e^{-\gamma}\rho(x)$, $x \in [0, \infty)$, where γ is Euler's constant, and ρ is the Dickman function, which we now describe. The Dickman function is the unique continuous function satisfying

$$\rho(x) = 1, \; x \in (0, 1],$$

and satisfying the differential-delay equation

$$x\rho'(x) + \rho(x - 1) = 0, \; x > 1.$$

By analyzing the Laplace transform of ρ, a rather short proof shows that $\int_0^{\infty} \rho(x)dx = e^{\gamma}$; thus $e^{-\gamma}\rho(x)$ is indeed a probability density on $[0, \infty)$. We will call this distribution the *Dickman distribution*. The distribution decays very rapidly; indeed, it is not hard to show that $\rho(s) \leq \frac{1}{\Gamma(s+1)}$. For an analysis of the Dickman function, see for example, [23] or [18].

Theorem 1.1 *Under both P_N and $P_N^{(k)}$, $k \geq 2$, the random variable $\frac{\log n}{\log N}$ converges weakly to the Dickman distribution.*

Remark 1.2 For $P_N^{(2)}$, Theorem 1.1 was first proved by Cellarosi and Sinai [6]. Their proof involved calculating characteristic functions and was quite tedious and long. Our short proof uses Laplace transforms and the asymptotic growth rate of the primes given by the Prime Number Theorem (henceforth PNT). After this paper was written, one of the authors of [13] pointed out to the present author that their paper also gives a simpler proof of the result in [6].

Using Theorem 1.1 we can recover a classical result from multiplicative number theory; namely,

Mertens' formula.

$$C_N = \sum_{n \in \Omega_N} \frac{1}{n} = \prod_{j=1}^{N}(1 - \frac{1}{p_j})^{-1} \sim e^{\gamma} \log N, \text{ as } N \to \infty. \tag{1.8}$$

(Traditionally the formula is written as $\prod_{p \le N}(1 - \frac{1}{p})^{-1} \sim e^{\gamma} \log N$, where the product is over all primes less than or equal to N. To show that the two are equivalent only requires the fact that $p_N = o(N^{1+\epsilon})$, for any $\epsilon > 0$.) A nice, alternative form of the formula is

$$\frac{\sum_{n \in \Omega_N} \frac{1}{n}}{\sum_{n=1}^{N} \frac{1}{n}} \sim e^{\gamma}.$$

Here is the derivation of Mertens' formula from Theorem 1.1. From the definition of P_N, we have $P_N(\frac{\log n}{\log N} \le 1) = \frac{1}{C_N} \sum_{n=1}^{N} \frac{1}{n}$. Thus, from Theorem 1.1, we have $\lim_{N \to \infty} \frac{1}{C_N} \sum_{n=1}^{N} \frac{1}{n} = \int_0^1 e^{-\gamma} \rho(x) dx = e^{-\gamma}$. Now (1.8) follows from this and the fact that $\sum_{n=1}^{N} \frac{1}{n} \sim \log N$.

A direct proof that $C_N \sim c \log N$, for some c, follows readily with the help of Mertens' second theorem (see (1.15)). The proof that the constant is e^{γ} is quite nontrivial. Of course, our proof of Mertens' formula via Theorem 1.1 uses the fact that $\int_0^{\infty} \rho(x) dx = e^{\gamma}$, but as noted, this result is obtained readily by analyzing the Laplace transform of ρ.

Why does the Dickman function arise? Our proof of Theorem 1.1 does not shed light on this question. However, in Sect. 3 we present a proof of the fact that if the limiting distribution of $\frac{\log n}{\log N}$ under P_N exists, then it must be the Dickman distribution. We believe that this is of independent interest, as it provides some intuition as to why the Dickman function arises. For more results in this spirit, see [21], which studies generalized Dickman distributions.

The Dickman function arises in probabilistic number theory in the context of so-called *smooth* numbers; that is, numbers all of whose prime divisors are "small." Let $\Psi(x, y)$ denote the number of positive integers less than or equal to x with no prime divisors greater than y. Numbers with no prime divisors greater than y are called *y-smooth* numbers. Then for $s \ge 1$, $\Psi(N, N^{\frac{1}{s}}) \sim N\rho(s)$, as $N \to \infty$. This result was first proved by Dickman in 1930 [7], whence the name of the function, with later refinements by de Bruijn [4]. (In particular, there are rather precise error terms.) See also [18] or [23]. Let $p^+(n)$ denote the largest prime divisor of n. Then Dickman's result states that the random variable $\frac{\log N}{\log p^+(n)}$ on the probability space $[N]$ with the uniform distribution converges weakly in distribution as $N \to \infty$ to the distribution whose distribution function is $1 - \rho(s)$, $s \ge 1$, and whose density is $-\rho'(s) = \frac{\rho(s-1)}{s}$, $s \ge 1$. Since $\frac{\log n}{\log N}$ on the probability space $[N]$ with the uniform distribution converges weakly in distribution to 1 as $N \to \infty$, an equivalent statement of Dickman's result is that the random variable $\frac{\log n}{\log p^+(n)}$ on the probability space $[N]$ with the uniform distribution converges weakly in distribution as $N \to \infty$ to the distribution whose distribution function is $1 - \rho(s)$, $s \ge 1$, For later use, we state this as follows in terms of the natural density:

$$D_{\mathrm{nat}}(\{n \in \mathbb{N} : p^+(n) \le n^{\frac{1}{s}}\}) = D_{\mathrm{nat}}(\{n \in \mathbb{N} : \frac{\log n}{\log p^+(n)} \ge s\}) = \rho(s), \ s \ge 1.$$

(1.9)

We will call $\{n \in \mathbb{N} : p^+(n) \le n^{\frac{1}{s}}\}$ the set of $n^{\frac{1}{s}}$-smooth numbers.

The standard number-theoretic proof of Dickman's result is via induction. It can be checked that this inductive proof also works to obtain a corresponding result for k-free integers. Thus,

$$D_{\mathrm{nat}}(\{n \in \mathbb{N} : p^+(n) \le n^{\frac{1}{s}}\}|S_k) = D_{\mathrm{nat}}(\{n \in \mathbb{N} : \frac{\log n}{\log p^+(n)} \ge s\}|S_k) = \rho(s),$$

for $s \ge 1$ and $k \ge 2$.

(1.10)

Proposition 1.3 shows that D_{nat} and $D_{\text{log-indep}}$ coincide on a certain natural algebra of sets. We will prove that they disagree on the sets appearing in (1.9) or (1.10); namely on the sets of $n^{\frac{1}{s}}$-smooth numbers, $s > 1$, and on the intersection of such a set with the set of k-free numbers, S_k, $k \ge 2$.

Theorem 1.2 *Under both P_N and $P_N^{(k)}$ the random variable $\frac{\log n}{\log p^+(n)}$ converges weakly as $N \to \infty$ to $D + 1$, where D has the Dickman distribution; that is,*

$$D_{\text{log-indep}}(\{n \in \mathbb{N} : p^+(n) \le n^{\frac{1}{s}}\}) = D_{\text{log-indep}}(\{n \in \mathbb{N} : \frac{\log n}{\log p^+(n)} \ge s\}) =$$

$$e^{-\gamma} \int_{s-1}^{\infty} \rho(x)dx, \ s \ge 1;$$

(1.11)

$$D_{\text{log-indep}}(\{n \in \mathbb{N} : p^+(n) \le n^{\frac{1}{s}}\}|S_k) = D_{\text{log-indep}}(\{n \in \mathbb{N} : \frac{\log n}{\log p^+(n)} \ge s\}|S_k) =$$

$$e^{-\gamma} \int_{s-1}^{\infty} \rho(x)dx, \ s \ge 1, k \ge 2.$$

Remark 1.3 Recalling that whenever the natural density exists, the logarithmic one does too and they are equal, it follows from (1.9) that $D_{\log}(\{n \in \mathbb{N} : p^+(n) \le n^{\frac{1}{s}}\}) = \rho(s)$. Since, as we've noted, the weights used in calculating the densities $D_{\log}$ and $D_{\text{log-indep}}$ have the same profile, but the sequences of subsets of $\mathbb{N}$ over which the limits are taken, namely $\{[N]\}_{N=1}^{\infty}$ and $\{\Omega_N\}_{N=1}^{\infty}$, are different, and since the integers in $[N]$ and in Ω_N are constructed from the same set $\{p_j\}_{j=1}^{N}$ of primes, and $[N] \subset \Omega_N$, intuition suggests that

$$\rho(s) \le e^{-\gamma} \int_{s-1}^{\infty} \rho(x)dx, \ s \ge 1;$$

(1.12)

that is, that under $D_{\text{log-indep}}$, $n^{\frac{1}{s}}$-smooth numbers are more likely than under D_{nat}. And indeed this is the case. Letting $H(s) = e^{-\gamma} \int_{s-1}^{\infty} \rho(x)dx - \rho(s)$, we have $H(1) =$

$H(\infty) = 0$. Differentiating H, and using the differential-delay equation satisfied by ρ, one has $H'(s) = -e^{-\gamma}\rho(s-1) - p'(s) = \rho(s-1)(\frac{1}{s} - e^{-\gamma})$. Thus, $H'(s)$ vanishes only at $s = e^{\gamma}$. Differentiating again and again using the differential-delay equation, one finds that $H''(e^{\gamma}) < 0$; thus, $H(s) \geq 0$, for $s \geq 1$, proving (1.12).

We now consider integers all of whose prime divisors are "large." Let $\Phi(x, y)$ denote the number of positive integers less than or equal to x all of whose prime divisors are greater than or equal to y. Numbers with no prime divisors less than y are called y-*rough* numbers. The Buchstab function $\omega(s)$, defined for $s \geq 1$, is the unique continuous function satisfying

$$\omega(s) = \frac{1}{s}, \ 1 \leq s \leq 2,$$

and satisfying the differential-delay equation

$$(s\omega(s))' = \omega(s-1), \ s > 2.$$

In 1937, Buchstab proved [5] that for $s > 1$, $\Phi(N, N^{\frac{1}{s}}) \sim \frac{N s\omega(s)}{\log N}$ as $N \to \infty$; whence the name of the function. See also [18] or [23]. Let $p^-(n)$ denote the smallest prime divisor of n. Then Buchstab's result states that

$$\frac{|\{n \in [N] : p^-(n) \geq N^{\frac{1}{s}}\}|}{N} = \frac{|\{n \in [N] : \frac{\log N}{\log p^-(n)} \leq s\}|}{N} \sim \frac{s\omega(s)}{\log N}, \qquad (1.13)$$

for $s > 1$, as $N \to \infty$.

Since $\frac{|\{n \in [N]: \frac{\log N}{\log n} > 1 + \epsilon\}|}{N} = N^{-\epsilon}$, it follows that (1.13) is equivalent to

$$\frac{|\{n \in [N] : p^-(n) \geq n^{\frac{1}{s}}\}|}{N} = \frac{|\{n \in [N] : \frac{\log n}{\log p^-(n)} \leq s\}|}{N} \sim \frac{s\omega(s)}{\log N}, \qquad (1.14)$$

for $s > 1$, as $N \to \infty$.

One has $\lim_{s \to \infty} \omega(s) = e^{-\gamma}$, and the rate of convergence is super-exponential [23]. We will call $\{n \in [N] : p^-(n) \geq n^{\frac{1}{s}}\}$ the set of $n^{\frac{1}{s}}$-rough numbers. (We note that the probability that the shortest cycle of a uniformly random permutation of $[N]$ is larger or equal to $\frac{N}{s}$ decays asymptotically as $\frac{s\omega(s)}{N}$ [1].)

Note that (1.14) also holds for $s = 1$, since in this case (1.14) reduces to $\frac{\Pi(N)}{N} \sim \frac{1}{\log N}$; that is, it reduces to the PNT. Buchstab assumed the PNT in proving (1.13).

What is the asymptotic probability of a prime number under the sequence of measures used to construct the logarithmic density $D_{\log}$ and under the sequence $\{P_N\}_{N=1}^{\infty}$ used to construct the density $D_{\log\text{-indep}}$? Mertens' second theorem states that

$$\sum_{p \leq N} \frac{1}{p} = \log \log N + M_0 + O(\frac{1}{\log N}), \qquad (1.15)$$

where the summation is over primes p, and where M_0 is called the Meissel-Mertens constant [19]. By the PNT, $p_N \sim N \log N$, thus by Mertens' second theorem,

$$\sum_{j=1}^{N} \frac{1}{p_j} \sim \log\log(N \log N) \sim \log\log N. \tag{1.16}$$

From (1.15) we conclude that for the sequence of measures used to construct the logarithmic density $D_{\log}$, the probability of a prime is

$$\frac{1}{\log N} \sum_{p \leq N} \frac{1}{p} \sim \frac{\log\log N}{\log N}. \tag{1.17}$$

Since

$$P_N(\{n \in \Omega_N : n \text{ is prime}\}) = \frac{1}{C_N} \sum_{j=1}^{N} \frac{1}{p_j},$$

from (1.16) and Mertens formula given in (1.8), we conclude that for the sequence $\{P_N\}_{N=1}^{\infty}$ use to construct the density $D_{\text{log-indep}}$, the probability of a prime satisfies

$$P_N(\{n \in \Omega_N : n \text{ is prime}\}) \sim \frac{e^{-\gamma} \log\log N}{\log N}. \tag{1.18}$$

From (1.17) and (1.18) it is clear that (1.14) cannot hold when the sequence of uniform measures on $[N]$, $N = 1, 2, \ldots$, appearing on the left hand side there is replaced either by the sequence of measures used to calculate the logarithmic density $D_{\log}$ or by the sequence $\{P_N\}_{N=1}^{\infty}$ used to calculate the density $D_{\text{log-indep}}$. However, letting

$$a_s(n) = \begin{cases} 1, & p^-(n) \geq n^{\frac{1}{s}}, \\ 0, & \text{otherwise}, \end{cases}$$

and $A_s(t) = \sum_{j=1}^{[t]} a_s(j)$, $t \geq 1$, a summation by parts gives

$$\sum_{n \leq N : p^-(n) \geq n^{\frac{1}{s}}} \frac{1}{n} = \sum_{n=1}^{N} \frac{a_s(n)}{n} = \frac{A_s(N)}{N} + \int_1^N \frac{A_s(t)}{t^2} dt. \tag{1.19}$$

By (1.14), $\frac{A_s(t)}{t} \sim \frac{s\omega(s)}{\log t}$ as $t \to \infty$; thus from (1.19) we have

$$\frac{1}{\log N} \sum_{n \leq N : p^-(n) \geq n^{\frac{1}{s}}} \frac{1}{n} \sim \log\log N \frac{s\omega(s)}{\log N}.$$

That is, modulo the change necessitated by comparing (1.17) to the PNT, Buchstab's result on $n^{\frac{1}{s}}$-rough numbers for the uniform measure in (1.14) carries over to the measures used in the construction of the logarithmic density.

Modulo the change necessitated by comparing (1.18) to the PNT, does Buchstab's result on $n^{\frac{1}{s}}$-rough numbers also carry over to the measures $\{P_N\}_{N=1}^\infty$ used in the construction of the density $D_{\text{log-indep}}$? Since (1.9) and (1.11) show that the *positive* densities with respect D_{nat} and $D_{\text{log-indep}}$ of the $n^{\frac{1}{s}}$-smooth sets $\{n \in \mathbb{N} : p^+(n) \le n^{\frac{1}{s}}\}$ do not coincide, it is interesting to discover that the answer is indeed affirmative.

Theorem 1.3 *For $s \ge 1$,*

$$P_N(\{n \in [N] : p^-(n) \ge n^{\frac{1}{s}}\}) = P_N\left(\frac{\log n}{\log p^-(n)} \le s\right) \sim (e^{-\gamma} \log \log N)\frac{s\omega(s)}{\log N},$$

as $N \to \infty$.

$$(1.20)$$

Recalling the definition of the Buchstab function, note that $V(s) \equiv s\omega(s)$ is the unique continuous function satisfying $V(s) = 1, 1 \le s \le 2$, and $V'(s) = \frac{V(s-1)}{s-1}$, for $s > 2$. In the proof of Theorem 1.3, we actually show that (1.20) holds with $s\omega(s)$ on the right hand side replaced by

$$v(s) \equiv \sum_{L=1}^{[s]} \Lambda_L(s),$$

where

$$\Lambda_1(s) = 1, \ s \ge 1;$$

$$\Lambda_2(s) = \int_1^{s-1} \frac{du_1}{u_1} = \log(s-1), \ s \ge 2;$$

$$\Lambda_L(s) = \int_{L-1}^{s-1} \int_{L-2}^{u_{L-1}-1} \cdots \int_1^{u_2-1} \prod_{j=1}^{L-1} \frac{du_j}{u_j}, \ s \ge L \ge 3.$$

$$(1.21)$$

Now $\Lambda'_L(s) = \frac{1}{s-1}\Lambda_{L-1}(s-1)$, for $s \ge L \ge 2$, while of course $\Lambda'_1(s) = 0$. Thus, $v(s) = 1$, for $1 \le s \le 2$ and $v'(s) = \frac{v(s-1)}{s-1}$, for $s > 2$. This proves the following result.

Proposition 1.4

$$s\omega(s) = 1 + \log(s-1) + \sum_{L=3}^{[s]} \int_{L-1}^{s-1} \int_{L-2}^{u_{L-1}-1} \cdots \int_1^{u_2-1} \prod_{j=1}^{L-1} \frac{du_j}{u_j}, \ s \ge 3. \quad (1.22)$$

The representation of the Buchstab function ω in (1.22) seems to be new. It is simpler than the following known representation [1, 15]:

$$s\omega(s) = 1 + \sum_{L=2}^{[s]} \frac{1}{L!} \int_{\substack{\frac{1}{s} \le y_j \le 1 \\ \frac{1}{s} \le 1-(y_1+y_2+\cdots+y_{L-1}) \le 1)}} \frac{1}{1-(y_1+y_2+\cdots+y_{L-1})} \prod_{j=1}^{L-1} \frac{dy_j}{y_j}.$$

Since $\lim_{s \to \infty} \omega(s) = e^{-\gamma}$, we also obtain what seems to be yet another representation of Euler's constant:

$$e^{-\gamma} = \lim_{N \to \infty} \frac{1}{N} \sum_{L=3}^{N} \int_{L-1}^{N} \int_{L-2}^{u_{L-1}-1} \cdots \int_{1}^{u_2-1} \prod_{j=1}^{L-1} \frac{du_j}{u_j}.$$

We conclude this introduction with some additional comments regarding Dickman's classical theorem that $\Psi(N, N^{\frac{1}{s}}) \sim N\rho(s)$, or it slightly refined version (1.9). The Dickman distribution is the distribution with density $e^{-\gamma}\rho(x)$. As is known, and as follows from the work in Sect. 3, if D is a random variable with the Dickman distribution, then

$$D \overset{dist}{=} D'U + U, \quad U \overset{dist}{=} \text{Unif}([0,1]), \quad D' \overset{dist}{=} D, \quad U \text{ and } D' \text{ independent.} \quad (1.23)$$

Now equivalent to (1.9) is the statement that $\frac{\log p^+(n)}{\log n}$ on $[N]$ with the uniform distribution converges weakly in distribution as $N \to \infty$ to the distribution whose distribution function is $\rho(\frac{1}{s})$, $s \in [0,1]$. The corresponding density function is then $\frac{-\rho'(\frac{1}{s})}{s^2} = \frac{1}{s}\rho(\frac{1}{s}-1)$. In the spirit of (1.23), it has been shown that if $\hat{D}$ denotes a random variable with this distribution, then

$$\hat{D} \overset{dist}{=} \max(1-U, \hat{D}U), \quad U \overset{dist}{=} \text{Unif}([0,1]), \quad U \text{ and } \hat{D} \text{ independent.}$$

In light of the comparison between (1.23) and the above equation, this distribution has been dubbed the *max-Dickman distribution* [20]. This distribution is the first coordinate of the Poisson-Dirichlet distribution on the infinite simplex $\{x = (x_1, x_2, \ldots) : x_i \ge 0, \sum_{i=1}^{\infty} x_i = 1\}$. The Poisson-Dirichlet distribution can be defined as the decreasing order statistics of the GEM distribution, where the GEM distribution is the "stick-breaking" distribution: let $\{U_n\}_{n=1}^{\infty}$ be IID uniform variables on $[0,1]$; let $Y_1 = U_1$, and let $Y_n = U_n \prod_{r=1}^{n-1}(1-U_r)$, $n \ge 2$; then $(Y_1, Y_2, \ldots)$ has the GEM distribution. The n-dimensional density function for the distribution of the first n coordinates of the Poisson-Dirichlet distribution is given by

$$f^{(n)}(s_1, s_2, \ldots, s_n) = \frac{1}{s_1 \cdots s_n} \rho(\frac{1-s_1-\cdots-s_n}{s_n}),$$

$$\text{for } 0 < s_n < \cdots < s_1 < 1 \text{ and } \sum_{j=1}^{n} s_j < 1.$$

Let $p_j^+(n)$ denote the jth largest distinct prime divisor of n, with $p_j^+(n) = 1$ if n has fewer than j distinct prime divisors. In 1972 Billingsley [2] gave a probabilistic proof of the fact that $\frac{1}{\log n}(\log p_1^+(n), \log p_2^+(n), \ldots)$ on $[N]$ with the uniform distribution converges weakly in distribution as $N \to \infty$ to the Poisson-Dirichlet distribution. However, he did not identify it as such as the theory of the Poisson-Dirichlet distribution had not yet been developed. (We note that the random vector consisting of the lengths of the cycles of a uniformly random permutation of $[N]$, arranged in decreasing order, when normalized by dividing their lengths by N, also converges as $N \to \infty$ to the Poisson-Dirichlet distribution [1].)

The rest of the paper is organized as follows. We prove Proposition 1.3 in Sect. 2. In Sect. 3 we prove that if the limiting distribution of $\frac{\log n}{\log N}$ under P_N exists, then it must be the Dickman distribution. The proofs of Theorems 1.1, 1.2, 1.3 are given successively in Sects. 4, 5, 6.

2 Proof of Proposition 3

For the proof of the proposition we need the following result which is obviously known; however, as we were unable to find it in a number theory text, we supply a proof in the appendix.

Proposition 2.1 *For $1 \leq l < k$,*

$$D_{nat}(\beta_{p_j} \geq l | S_k) \equiv \frac{D_{nat}(\{\beta_{p_j} \geq l\} \cap S_k)}{D_{nat}(S_k)} = \frac{(\frac{1}{p_j})^l - (\frac{1}{p_j})^k}{1 - (\frac{1}{p_j})^k}. \tag{2.1}$$

Remark 2.1 When $k = 2$ and $l = 1$, (2.1) becomes $D_{nat}(\beta_{p_j} \geq 1 | S_2) = \frac{1}{1+p_j}$. That is, among square-free numbers, the natural density of those divisible by the prime p_j is $\frac{1}{p_j+1}$.

Proof of Proposition 1.3. In light of Proposition 1.1, it follows immediately that for $l \leq N$, the random vector $\{\beta_{p_j}\}_{j=1}^l$ under P_N has the distribution of $\{X_{p_j}\}_{j=1}^l$ under P, this latter distribution being the weak limit as $N \to \infty$ of the distribution of $\{\beta_{p_j}\}_{j=1}^l$ on $[N]$ with the uniform distribution. From this it follows that $D_{\text{log-indep}}$ and D_{nat} coincide on the algebra of sets generated by the inverse images of the $\{\beta_{p_j}\}_{j=1}^\infty$.

It is well-known that $D_{nat}(S_k) = \frac{1}{\zeta(k)}$, where $\zeta(s) = \sum_{n=1}^\infty \frac{1}{n^s}$ is the Riemann zeta function [22]. On the other hand, by Proposition 1.1 we have

$$P_N(S_k) = P(X_{p_j} < k, \ j \in [N]) = \prod_{j=1}^N P(X_j < k) = \prod_{j=1}^N (1 - \frac{1}{p_j^k}),$$

and so by the Euler product formula we conclude that

$$D_{\text{log-indep}}(S_k) = \lim_{N \to \infty} P_N(S_k) = \lim_{N \to \infty} \prod_{j=1}^{N}(1 - \frac{1}{p_j^k}) = \frac{1}{\zeta(k)}.$$

Thus, the two densities coincide on the algebra generated by $\{S_k\}_{k=2}^{\infty}$.

Also, for $j \leq N$, $k \geq 2$ and $l < k$, we have

$$P_N^{(k)}(\beta_{p_j} \geq l) = P_N(\beta_{p_j} \geq l | S_k) = P\left(X_{p_j} \geq l | X_{p_i} < k, i = 1, \ldots, N\right) =$$

$$P(X_{p_j} \geq l | X_{p_j} < k) = \frac{\sum_{i=l}^{k-1}(\frac{1}{p_j})^i(1 - \frac{1}{p_j})}{\sum_{i=0}^{k-1}(\frac{1}{p_j})^i(1 - \frac{1}{p_j})} = \frac{(\frac{1}{p_j})^l - (\frac{1}{p_j})^k}{1 - (\frac{1}{p_j})^k}.$$

Thus, $D_{\text{log-indep}}(\beta_{p_j} \geq l | S_k) \equiv \frac{D_{\text{log-indep}}(\{\beta_{p_j} \geq l\} \cap S_k)}{D_{\text{log-indep}}(S_k)} = \frac{(\frac{1}{p_j})^l - (\frac{1}{p_j})^k}{1 - (\frac{1}{p_j})^k}$. Recalling Proposition 2.1, we conclude that the two densities indeed coincide on the algebra generated by the inverse images of $\{\beta_{p_j}\}_{j=1}^{\infty}$ and the sets $\{S_k\}_{k=2}^{\infty}$. □

3   If the Limiting Distribution Exists, It Must Be Dickman

In this section we present a proof of the following theorem, independent of our proof of Theorem 1.1,

Theorem 3.1 *If the limiting distribution of $\frac{\log n}{\log N}$ under P_N exists, then it must be the Dickman distribution.*

Proof Let

$$J_N^+ = \max\{j \in [N] : X_{p_j} \neq 0\},$$

with $\max \emptyset \equiv 0$. By Proposition 1.1, the distribution of $\frac{\log n}{\log N}$ under P_N is equal to the distribution of

$$D_N \equiv \frac{1}{\log N} \sum_{n=1}^{N} X_{p_j} \log p_j = \left(\frac{\log J_N^+}{\log N}\right) \frac{1}{\log J_N^+} \sum_{j=1}^{J_N^+ - 1} X_{p_j} \log p_j +$$

$$X_{p_{J_N^+}} \frac{\log p_{J_N^+}}{\log N}, \tag{3.1}$$

where, of course, the sum on the right hand side above is interpreted as equal to 0 if $J_N^+ \leq 1$, and where we define $p_0 = 1$. Our assumption is that $\{D_N\}_{N=1}^{\infty}$ converges weakly to some distribution. Since $P(J_N^+ \leq j) = \prod_{m=j+1}^{N}(1 - \frac{1}{p_m})$, we have $J_N^+ \to \infty$ a.s. as $N \to \infty$. Also, by the independence of $\{X_{p_j}\}_{j=1}^{\infty}$, we have $\sum_{j=1}^{J_N^+ - 1} X_{p_j} \log p_j | \{J_N^+ = j_0\} \stackrel{\text{dist}}{=} \sum_{j=1}^{j_0-1} X_{p_j} \log p_j$. Thus, $\frac{1}{\log J_N^+} \sum_{j=1}^{J_N^+ - 1} X_{p_j} \log p_j$ converges weakly to the same distribution. Using no more than the weak form of

Merten's formula (namely, $\prod_{j=1}^{N}(1 - \frac{1}{p_j})^{-1} \sim c \log N$, for some c) for the asymptotic equivalence below, we have for $0 < x < 1$,

$$P(\frac{\log J_N^+}{\log N} \le x) = P(J_N^+ \le N^x) = \prod_{j=[N^x+1]}^{N} (1 - \frac{1}{p_j}) \sim \frac{\log N^x}{\log N} = x. \quad (3.2)$$

Using only the fact that $p_j = o(j^{(1+\epsilon)})$, for any $\epsilon > 0$, it follows that (3.2) also holds with $\frac{\log J_N^+}{\log N}$ replaced by $\frac{\log p_{J_N^+}}{\log N}$. Note that $X_{p_{J_N^+}}$ conditioned on $\{J_N^+ = j_0\}$ is distributed as $X_{p_{j_0}}$ conditioned on $\{X_{p_{j_0}} \ge 1\}$. A trivial calculation shows that the conditional distribution of X_{p_j}, conditioned on $X_{p_j} \ge 1$, converges weakly to 1 as $j \to \infty$. From the above facts and (3.1) it follows that if D denotes a random variable distributed according to the limiting distribution of $\{D_N\}_{N=1}^{\infty}$, then

$$D \overset{\text{dist}}{=} D'U + U, \quad U \overset{dist}{=} \text{Unif}([0, 1]), \quad D' \overset{\text{dist}}{=} D, \quad U \text{ and } D' \text{ independent}.$$

From this, it is a calculus exercise to show that D has a continuous density f, that f is equal to some constant c on $(0, 1]$, and that f satisfies the differential-delay equation satisfied by the Dickman function ρ on $x > 1$. (See, for example, [21].) Thus $f = c\rho$. Since f is a density and since $\int_0^{\infty} \rho(x)dx = e^{\gamma}$, it follows that the density of D is $e^{-\gamma}\rho$.

4   Proof of Theorem 1.1

We first prove the theorem for P_N. Let E_N denote the expectation with respect to P_N. Using Proposition 1.1, we have

$$E_N \frac{\log n}{\log N} = \frac{1}{\log N} \sum_{j=1}^{N} E X_{p_j} \log p_j = \frac{1}{\log N} \sum_{j=1}^{N} \frac{\log p_j}{p_j - 1}.$$

Mertens' first theorem [19] states that $\sum_{p \le N} \frac{\log p}{p} \sim \log N$, where the sum is over all primes less than or equal to N. Thus, using nothing more than the trivial bound $p_N \le N^k$, for some k, it follows that $\{E_N \frac{\log n}{\log N}\}_{N=1}^{\infty}$ is bounded, and therefore that the distributions of the nonnegative random variables $\{\frac{\log n}{\log N}\}_{N=1}^{\infty}$ under $\{P_N\}_{N=1}^{\infty}$ are tight. In the next paragraph we will prove that their Laplace transforms converge to $\exp(-\int_0^1 \frac{1-e^{-tx}}{x}dx)$. This proves that the distributions converge weakly. By the argument in the paragraph containing (3.1), it then follows that the limiting distribution is the Dickman distribution. Alternatively, the above function is known to be the Laplace transform of the Dickman distribution [18, 23].

By Proposition 1.1, we have for $t \ge 0$,

$$E_N \exp(-t\frac{\log n}{\log N}) = E \exp(-\frac{t}{\log N}\sum_{j=1}^{N} X_{p_j}\log p_j) = \prod_{j=1}^{N} E \exp(-\frac{t\log p_j}{\log N}X_{p_j}).$$
$$(4.1)$$

For $s \geq 0$,

$$E \exp(-sX_{p_j}) = \sum_{k=0}^{\infty} e^{-sk}(\frac{1}{p_j})^k(1 - \frac{1}{p_j}) = (1 - \frac{1}{p_j})\frac{1}{1 - \frac{e^{-s}}{p_j}} = \frac{1}{1 + \frac{1-e^{-s}}{p_j - 1}}.$$
$$(4.2)$$

From (4.1) and (4.2) we have

$$\log E_N \exp(-t\frac{\log n}{\log N}) = -\sum_{j=1}^{N}\log\left(1 + \frac{1 - \exp(-t\frac{\log p_j}{\log N})}{p_j - 1}\right). \qquad (4.3)$$

Now $x - \frac{x^2}{2} \leq \log(1 + x) \leq x$, for $x \geq 0$, and by the bounded convergence theorem, $\lim_{N\to\infty}\sum_{j=1}^{N}\left(\frac{1-\exp(-t\frac{\log p_j}{\log N})}{p_j-1}\right)^2 = 0$; thus,

$$\lim_{N\to\infty}\log E_N \exp(-t\frac{\log n}{\log N}) = -\lim_{N\to\infty}\sum_{j=1}^{N}\frac{1 - \exp(-t\frac{\log p_j}{\log N})}{p_j - 1}. \qquad (4.4)$$

Let $x_j^{(N)} = \frac{\log p_j}{\log N}$ and $\Delta_j^{(N)} = x_{j+1}^{(N)} - x_j^{(N)}$. By the PNT, $p_j \sim j\log j$, as $j \to \infty$; thus

$$\log p_{j+1} - \log p_j \sim \log\frac{(j+1)\log(j+1)}{j\log j} = \log\left((1 + \frac{1}{j})(1 + \frac{\log(1 + \frac{1}{j})}{\log j})\right) \sim \frac{1}{j} \sim \frac{\log p_j}{p_j}.$$
$$(4.5)$$

Consequently,

$$\Delta_j^{(N)} \sim \frac{\log p_j}{p_j\log N}, \quad \text{uniformly as } j, N \to \infty. \qquad (4.6)$$

Note also that

$$\lim_{N\to\infty} x_1^{(N)} = 0, \quad \lim_{N\to\infty} x_N^{(N)} = 1. \qquad (4.7)$$

We rewrite the summand on the right hand side of (4.4) as

$$\sum_{j=1}^{N}\frac{1 - \exp(-t\frac{\log p_j}{\log N})}{p_j - 1} = \sum_{j=1}^{N}\frac{1 - \exp(-t\frac{\log p_j}{\log N})}{\frac{\log p_j}{\log N}}\frac{\log p_j}{(p_j - 1)\log N} = \\ \sum_{j=1}^{N}\frac{1 - \exp(-tx_j^{(N)})}{x_j^{(N)}}\frac{\log p_j}{(p_j - 1)\log N}.$$
$$(4.8)$$

From (4.6)-(4.8) along with (4.4) we conclude that

$$\lim_{N\to\infty} E_N \exp(-t\frac{\log n}{\log N}) = \exp(-\int_0^1 \frac{1 - e^{-tx}}{x} dx).$$ (4.9)

This completes the proof of the theorem for P_N.

We now turn to $P_N^{(k)}$. Let $E_N^{(k)}$ denote the expectation with respect to $P_N^{(k)}$. By Proposition 1.2,

$$E_N^{(k)} \exp(-t\frac{\log n}{\log N}) = E \exp(-\frac{t}{\log N} \sum_{j=1}^N X_{p_j}^{(k)} \log p_j) = \prod_{j=1}^N E \exp(-\frac{t \log p_j}{\log N} X_{p_j}^{(k)}).$$ (4.10)

For $s \geq 0$,

$$E \exp(-s X_{p_j}^{(k)}) = \sum_{l=0}^{k-1} e^{-sl} (\frac{1}{p_j})^l \frac{1 - \frac{1}{p_j}}{1 - (\frac{1}{p_j})^k} = \frac{1 - \frac{1}{p_j}}{1 - (\frac{1}{p_j})^k} \frac{1 - (\frac{e^{-s}}{p_j})^k}{1 - \frac{e^{-s}}{p_j}}.$$ (4.11)

Comparing the equality between the first and third expressions in (4.2) with (4.11), we have

$$E \exp(-s X_{p_j}^{(k)}) = \frac{1 - (\frac{e^{-s}}{p_j})^k}{1 - (\frac{1}{p_j})^k} E \exp(-s X_{p_j}) = \left(1 + \frac{(\frac{1}{p_j})^k(1 - e^{-sk})}{1 - (\frac{1}{p_j})^k}\right) E \exp(-s X_{p_j}).$$ (4.12)

Thus, from (4.1), (4.10) and (4.12) we have

$$E_N^{(k)} \exp(-t\frac{\log n}{\log N}) = E_N \exp(-t\frac{\log n}{\log N}) \prod_{j=1}^N \left(1 + \frac{(\frac{1}{p_j})^k(1 - \exp(-\frac{kt \log p_j}{\log N}))}{1 - (\frac{1}{p_j})^k}\right).$$ (4.13)

By the bounded convergence theorem,

$$\lim_{N\to\infty} \sum_{j=1}^N \frac{(\frac{1}{p_j})^k(1 - \exp(-\frac{kt \log p_j}{\log N}))}{1 - (\frac{1}{p_j})^k} = 0.$$ (4.14)

Thus, from (4.9), (4.13) and (4.14), we conclude that

$$\lim_{N\to\infty} E_N^{(k)} \exp(-t\frac{\log n}{\log N}) = \exp(-\int_0^1 \frac{1 - e^{-tx}}{x} dx).$$

□

5 Proof of Theorem 1.2

We prove the theorem for P_N; the proof for $P_N^{(k)}$ is done analogously. For definiteness and convenience, we define $\frac{\log n}{\log p^+(n)}|_{n=1} = 0$. Let

$$J_N^+ = \max\{j \in [N] : X_{p_j} \neq 0\},$$

with max $\emptyset$ defined to be 0. By Proposition 1.1, $\frac{\log n}{\log p^+(n)}$ under P_N is equal in distribution to $\frac{1_{\{J_N^+ \neq 0\}}}{\log p_{J_N^+}} \sum_{j=1}^N X_{p_j} \log p_j$. On $\{J_N^+ \neq 0\}$, we write

$$\frac{1}{\log p_{J_N^+}} \sum_{j=1}^N X_{p_j} \log p_j = \frac{1}{\log p_{J_N^+}} \sum_{j=1}^{J_N^+ - 1} X_{p_j} \log p_j + X_{p_{J_N^+}}.$$

As noted in the paragraph containing (3.1), $J_N^+ \to \infty$ a.s. as $N \to \infty$. Also, by the independence of $\{X_{p_j}\}_{j=1}^\infty$, we have $\sum_{j=1}^{J_N^+ - 1} X_{p_j} \log p_j | \{J_N^+ = j_0\} \overset{\text{dist}}{=} \sum_{j=1}^{j_0 - 1} X_{p_j} \log p_j$. Thus, it follows from Theorem 1.1 that $\frac{1}{\log J_N^+} \sum_{j=1}^{J_N^+ - 1} X_{p_j} \log p_j$ converges weakly to the Dickman distribution. By the PNT, $p_{J_N^+} \sim J_N^+ \log J_N^+$; thus also $\frac{1}{\log p_{J_N^+}} \sum_{j=1}^{J_N^+ - 1} X_{p_j} \log p_j$ a.s. converges weakly to the Dickman distribution.

Note that $X_{p_{J_N^+}}$ conditioned on $\{J_N^+ = j_0\}$ is distributed as $X_{p_{j_0}}$ conditioned on $\{X_{p_{j_0}} \geq 1\}$. A trivial calculation shows that $X_{p_{j_0}}$ conditioned on $\{X_{p_{j_0}} \geq 1\}$ converges weakly to 1 as $j_0 \to \infty$; thus, $X_{p_{j_N}}$ converges weakly to 1. Consequently, $\frac{\log n}{\log p^+(n)}$ under P_N converges weakly to $D + 1$ as $N \to \infty$. $\qquad\square$

6 Proof of Theorem 1.3

As noted after the statement of the theorem, we will prove (1.20) with $s\omega(s)$ replaced by $\sum_{L=1}^{[s]} \Lambda_L(s)$, where Λ_L is as in (1.21). That is, we will prove that

$$P_N\left(\frac{\log n}{\log p^+(n)} \leq s\right) \sim (e^{-\gamma} \log \log N)^{\frac{\sum_{L=1}^{[s]} \Lambda_L(s)}{\log N}}, \quad s \geq 1. \qquad (6.1)$$

We will first prove (6.1) for $s \in [1, 2]$, then for $s \in [2, 3]$, and then for $s \in [3, 4]$. After treating these three particular cases, an inductive argument for the general case of $s \in [L, L+1]$ will be explained succinctly.

For definiteness and convenience, we define $\frac{\log n}{\log p^-(n)}|_{n=1} = 0$. Of course, $\frac{\log n}{\log p^-(n)} \geq 1$, for $n \geq 2$. Let

$$J_N^- = \min\{j \in [N] : X_{p_j} \neq 0\},$$

with min ∅ defined to be 0. Note that by (1.8),

$$P(\frac{\log n}{\log p^-(n)} < 1) = P(J_N^- = 0) = C_N^{-1} \sim \frac{e^{-\gamma}}{\log N}. \tag{6.2}$$

By Proposition 1.1, $\frac{\log n}{\log p^-(n)}$ under P_N is equal in distribution to $\frac{1_{\{J_N^- \neq 0\}}}{\log p_{J_N^-}} \sum_{j=1}^N X_{p_j} \log p_j$. Thus, we have

$$P_N(L \leq \frac{\log n}{\log p^-(n)} \leq s) =$$

$$\sum_{a=1}^N P\left(L \log p_a \leq \sum_{j=a}^N X_{p_j} \log p_j \leq s \log p_a | J_N^- = a\right) P(J_N^- = a), \text{ for } L \in \mathbb{N},$$
$$\tag{6.3}$$

and

$$P(J_N^- = a) = \frac{1}{p_a} \prod_{j=1}^{a-1} (1 - \frac{1}{p_j}). \tag{6.4}$$

Under the conditioning $\{J_N^- = a\}$, the random variables $\{X_{p_j}\}_{j=a}^N$ are still independent, and for $j > a$, X_{p_j} is distributed as before, namely according to $\text{Geom}(1 - \frac{1}{p_j})$; however X_{p_a} is now distributed as a $\text{Geom}(1 - \frac{1}{p_a})$ random variable conditioned to be positive.

Consider first $L = 1$ and $s \in [1, 2]$. For $s \neq 2$, the inequality $\log p_a \leq \sum_{j=a}^N X_{p_j} \log p_j \leq s \log p_a$ in (6.3) under the conditional probability $P(\cdot | J_N^- = a)$ will hold if and only if $X_{p_a} = 1$ and $X_{p_j} = 0$, for $a + 1 \leq j \leq N$. For $s = 2$ it will hold if and only if X_{p_a} is equal to either 1 or 2 and $X_{p_j} = 0$, for $a + 1 \leq j \leq N$. Thus, we have

$$P\left(\log p_a \leq \sum_{j=a}^N X_{p_j} \log p_j \leq s \log p_a | J_N^- = a\right) =$$
$$\tag{6.5}$$
$$\begin{cases} \prod_{j=a}^N (1 - \frac{1}{p_j}), & s \in [1, 2); \\ \prod_{j=a}^N (1 - \frac{1}{p_j}) + \frac{1}{p_a} \prod_{j=a}^N (1 - \frac{1}{p_j}), & s = 2. \end{cases}$$

From (6.2)–(6.5), along with (1.8) and (1.16) and the fact that $\Lambda_1(s) \equiv 1$ for $s \in [1, 2]$, we obtain

$$P_N(\frac{\log n}{\log p^-(n)} \leq s) \sim C_N^{-1} \sum_{a=1}^N \frac{1}{p_a} \sim (e^{-\gamma} \log \log N) \frac{\Lambda_1(s)}{\log N}, \quad s \in [1, 2]. \tag{6.6}$$

Now consider $L = 2$ and $s \in [2, 3]$. Let

$$J_{a,1}(s) = \max\{j : p_j \leq p_a^{s-1}\}.$$

(Note that $J_{a,1}(s) \geq a$, for $s \geq 2$.) Then for $s \in [2, 3)$, the inequality $2 \log p_a \leq \sum_{j=a}^{N} X_{p_j} \log p_j \leq s \log p_a$ in (6.3) under the conditional probability $P(\ \cdot\ |J_N^- = a)$ will hold if and only if either $X_{p_a} = 2$ and $X_{p_j} = 0$ for $a + 1 \leq j \leq N$, or $X_{p_a} = 1$, $X_{p_j} = 1$ for exactly one j satisfying $a + 1 \leq j \leq J_{a,1}(s) \wedge N$, and $X_{p_j} = 0$ for all other j satisfying $a + 1 \leq j \leq N$. Thus, we have

$$P\left(2 \log p_a \leq \sum_{j=a}^{N} X_{p_j} \log p_j \leq s \log p_a | J_N^- = a\right) =$$

$$\frac{1}{p_a} \prod_{j=a}^{N} (1 - \frac{1}{p_a}) + \sum_{l=a+1}^{J_{a,1}(s) \wedge N} \frac{1}{p_l} \prod_{j=a}^{N} (1 - \frac{1}{p_j}), \ s \in [2, 3), \tag{6.7}$$

where, of course, the sum on the right hand side above is interpreted as 0 if $J_{a,1}(s) = a$. For the case $s = 3$, there is also the possibility of $X_{p_a} = 3$ and $X_{p_j} = 0$ for $a + 1 \leq j \leq N$. The $P(\ \cdot\ |J_N^- = a)$-probability of this is $\frac{1}{p_a^2} \prod_{j=a}^{N} (1 - \frac{1}{p_a})$. Thus, with $s = 3$, (6.7) has the additional term $\frac{1}{p_a^2} \prod_{j=a}^{N} (1 - \frac{1}{p_a})$ on the right hand side. However, this term does not contribute to the leading order asymptotics. From (6.3), (6.4) and (6.7), we obtain

$$P_N(2 \leq \frac{\log n}{\log p^-(n)} \leq s) = C_N^{-1} \left(\sum_{a=1}^{N} \frac{1}{p_a^2} + \sum_{a=1}^{N} \frac{1}{p_a} \sum_{l=a+1}^{J_{a,1}(s) \wedge N} \frac{1}{p_l} \right), \ s \in [2, 3). \tag{6.8}$$

Since $p_a \sim a \log a$ as $a \to \infty$, it follows that

$$J_{a,1}(s) \log J_{a,1}(s) \sim (a \log a)^{s-1}, \ \text{as } a \to \infty. \tag{6.9}$$

Taking the logarithm of each side in (6.9), we obtain

$$\lim_{a \to \infty} \frac{\log J_{a,1}(s)}{\log a} = s - 1. \tag{6.10}$$

Using Mertens' second theorem in the form (1.15) along with the fact that $p_j \sim j \log j$, we have

$$\sum_{l=a+1}^{J_{a,1}(s)} \frac{1}{p_l} \sim \log \log \left(J_{a,1}(s) \log J_{a,1}(s)\right) - \log \log(a \log a) \sim \log \frac{\log J_{a,1}(s)}{\log a}, \ \text{as } a \to \infty, \tag{6.11}$$

and thus, by (6.10),

$$\lim_{a \to \infty} \sum_{l=a+1}^{J_{a,1}(s)} \frac{1}{p_l} = \log(s - 1). \tag{6.12}$$

Now choose any $b \in (0, \frac{1}{s})$. Then $(N^b \log N^b)^s < N$ for all large N. By (6.9),

$$J_{a,1}(s) \leq N, \text{ for } a \leq N^b \text{ and sufficiently large } N. \tag{6.13}$$

By Mertens' second theorem in the form (1.16), we have

$$\sum_{a=1}^{N} \frac{1}{p_a} = \sum_{a=1}^{N^b} \frac{1}{p_a} + O(1) \sim \log \log N. \tag{6.14}$$

From (6.12)-(6.14), we obtain

$$\sum_{a=1}^{N} \frac{1}{p_a} \sum_{l=a+1}^{J_{a,1}(s) \wedge N} \frac{1}{p_l} \sim \sum_{a=1}^{N^b} \frac{1}{p_a} \sum_{l=a+1}^{J_{a,1}(s)} \frac{1}{p_l} \sim (\log \log N) \log(s-1). \tag{6.15}$$

Recalling the asymptotic behavior of C_N, recalling from (1.21) that $\Lambda_2(s) = \log(s-1)$ for $s \geq 2$, and using (6.8) and (6.15), we conclude that

$$P_N\left(2 \leq \frac{\log n}{\log p^-(n)} \leq s\right) \sim (e^{-\gamma} \log \log N) \frac{\Lambda_2(s)}{\log N}, \quad s \in [2, 3], \tag{6.16}$$

where the inclusion of the right endpoint $s = 3$ follows from the remarks made after (6.7). From (6.6) with $s = 2$ and (6.16), along with the fact that $\Lambda_1(s) \equiv 1$, we obtain

$$P_N\left(\frac{\log n}{\log p^-(n)} \leq s\right) \sim (e^{-\gamma} \log \log N) \frac{\Lambda_1(s) + \Lambda_2(s)}{\log N}, \quad s \in [2, 3]. \tag{6.17}$$

Now consider $L = 3$ and $s \in [3, 4]$. In fact we will work with $s \in [3, 4)$ since the case $s = 4$ is slightly different but leads to the same asymptotics, similar to the remarks after (6.7). Then the inequality $3 \log p_a \leq \sum_{j=a}^{N} X_{p_j} \log p_j \leq s \log p_a$ in (6.3) under the conditional probability $P(\ \cdot\ |J_N^- = a)$ will hold if and only if one of the following four situations obtains:

(1) $X_{p_a} = 3; X_{p_j} = 0$, for $a + 1 \leq j \leq N$.

(2) $X_{p_a} = 2; X_{p_j} = 1$ for exactly one j satisfying $a + 1 \leq j \leq J_{a,1}(s-1) \wedge N$; $X_{p_j} = 0$ for all other j satisfying $a + 1 \leq j \leq N$.

(3) $X_{p_a} = 1; X_{p_j} = 1$ for exactly one j satisfying $J_{a,1}(3) < j \leq J_{a,1}(s) \wedge N$; $X_{p_j} = 0$ for all other j satisfying $a + 1 \leq j \leq N$.

(4) $X_{p_a} = 1$; there exist j_1, j_2, satisfying $a + 1 \leq j_1 \leq j_2 \leq N$ and $p_{j_1} p_{j_2} \leq p_a^{s-1}$, such that $X_{j_1} = X_{j_2} = 1$, if $j_1 \neq j_2$ and $X_{j_1} = 2$ if $j_1 = j_2$; $X_{p_j} = 0$ for all other j satisfying $a + 1 \leq j \leq N$.

$$\tag{6.18}$$

Because $\sum_{a=1}^{\infty} \frac{1}{p_a^2} < \infty$, the probabilities from situations (1) and (2) in (6.18) do not contribute to the leading order asymptotics of $P_N(3 \leq \frac{\log n}{\log p^-(n)} \leq s)$, just as in the case $L = 2$ and $s \in [2, 3)$, the probability from the case $X_{p_a} = 2$ did not contribute to the leading order asymptotics there. (The contribution there from the case $X_{p_a} = 2$ is the term $C_N^{-1} \sum_{a=1}^{N} \frac{1}{p_a^2}$ in (6.8).)

The analysis of the contribution from situation (3) in (6.18) follows the same line of analysis as above when $L = 2$ and $s \in [2, 3)$ for the case $X_{p_a} = 1$, $X_{p_j} = 1$ for exactly one j satisfying $a + 1 \leq j \leq J_{a,1}(s) \wedge N$, and $X_{p_j} = 0$ for all other j satisfying $a + 1 \leq j \leq N$. The difference is that there one had $X_{p_j} = 1$ for exactly one j satisfying $a + 1 \leq j \leq J_{a,1}(s) \wedge N$, while here one has $X_{p_j} = 1$ for exactly one j satisfying $J_{a,1}(3) < j \leq J_{a,1}(s) \wedge N$. Thus, whereas the corresponding contribution there was the term $\sum_{a=1}^{N} \frac{1}{p_a} \sum_{l=a+1}^{J_{a,1}(s) \wedge N} \frac{1}{p_l}$ in (6.8), the contribution here will be $\sum_{a=1}^{N} \frac{1}{p_a} \sum_{l=J_{a,1}(3)+1}^{J_{a,1}(s) \wedge N} \frac{1}{p_l}$. Similar to (6.11), we have $\sum_{l=J_{a,1}(3)+1}^{J_{a,1}(s) \wedge N} \frac{1}{p_l} \sim \log \frac{\log J_{a,1}(s)}{\log J_{a,1}(3)}$, and from (6.10) we have $\lim_{a \to \infty} \frac{\log J_{a,1}(s)}{\log J_{a,1}(3)} = \frac{s-1}{3-1} = \frac{s-1}{2}$. Thus, similar to (6.15), we obtain

$$
\sum_{a=1}^{N} \frac{1}{p_a} \sum_{l=J_{a,1}(3)+1}^{J_{a,1}(s) \wedge N} \frac{1}{p_l} \sim \sum_{a=1}^{N^b} \frac{1}{p_a} \sum_{l=J_{a,1}(3)+1}^{J_{a,1}(s)} \frac{1}{p_l} \sim (\log \log N)\big(\log(s-1) - \log 2\big).
$$
(6.19)

And finally, similar to (6.16), the contribution to $P_N (3 \leq \frac{\log n}{\log p^-(n)} \leq s)$ from situation (3), which we denote by $\rho_3(s)$, satisfies

$$
\rho_3(s) \sim (e^{-\gamma} \log \log N) \frac{\Lambda_2(s) - \Lambda_2(3)}{\log N}, \quad s \in [3, 4],
$$
(6.20)

where the inclusion of the right endpoint $s = 4$ follows from the remarks made at the beginning of the treatment of the case $s \in [3, 4]$.

We know analyze the contribution from situation (4) in (6.18). From (6.3) and (6.4), the contribution to $P_N (3 \leq \frac{\log n}{\log p^-(n)} \leq s)$ from situation (4), which we will denote by $\rho_4(s)$, is

$$
\rho_4(s) = C_N^{-1} \sum_{a=1}^{N} \frac{1}{p_a} \sum_{\substack{a+1 \leq j_1 \leq j_2 \leq N \\ p_{j_1} p_{j_2} \leq p_a^{s-1}}} \frac{1}{p_{j_1} p_{j_2}}.
$$
(6.21)

Define

$$
J_a(s, j_1) = \max\{j : p_j \leq \frac{p_a^{s-1}}{p_{j_1}}\}, \quad J_{a,2}(s) = \max\{j : p_j^2 \leq p_a^{s-1}\}.
$$

Then

$$
\rho_4(s) = C_N^{-1} \sum_{a=1}^{N} \frac{1}{p_a} \sum_{j_1=a+1}^{J_{a,2}(s) \wedge N} \frac{1}{p_{j_1}} \sum_{j_2=j_1}^{J_a(s, j_1) \wedge N} \frac{1}{p_{j_2}}.
$$
(6.22)

Since $p_j \sim j \log j$, it follows that $J_{a,2}(s) \log J_{a,2}(s) \sim (a \log a)^{\frac{s-1}{2}}$, as $a \to \infty$. Taking the logarithm of both sides above, it follows that $\log J_{a,2}(s) \sim \frac{s-1}{2} \log a$ as $a \to \infty$. Thus

$$J_{a,2}(s) \sim \frac{2}{s-1} a^{\frac{s-1}{2}} (\log a)^{\frac{s-3}{2}}, \quad \text{as } a \to \infty. \tag{6.23}$$

Consider now $J_a(s, j_1)$, for $a + 1 \leq j_1 \leq J_{a,2}(s)$. Similarly as in the above paragraph, it follows that $J_a(s, j_1) \log J_a(s, j_1) \sim \frac{(a \log a)^{s-1}}{j_1 \log j_1}$, as $j_1, a \to \infty$. Since $j_1 \leq J_{a,2}(s)$, it follows from (6.23) that $j_1 = o(a^{s-1})$. Thus, taking the logarithm of both sides above, we have

$$\log J_a(s, j_1) \sim (s-1) \log a - \log j_1, \quad \text{as } j_1, a \to \infty. \tag{6.24}$$

Therefore,

$$J_a(s, j_1) \sim \frac{a^{s-1} (\log a)^{s-1}}{j_1 \log j_1 \big((s-1) \log a - \log j_1 \big)}, \quad \text{as } j_1, a \to \infty. \tag{6.25}$$

In light of (6.23) and (6.25), we can choose $b \in (0, 1)$, depending on s, such that $J_a(s, j_1) \leq N$ and $J_{a,2}(s) \leq N$, for all $a \leq N^b$ and all sufficiently large N. Thus, from (6.14) and (6.22), we have, similar to the first asymptotic equivalence in (6.15),

$$\rho_4(s) \sim C_N^{-1} \sum_{a=1}^{N^b} \frac{1}{p_a} \sum_{j_1=a+1}^{J_{a,2}(s)} \frac{1}{p_{j_1}} \sum_{j_2=j_1}^{J_a(s,j_1)} \frac{1}{p_{j_2}}. \tag{6.26}$$

By (1.16) and (6.24), we have

$$\sum_{j_2=j_1}^{J_a(s,j_1)} \frac{1}{p_{j_2}} \sim \log \frac{\log J_a(s, j_1)}{\log j_1} \sim \log \frac{(s-1) \log a - \log j_1}{\log j_1} = \log \Big((s-1) \frac{\log a}{\log j_1} - 1 \Big),$$

as $j_1, a \to \infty$.

$$\tag{6.27}$$

Using (6.27), (6.23) and the fact that $p_j \sim j \log j$ as $j \to \infty$, we have

$$\sum_{j_1=a+1}^{J_{a,2}(s)} \frac{1}{p_{j_1}} \sum_{j_2=j_1}^{J_a(s,j_1)} \frac{1}{p_{j_2}} \sim \sum_{j_1=a+1}^{J_{a,2}(s)} \frac{1}{j_1 \log j_1} \log \Big((s-1) \frac{\log a}{\log j_1} - 1 \Big) \sim$$

$$\int_a^{J_{a,2}(s)} \frac{1}{x \log x} \log \Big((s-1) \frac{\log a}{\log x} - 1 \Big) dx \sim \tag{6.28}$$

$$\int_a^{a^{\frac{s-1}{2}}} \frac{1}{x \log x} \log \Big((s-1) \frac{\log a}{\log x} - 1 \Big) dx, \quad \text{as } a \to \infty,$$

where the final asymptotic equivalence follows from the iterated logarithmic growth rate of the indefinite integral of the integrand appearing in the equation. Making the substitution

$$x = a^{\frac{s-1}{u_2}}$$

reveals that the second integral in (6.28) is in fact independent of a. We obtain

$$\int_a^{a^{\frac{s-1}{2}}} \frac{1}{x \log x} \log \left((s-1) \frac{\log a}{\log x} - 1 \right) dx = \int_2^{s-1} \frac{du_2}{u_2} \log(u_2 - 1) =$$
$$\int_2^{s-1} \int_1^{u_2-1} \frac{du_1}{u_1} \frac{du_2}{u_2} = \Lambda_3(s). \tag{6.29}$$

From (6.28) and (6.29), we conclude that

$$\lim_{a \to \infty} \sum_{j_1=a+1}^{J_{a,2}(s)} \frac{1}{p_{j_1}} \sum_{j_2=j_1}^{J_a(s,j_1)} \frac{1}{p_{j_2}} = \Lambda_3(s). \tag{6.30}$$

Thus, recalling the asymptotic behavior of C_N, from (6.30), (6.26) and (6.14) we conclude that

$$\rho_4(s) \sim (e^{-\gamma} \log \log N) \frac{\Lambda_3(s)}{\log N}, \quad s \in [3, 4], \tag{6.31}$$

where the inclusion of the right endpoint $s = 4$ follows from the remarks made at the beginning of the treatment of the case $s \in [3, 4]$. From (6.20) and (6.31), we conclude that

$$P_N(3 \le \frac{\log n}{\log p^-(n)} \le s) \sim (e^{-\gamma} \log \log N) \frac{(\Lambda_2(s) - \Lambda_2(3)) + \Lambda_3(s)}{\log N}, \quad s \in [3, 4]. \tag{6.32}$$

From (6.17) with $s = 3$ and (6.32), and recalling that $\Lambda_1(s) \equiv 1$, we have

$$P_N(\frac{\log n}{\log p^-(n)} \le s) \sim (e^{-\gamma} \log \log N) \frac{\Lambda_1(s) + \Lambda_2(s) + \Lambda_3(s)}{\log N}, \quad s \in [3, 4].$$

We now consider the general case that $s \in [L, L + 1]$. By induction, we have

$$P_N(\frac{\log n}{\log p^-(n)} \le s) \sim (e^{-\gamma} \log \log N) \frac{\sum_{l=1}^{[s]} \Lambda_l(s)}{\log N}, \quad s \le L. \tag{6.33}$$

Making a list similar to (6.18), and analyzing the situations as was done there, one concludes that the situations with $X_{p_a} \ge 2$ do not contribute to the leading order asymptotics of $P_N(L \le \frac{\log n}{\log p^-(n)} \le s)$, while the situations with $X_{p_a} = 1$ do contribute. When $X_{p_a} = 1$, we obtain $L - 1$ situations, with all but one of them of the form already treated in the case of $L - 1$. (In (6.18), where $L = 3$, there were 2 such situations—labeled there (3) and (4), and one of them, namely (3), was of the form already treated for $L = 2$.) Thus, by induction and by the argument used to show that the contribution from situation (3) in (6.18) is as it appears in (6.20), these terms will give asymptotic contributions

$$(e^{-\gamma} \log \log N) \frac{\Lambda_1(s) - \Lambda_1(L)}{\log N}, \ldots, (e^{-\gamma} \log \log N) \frac{\Lambda_{L-1}(s) - \Lambda_{L-1}(L)}{\log N}.$$

$$(6.34)$$

We now look at the new situation that arises; namely the one in which $X_{p_a} = 1$ and there exist $j_1, \ldots, j_{L-1}$ satisfying $a + 1 \le j_1 \le \cdots \le j_{L-1} \le N$ and $\prod_{i=1}^{L-1} p_{j_i} \le p_a^{s-1}$, such that for $j \in \{a+1, \ldots, N\}$, X_j is equal to the number of times j appears among the $\{j_i\}_{i=1}^{L-1}$. From (6.3) and (6.4), the contribution to $P_N(L \le \frac{\log n}{\log p^-(n)} \le s)$ from this situation, similar to (6.21) in the case $L = 3$, is

$$C_N^{-1} \sum_{a=1}^{N} \frac{1}{p_a} \sum_{\substack{a+1 \le j_1 \le \cdots \le j_{L-1} \le N \\ \prod_{i=1}^{L-1} p_{j_i} \le p_a^{s-1}}} \frac{1}{\prod_{i=1}^{L-1} p_{j_i}}.$$

$$(6.35)$$

An analysis analogous to that implemented between (6.21) and (6.30) gives

$$\lim_{a \to \infty} \sum_{\substack{a+1 \le j_1 \le \cdots \le j_{L-1} \le N \\ \prod_{i=1}^{L-1} p_{j_i} \le p_a^{s-1}}} \frac{1}{\prod_{i=1}^{L-1} p_{j_i}} = \int_{L-1}^{s-1} \int_{L-2}^{u_{L-1}-1} \cdots \int_{1}^{u_2-1} \prod_{j=1}^{L-1} \frac{du_j}{u_j} = \Lambda_L(s).$$

$$(6.36)$$

From (6.35) and (6.36) it follows that the contribution to the leading order asymptotics of $P_N(L \le \frac{\log n}{\log p^-(n)} \le s)$ from this situation is $(e^{-\gamma} \log \log N) \frac{\Lambda_L(s)}{\log N}$. We conclude from this and (6.34) that

$$P_N(L \le \frac{\log n}{\log p^-(n)} \le s) \sim (e^{-\gamma} \log \log N) \frac{\Lambda_L(s) + \sum_{l=1}^{L-1} (\Lambda_l(s) - \Lambda_l(L))}{\log N}, \quad s \in [L, L+1].$$

$$(6.37)$$

From (6.33) with $s = L$ and from (6.37), we conclude that

$$P_N(L \frac{\log n}{\log p^-(n)} \le s) \sim (e^{-\gamma} \log \log N) \frac{\sum_{l=1}^{L} \Lambda_l(s)}{\log N}, \quad s \in [L, L+1].$$

This completes the proof of (6.1). $\qquad\square$

Appendix: Proof of Proposition 2.1

For notational convenience, we will work with p instead of p_j. The proof is via the inclusion-exclusion principle along with the fact that $D_{\mathrm{nat}}(S_k) = \frac{1}{\zeta(k)}$, where S_k denotes the k-free integers, as was noted with a reference in the proof of Proposition 1.3. Recall that $1 \le l < k$. We have

$$I_N \equiv |\{n : p^l | n, n \le N, n \in S_k\}| = |\{n_1 : n_1 \le [\frac{N}{p^l}], n_1 \in S_k, p^{k-l} \nmid n_1\}| =$$

$$|\{n_1 : n_1 \le [\frac{N}{p^l}], n_1 \in S_k\}| - |\{n_1 : n_1 \le [\frac{N}{p^l}], n_1 \in S_k, p^{k-l} | n_1\}| \equiv I_{N,1} - I_{N,2}.$$

Similarly,

$$I_{N,2} = |\{n_2 : n_2 \le [\frac{N}{p^k}], n_2 \in S_k, p^l \nmid n_2\}| =$$

$$|\{n_2 : n_2 \le [\frac{N}{p^k}], n_2 \in S_k\}| - |\{n_2 : n_2 \le [\frac{N}{p^k}], n_2 \in S_k, p^l | n_2\}| \equiv I_{N,3} - I_{N,4},$$

and

$$I_{N,4} = |\{n_3 : n_3 \le [\frac{N}{p^{k+l}}], n_3 \in S_k, p^{k-l} \nmid n_3\}| =$$

$$|\{n_3 : n_3 \le [\frac{N}{p^{k+l}}], n_3 \in S_k\}| - |\{n_3 : n_3 \le [\frac{N}{p^{k+l}}], n_3 \in S_k, p^{k-l} | n_3\}| \equiv I_{N,5} - I_{N,6}.$$

So up to this point, we have

$$I_N = I_{N,1} - I_{N,3} + I_{N,5} - I_{N,6}.$$

Now $\lim_{N\to\infty} \frac{I_{N,1}}{N} = \frac{1}{p^l} D_{\mathrm{nat}}(S_k) = \frac{1}{p^l \zeta(k)}$, $\lim_{N\to\infty} \frac{I_{N,3}}{N} = \frac{1}{p^k} D_{\mathrm{nat}}(S_k) = \frac{1}{p^k \zeta(k)}$ and $\lim_{N\to\infty} \frac{I_{N,5}}{N} = \frac{1}{p^{k+l}} D_{\mathrm{nat}}(S_k) = \frac{1}{p^{k+l}\zeta(k)}$. Continuing this process of inclusion-exclusion, we have

$$I_N = \sum_{m=0}^{\infty} I_{N,4m+1} - \sum_{m=0}^{\infty} I_{N,4m+3},$$

where for each N only a finite number of the summands above are non-zero. Now

$$\lim_{N\to\infty} \frac{I_{N,4m+1}}{N} = \frac{1}{p^{mk+l}\zeta(k)}, \quad m = 0, 1, \ldots,$$

and

$$\lim_{N\to\infty} \frac{I_{N,4m+3}}{N} = \frac{1}{p^{(m+1)k}\zeta(k)}, \quad m = 0, 1, \ldots.$$

From this we conclude that

$$D_{nat}(\beta_p \geq l, S_k) = \frac{1}{\zeta(k)} \sum_{m=0}^{\infty} \frac{1}{p^{mk+l}} - \frac{1}{\zeta(k)} \sum_{m=0}^{\infty} \frac{1}{p^{(m+1)k}} = \frac{1}{\zeta(k)} \frac{\frac{1}{p^l} - \frac{1}{p^k}}{1 - \frac{1}{p^k}}.$$

Thus, $D_{nat}(\beta_p \geq l | S_k) = \frac{\frac{1}{p^l} - \frac{1}{p^k}}{1 - \frac{1}{p^k}}$. $\qquad\square$

References

1. Arratia, R., Barbour, A., Tavaré, S.: Logarithmic combinatorial structures: a probabilistic approach. EMS Monographs in Mathematics. European Mathematical Society, EMS, Zurich (2003)
2. Billingsley, P.: On the distribution of large prime divisors, collection of articles dedicated to the memory of Alfréd Rényi I. Period. Math. Hungar. **2**, 283–289 (1972)
3. Billingsley, P.: The probability theory of additive arithmetic functions. Ann. Probab. **2**, 749–791 (1974)
4. de Bruijn, N.: On the number of positive integers $\leq x$ and free of prime factors $> y$. Nederl. Acad. Wetensch. Proc. Ser. A. **54**, 50–60 (1951)
5. Buchstab, A.: An asymptotic estimation of a general number-theoretic function. Mat. Sbornik **44**, 1239–1246 (1937)
6. Cellarosi, F., Sinai, Y.: Non-standard limit theorems in number theory. In: Prokhorov and Contemporary Probability Theory, vol. 33, pp. 197–213. Springer Proceedings in Mathematics & Statistics. Springer, Heidelberg (2013)
7. Dickman, K.: On the frequency of numbers containing prime factors of a certain relative magnitude. Ark. Math. Astr. Fys. **22**, 1–14 (1930)
8. Durrett, R.: Probability Theory and Examples, 3rd edn. Brooks/Cole, Belmont, CA (2005)
9. Elliott, P.D.T.A.: Probabilistic Number Theory. I. Mean-Value Theorems. Grundlehren der Mathematischen Wissenschaften, vol. 239. Springer, New York (1979)
10. Elliott, P.D.T.A.: Probabilistic Number Theory. II. Central Limit Theorems. Grundlehren der Mathematischen Wissenschaften, vol. 240. Springer, New York (1980)
11. Erdös, P., Wintner, A.: Additive arithmetical functions and statistical independence. Am. J. Math. **61**, 713–721 (1939)
12. Erdös, P., Kac, M.: The Gaussian law of errors in the theory of additive number theoretic functions. Am. J. Math. **62**, 738–742 (1940)
13. Giuliano, R., Macci, C.: Asymptotic results for weighted means of random variables which converge to a Dickman distribution, and some number theoretical applications. ESAIM Probab. Stat. **19**, 395–413 (2015)
14. Hardy, G.H., Ramanujan, S.: Proof that almost all numbers n are composed of about $\log \log n$ prime factors. Proc. London Math. Soc. **16**, 242–243 (1917). Collected Papers of Srinivasa Ramanujan. AMS Chelsea Publications, Providence, RI (2000)
15. Lagarias, J.: Euler's constant: Euler's work and modern developments. Bull. Am. Math. Soc. (N.S.) **50**, 527–628 (2013)
16. Mehrdad, B., Zhu, L.: Moderate and large deviations for the Erdös-Kac theorem. Q. J. Math. **67**, 147–160 (2016)
17. Mehrdad, B., Zhu, L.: Limit theorems for empirical density of greatest common divisors. In: Mathematical Proceedings of the Cambridge Philosophical Society, vol. 161, pp. 517–533 (2016)
18. Montgomery, H., Vaughan, R.: Multiplicative Number Theory. I. Classical Theory. Cambridge Studies in Advanced Mathematics, vol. 97. Cambridge University Press, Cambridge (2007)

19. Nathanson, M.: Elementary Methods in Number Theory, Graduate Texts in Mathematics, vol. 195. Springer, New York (2000)
20. Penrose, M., Wade, A.: Random minimal directed spanning trees and Dickman-type distributions. Adv. Appl. Probab. **36**, 691–714 (2004)
21. Pinsky, R.: On the strange domain of attraction to generalized Dickman distributions for sums of independent random variables. Electron. J. Probab. **23** (2018). (Paper No. 3, 17 pp)
22. Shapiro, H.: Introduction to the Theory of Numbers. Wiley, New York (1983)
23. Tenenbaum, G.: Introduction to Analytic and Probabilistic Number Theory. Cambridge Studies in Advanced Mathematics. Cambridge University Press, Cambridge (1995)
24. Turán, P.: On a theorem of Hardy and Ramanujan. J. Lond. Math. Soc. **9**, 274–276 (1934)

Printed in the United States
By Bookmasters